Recommendations on Piling (EA-Pfähle)

WILEY
Ernst & Sohn
A Wiley Brand

DGGT
Deutsche Gesellschaft für Geotechnik e. V.
German Geotechnical Society

Recommendations on Piling

(EA-Pfähle)

Edited by
German Geotechnical Society
(Deutsche Gesellschaft für Geotechnik e.V.)

WILEY
Ernst & Sohn
A Wiley Brand

DGGT
Deutsche Gesellschaft
für Geotechnik e. V.
German Geotechnical Society

AK 2.1 Piling Committee of the German Geotechnical Society
(*Deutsche Gesellschaft für Geotechnik e.V.*)

Chairman until June 2013
Univ.-Prof.(em) Dr.-Ing. Hans-Georg Kempfert
Potosistraße 27
22587 Hamburg, Germany
kempfert@kup-geotechnik.de

Chairman (from July 2013)
Uni.-Prof. Dr.-Ing. habil. Christian Moormann
University of Stuttgart
Institute of Geotechnical Engineering
Pfaffenwaldring 35
70569 Stuttgart
christian.moormann@igs.uni-stuttgart.de

Library of Congress Card No.:
applied for
British Library Cataloguing-in-Publication Data
A catalogue record for this book is available from the British Library.
Bibliographic information published by the Deutsche Nationalbibliothek
The Deutsche Nationalbibliothek lists this publication in the Deutsche Nationalbibliografie; detailed bibliographic data are available on the Internet at <http://dnb.d-nb.de>.

Cover design: Design Pur GmbH
Production management: pp030 – Produktionsbüro Heike Praetor, Berlin
Typesetting: Beltz Bad Langensalza GmbH, Bad Langensalza
Printing and binding: betz Druck GmbH, Darmstadt

Printed in the Federal Republic of Germany.
Printed on acid-free paper.

Print ISBN: 978-3-433-03018-9
ePDF ISBN: 978-3-433-60414-4
ePub ISBN: 978-3-433-60413-7
mobi ISBN: 978-3-433-60412-0
oBook ISBN: 978-3-433-60411-3

Members of the AK 2.1 Piling Committee of the German Geotechnical Society

At the time of publication of these recommendations the Piling Committee consisted of the following members:

Univ.-Prof. (em) Dr.-Ing. H.-G. Kempfert, Hamburg (Chairman)
Dr.-Ing. W.-R. Linder, Essen (Deputy Chairman)
Dipl.-Ing. B. Böhle, Essen
Dipl.-Ing. W. Brieke, Düsseldorf
Dipl.-Ing. G. Dausch, Mannheim
Dipl.-Ing. E. Dornecker, Karlsruhe
Dipl.-Ing. A. Ellner, Nürnberg
Dipl.-Ing. M. Glimm, Hamburg
Dipl.-Ing. R. Jörger, Wiesbaden
Dr.-Ing. O. Klingmüller, Mannheim
Dipl.-Ing. O. Krist, Munich
Univ.-Prof. Dr.-Ing. habil. Chr. Moormann, Stuttgart
Dr.-Ing. K. Morgen, Hamburg
Prof. Dr.-Ing. D. Placzek, Essen
Prof. Dr.-Ing. B. Plaßmann, Mainz
Dipl.-Ing. U. Plohmann, Eggenstein
Dr.-Ing. habil. K. Röder, Leipzig
Dr.-Ing. P. Schwarz, Munich
Dr.-Ing. W. Schwarz, Schrobenhausen
Dr.-Ing. S. Weihrauch, Hamburg
Dipl.-Ing. J. Voth, Hamburg

Further members of the Piling committee since the 1st German edition was published (2007) were:

Dipl.-Ing. W. Körner, Hamburg
Dr.-Ing. H.G. Schmidt, Ladenburg

The following subcommittees were involved in the compilation of the Pile Recommendations:

'Dynamic Pile Testing' subcommittee: Sections 10 and 12 and Appendix C

Dr.-Ing. O. Klingmüller, Mannheim (Convenor)
Dipl.-Ing. A. Beneke, Achim
Dr.-Ing. U. Ernst, Nuremberg
Dipl.-Ing. J. Fischer, Braunschweig
Dr.-Ing. M. Fritsch, Braunschweig
Dipl.-Ing. P. Grud, Denmark

Dipl.-Ing. G. Kainrath, Austria
Dr.-Ing. F. Kirsch, Berlin
P. Middendorp, MSc, Netherlands
Dr. rer.nat. E. Niederleithinger, Berlin
Dr. F. Rausche, USA
Prof. Dr.-Ing. W. Rücker, Berlin
Dr.-Ing. M. Schallert, Mannheim
D. Schau, Büdelsdorf
Dr.-Ing. W. Schwarz, Schrobenhausen
Univ.-Prof. Dr.-Ing. J. Stahlmann, Braunschweig
R. Skov, MSc, Denmark
Dr.-Ing. G. Ulrich, Leutkirch
Dr.-Ing. B. Wienholz, Oldenburg

'Piled Raft and Pile Group Foundations' subcommittee: Involved in Sections 8.1.1, 8.2.1, 8.3.1 and 8.4.1 (until 2007)

Prof. Dr.-Ing. Th. Richter, Berlin (Convenor)
Dipl.-Ing. U. Barth, Mannheim
Univ.-Prof. Dr.-Ing. R. Katzenbach, Darmstadt
Univ.-Prof. Dr.-Ing. H.-G. Kempfert, Kassel
Prof. Dr.-Ing. B. Lutz, Berlin
Dr.-Ing. Y. El-Mossallamy, Darmstadt
Dr.-Ing. H. Wahrmund, Cologne
Univ.-Prof. Dr.-Ing. habil. Chr. Moormann, Stuttgart

AK 1.4 'Soil Dynamics' and AK 2.1 'Piles' joint subcommittee: Section 13 and Appendix D

Univ.-Prof. Dr.-Ing. habil. S. Savidis, Berlin (Chairman AK 1.4)
Univ.-Prof. (em) Dr.-Ing. H.-G. Kempfert, Hamburg (Chairman AK 2.1)
Univ.-Prof. Dr.-Ing. M. Achmus, Hannover
Dr.-Ing. J. Dührkop, Hamburg
Dr.-Ing. H.-G. Hartmann, Frankfurt
Dr.-Ing. U. Hartwig, Stuttgart
Dr.-Ing. F. Kirsch, Berlin
PD Dr.-Ing. habil. K. Lesny, Essen
Dr.-Ing. F. Rackwitz, Berlin
Prof. Dr.-Ing. Th. Richter, Berlin
Dr.-Ing. P. Schwarz, München
Dr.-Ing. E. Tasan, Berlin
Dr.-Ing. S. Thomas, Kassel
Dr.-Ing. S. Weihrauch, Hamburg
Dr.-Ing. J. Wiemann, Hamburg

Preface of the English Version of the Recommendations of the Piling Committee of the German Geotechnical Society

(1) The present English version of the Recommendations of the Piling Committee of the German Geotechnical Society (Empfehlungen des Arbeitskreises „Pfähle“ – EA-Pfähle), is with a few exceptions a direct and complete translation of the 2nd edition 2012 of the German original. Even in case that some issues and, in particular, references might internationally not be known as in Germany, this approach was taken in order not to permit differing provisions between the German and this English version, if the latter is used in German speaking countries. It might also be possible that contractual or legal problems could arise between German and English speaking users if they could refer to a not identical translation.

(2) As outlined in Section 1, Germany has published the German language versions of the European design (e.g. Eurocode 7) and execution standards (e.g. EN 1536, EN 12699, etc.). So was EN 1997-1:2004-11: Eurocode 7: Geotechnical design – Part 1: General rules published in Germany as DIN EN 1997-1:2009-09: Eurocode 7: Entwurf, Berechnung und Bemessung in der Geotechnik – Teil 1: Allgemeine Regeln.

(3) For application in Germany, the standard DIN EN 1997-1:2009-09 was complemented with additional rules by DIN 1054:2010-12: Baugrund – Sicherheitsnachweise im Erd- und Grundbau – Ergänzende Regelungen zu DIN EN 1997-1 (Ground – Verification of the safety of earthworks and foundations) and by the National Annex EN 1997-1/NA:2010-12: Nationaler Anhang – National festgelegte Parameter – Eurocode 7: Entwurf, Berechnung und Bemessung in der Geotechnik – Teil 1 (National Annex – Nationally determined parameters – Eurocode 7: Geotechnical design – Part 1: General rules). The three documents were implemented together and their application is binding in Germany.

(4) Complementing the European execution standards, national specifications were published and implemented in Germany under the heading “DIN SPEC”.

(5) DIN 1054:2010-12: Baugrund – Sicherheitsnachweise im Erd- und Grundbau – Ergänzende Regelungen zu DIN EN 1997-1 (Ground – Verification of the safety of earthworks and foundations) frequently refers to the Recommendations of the Piling Committee of the German Geotechnical Society (EA-Pfähle) e.g. for the application of values based on well established experience for pile resistances, for the static and dynamic load testing, concerning pile resistances under cyclic loads, etc.

(6) Where this English version of the Piling Recommendations makes reference to the above mentioned German standards DIN 1054 or the DIN SPEC,

the English speaking user should know, that all technical contents of the German complementing standards to Eurocode EC 7 and the European execution standards are also covered in these Recommendations.

(7) This means that the user of the Piling Recommendations (EA-Pfähle), when consequently applying, also fulfils the provisions of the complementing standards implemented in Germany.

(8) Technical terms and symbols used in these Recommendations, if in addition to those of DIN EN 1997-1 (Eurocode EC 7-1), are listed in Annex A1 and, as necessary, explained in this English version.

(9) In addition, the text can contain notes to supply the user with additional explanations for technical issues which are common in Germany or making reference to international procedures.

(10) The Piling Recommendations are technical rules representing good and generally accepted codes of practice. They are the result of technical and scientific co-operation on honorary basis of the members of the Piling Committee of the German Geotechnical Society.

Preface of the 2nd German edition

(1) Germany has a long tradition of standardisation with regard to the execution and design of piled foundations and individual pile systems. The German 'Piles' standardisation committee (DIN NA 005-05-07 AA) and the German Geotechnical Society's AK 2.1 'Piles' Working Group (hereafter called as the Piling Committee), have cooperated on these topics for many years, with members sitting in both bodies. In recent decades this joint committee has compiled the piling standards DIN 4026 (driven piles), DIN 4014 (bored piles) and DIN 4128 (grouted piles), and the piling section of DIN 1054 (Section 5), and adapted the individual editions to meet current best practice.

(2) Since European standardisation began at the end of the 1980s, the committee has been tasked with accompanying the development of the European execution standards EN 1536 (bored piles), EN 12699 (displacement piles) and EN 14199 (micropiles) as a national mirror committee. It was also tasked to supervise the edition of the German language versions of the execution standards, namely DIN EN 1536, DIN EN 12699 and DIN EN 14199. In terms of pile analysis and design, the committee focused on the piling sections of DIN 1054:2005-01 and DIN 1054:2010-12, paying particular attention to the partial safety factor approach.

(3) To supplement this work, the Piling Committee decided to produce summary recommendations for pile analysis and design, of which the first edition was published 2007 as 'EA-Pfähle' and which are presented in the second edition. The 'EA-Pfähle' sees itself in the tradition of similar DGGT (German Geotechnical Society) recommendations such as EAB (Empfehlungen Arbeitsausschuss Baugruben – Recommendations on Excavations), EBGEO (Empfehlungen für den Entwurf und die Berechnung von Erdkörpern mit Bewehrungen aus Geokunststoffen – Recommendations for Design and Analysis of Earth Structures using Geosynthetic Reinforcements), etc., which are now well-established as best practice regulations.

(4) With the publication of the Eurocode standards handbooks for structural engineering and the handbook for Eurocode 7, Geotechnical Design – Part 1: General Rules (1st edition 2011), European standardisation in this field has, for the time being, come to a conclusion. The standards were implemented as binding building regulations in Germany with the cut-off date 01.07.2012. The handbook Eurocode 7, Geotechnical Design – Part 1: General Rules, contains as summary DIN EN 1997-1:2009-09 (Eurocode EC 7-1: Geotechnical Design – Part 1: General Rules), DIN 1054:2010-12 (Ground – Verification of the Safety of Earthworks and Foundations – Supplementary Rules to DIN EN 1997-1) and DIN EN 1997-1/NA:2010-12 (National Annex).

(5) DIN 1054:2010-12 refers to 'EA-Pfähle' at various points dealing with pile analysis and design, e.g. for the tabled values for pile resistances based on

well established experience. For formal reasons DIN 1054:2010-12 refers to the first edition of 'EA-Pfähle'; however, it is only with the this second edition that the technical link to the EC 7-1 Handbook [44] has been properly accomplished thematically. This is especially the case for analysis methods and for terminology with regard to the EC 7-1 Handbook [44], meaning that coherent and coordinated pile foundation analysis and design regulations are now available to the user.

(6) In addition, design related issues in terms of the European pile execution standards DIN EN 1536, DIN EN 12699 and DIN 14199 are also dealt with in 'EA-Pfähle'. A comparative classification of the pile systems used in construction practice (Section 2) simplifies correlation between the standards. Detailed regulations on static and dynamic pile testing (Sections 9 and 10), and quality assurance guidelines and methods (Sections 11 and 12) aim to promote technically high-quality pile construction.

(7) A new Section 13, not included in the first edition, deals with load-bearing behaviour and analysis methods for piles subjected to variable actions. Particular attention was paid to the load-bearing behaviour of piles under cyclic loads, such as often occur in foundations for wind turbines (offshore, onshore), but also in highways engineering, etc.

(8) Whereas the first edition of 'EA-Pfähle' was initially understood to be a draft, and at the time was only recommended for trial use, the draft character is no more applicable on this second edition. In the interim, the piling community has had the opportunity to test the provisions and pass on their comments and proposals to the Piling Committee. This is also the case for the new Section 13, which was previously published in a variety of organs. All statements on the first edition were dealt with by the Committee when compiling the second edition and, if justified, included. Appendices A5, A6 and D contain "informative" technical guidelines, not yet classified as best practice, but instead representing the current scientific view.

(9) In terms of the compulsory nature of the present recommendations 'EA-Pfähle', 2nd edition, the user is referred to the "Notes for the User" published in EAB (2006), 4th edition, Ernst & Sohn, which are applicable in a similar manner here.

(10) The German Geotechnical Society's Piling Committee (AK 2.1 'Piles' Working Group) asks to send any suggestions and correspondence concerning further development of the Recommendations to the Chairman of AK 2.1 (see Imprint for address).

Hamburg, 2012 *Hans-Georg Kempfert*

Inhaltsverzeichnis

1 Introduction to the Recommendations and their Application Principles

1.1 National and International Regulations for Piling Works

(1) Since the implementation of DIN EN 1997-1:2009-09: Eurocode 7: Geotechnical Design – Part 1: General Rules, pile analysis and design in Germany is governed by

- Section 7 of Eurocode EC 7-1 (Eurocode 7), in conjunction with
- DIN 1054:2010-12: Subsoil – Verification of the Safety of Earthworks and Foundations – Supplementary Rules to the German version DIN EN 1997-1, and the
- National Annex to EC 7-1, namely DIN EN 1997-1/NA:2010-12: National Annex – Nationally Determined Parameters – Eurocode 7: Geotechnical Design – Part 1: General Rules.

These three coordinated documents are summarised in the German Eurocode 7 Handbook, Volume 1 [44].

Note: In case amendments or corrections are made to the standards included in the German Eurocode 7 Handbook, Volume 1 [44], the changes must be adopted even when not yet incorporated in [44].

(2) In addition, the individual pile systems are governed by the following execution standards:

DIN EN 1536:	Execution of special geotechnical works – Bored piles.
DIN SPEC 18140:	Supplementary provisions to DIN EN 1536.
DIN EN 12699:	Execution of special geotechnical works – Displacement piles.
DIN SPEC 18538:	Supplementary provisions to DIN EN 12699.
DIN EN 14199:	Execution of special geotechnical works – Micropiles.
DIN SPEC 18539:	Supplementary provisions to DIN EN 14199.
DIN EN 12794:	Precast concrete products – Foundation piles.
DIN EN 1993-5:	Design of steel structures – Part 5: Piling.

(3) Because diaphragm wall elements are often employed in the same way as pile foundations, the respective execution standard must also be considered:

DIN EN 1538:	Execution of special geotechnical works – Diaphragm walls

in conjunction with:

DIN 4126:	Stability analysis of diaphragm walls.

Editors: German Geotechnical Society (*Deutsche Gesellschaft für Geotechnik e.V.*)

(4) In addition, several ISO standards are being compiled for a number of special topics relating to piles. They are however not likely to be implemented as building regulations in Germany. Currently, these include:

DIN EN ISO 22477-1:	Geotechnical investigation and testing – Testing of geotechnical structures – Part 1: Pile load test by static axially loaded compression.

On a national basis, the regulations in Section 9 should be adopted for static pile testing.

1.2 Types of Analysis and Limit States using the Partial Safety Factor Approach

1.2.1 New standards generation and their application to pile foundations

(1) By European Commission decision national building design and execution standards either already have been or will in future be replaced by European standards. Actually numerous European standards have been published for geotechnical design and execution of special geotechnical works.

(2) The European standards governing pile execution are listed in 1.1.

(3) Analysis and design of pile foundations is dealt with in the European standard DIN EN 1997-1: Geotechnical Design (Eurocode 7) in conjunction with DIN 1054 and DIN EN 1997-1/NA, see 1.1. These three standards were implemented by the German Building Authorities for use in Germany as of 01.07.2012.

(4) Until the time of implementation of the Eurocodes as binding building regulations a new generation of national standards using the partial safety factor approach served as temporary solution for all fields of structural engineering. The following standards in particular represented the governing standards for pile foundations:

DIN 1055-100:2001-03:	Basis of structural design;
DIN 1054:2005-01:	Verification of the safety of earthworks and foundations;
DIN 18800:1990-11:	Steel structures and;
DIN 1045-2:2001-07:	Concrete, reinforced and prestressed concrete structures – Part 2: Concrete – Specification, properties, production and conformity – Application rules for DIN EN 206-1.

(5) These Recommendations on Piling (EA-Pfähle) are based on the standards listed in 1.1 above and, for design in particular, on Eurocode EC 7-1, in conjunction with DIN 1054 and the NA as stipulated in 1.1 (1).

1.2.2 Actions, effects and resistances

(1) The partial safety factor approach has its origins in probability theory, as used to specify the requisite safety factors from a probabilistic perspective. In contrast to this, with the implementation of DIN 1054:2005-01, the new geotechnical standards generation follows a more pragmatic split of the previously common global safety factors into partial safety factors for actions or effects, and partial safety factors for resistances.

(2) The basis for stability analyses is represented by the characteristic values for actions and resistances. The characteristic value, represented by the index "k", is a value of which the specified probability is assumed not to be exceeded or fallen short of during the reference period, taking into consideration the design working life of the structure or the corresponding design situation. Characteristic values are generally specified based on testing, measurements, analyses or empiricism.

(3) If the "internal" or "external" pile capacity needs to be analysed, the effects at the pile head or at given depths are required:

- As action effects, e.g. normal force, shear force, bending moment;
- As stresses, e.g. compression, tension, bending stress, shear stress or equivalent stress.

In addition, further effects of actions can occur:

- As dynamic or cyclic loads;
- As change to the structural element, e.g. strain, deformation or crack width;
- As change in the position of isolated piles or pile groups, e.g. displacement, settlement, rotation.

(4) The cross-section and the internal resistance of the material are the governing factors in the design of individual components. Specific standards are to be applied for this purpose.

(5) The characteristic values of the effects are multiplied by partial safety factors, those of resistances are divided. The values acquired in this way are designated as design values of effects or resistances respectively and are characterised by the index "d". For stability analyses, different limit states are distinguished, also see 1.2.3, 1.2.4 and 3.1.1 (4).

(6) In addition to actions, design situations are also taken into consideration for pile analyses, similar to other structural elements. To this end the previous loading cases LC 1, LC 2 and LC 3, adopted for use in analysis according to DIN 1054:2005-01, have been converted to design situations for use in analyses after DIN EN 1997-1 (EC 7-1) and DIN 1054:2010-12, and DIN EN 1990 as follows:

- BS-P (persistent (design) situation);
- BS-T (transient (design) situation) and;
- BS-A (accidental (design) situation).

In addition, there is the seismic design situation BS-E. More detailed information is given in the EC 7-1 German Handbook [44] and in [133].

1.2.3 Limit states and national application of the EC 7-1 German Handbook

(1) The term "limit state" is used with two different meanings:

a) In soil mechanics, the state of the soil in which displacements of individual soil particles against each other are so great that the mobilisable shear strength achieves its greatest values in either the entire soil mass, or at least in the region of a failure plane, is known as the "limit state of plastic flow". It cannot become greater even if more movement occurs, but could become smaller. The limit state of plastic flow characterises the active earth pressure, passive earth pressure, bearing capacity, "external" pile failure, slope stability, and global stability.
b) A second limit state in the sense of the new safety factor approach is a state of the load-bearing structure where, if exceeded, the design requirements are no longer fulfilled.

(2) The following limit states are differentiated using the partial safety factor approach:

a) The ultimate limit state is a condition of the structure which, if exceeded, immediately leads to a calculated collapse or another form of failure. It is referred to as "ULS" in EC 7 and DIN 1054. Further distinguishing is described in (5).
b) The serviceability limit state is a condition of the structure which, if exceeded, no longer fulfils the conditions specified for its use. It is referred to as "SLS" in EC 7 and DIN 1054.

(3) For ultimate limit state analysis (ULS), Eurocode EC 7-1 provides three options. With one exception (see (6) and (9)), the supplementary rules of DIN 1054 for use in Germany are based on Analysis Method 2 of EC 7-1. The partial safety factors are applied to both, effects and resistances. To differentiate this from the other permitted scenario, in which the partial safety factors are not applied to the effects but to the actions, this procedure is designated as Analysis Method 2* in [133], also see the EC 7-1 Handbook [44].

(4) The National Annex to EC 7-1 and DIN 1054 represent a formal link between EC 7-1 and national German standards, see EC 7-1 German Handbook [44]. In DIN 1054 and the National Annex it is stated which of the possible analysis methods and partial safety factors are applicable in the respective national domains. In addition, the applicable, supplementary national codes may also be given. The supplementary national codes may not contradict EC 7-1. Moreover, the National Annex may not repeat information already given in EC 7-1.

(5) Eurocode EC 7-1 distinguishes the following subordinate limit states of the ultimate limit state (ULS):

a) EQU: Loss of equilibrium of the structure, regarded as a rigid body, or the ground. The designation is derived from “equilibrium”.
b) STR: Internal failure or very large deformations of the structure or its structural elements, where the strength of the materials governs the resistance. The designation is derived from “structural failure”.
c) GEO: Failure or very large deformation of the structure or the ground, where the strength of the soil or rock governs the resistance. The designation is derived from “geotechnical failure”.
d) UPL: Loss of equilibrium of the structure or the ground due to buoyancy or water pressure. The designation is derived from “uplift”.
e) HYD: Hydraulic ground failure, internal erosion of or piping in the ground, caused by a flow gradient. The designation is derived from “hydraulic failure”.

(6) In the terminology of the EC 7-1 German Handbook [44] the GEO limit state is sub-divided into GEO-2 and GEO-3:

a) GEO-2: Failure or very large deformation of the ground in conjunction with the calculation of the action effects and the dimensions, i.e. when utilising shear strength for passive earth pressure, for sliding resistance and bearing resistance, and when analysing deep slide surface stability for anchored retaining walls, and for base resistance and skin friction of pile foundations. The GEO-2 limit state calculation follows Analysis Method 2*, see (3), as outlined in the EC 7-1 German Handbook [44].
b) GEO-3: Failure or very large deformation of the ground in conjunction with the analysis of the overall stability, i.e. when utilising the shear strength for analysis of the safety against slope failure and global failure and, normally, when analysing slope stabilisation measures, including consideration of structural elements, e.g. anchors or piles. The GEO-3 limit state calculation follows Analysis Method 3 as outlined in the EC 7-1 German Handbook [44].

(7) The EQU, UPL and HYD limit states describe the loss of static equilibrium. These include:

a) Safety analysis against overturning (EQU);
b) Safety analysis against buoyancy or uplift, e.g. of a tension pile group (UPL);
c) Safety analysis against hydraulic heave (HYD).

(8) The limit states EQU, UPL and HYD involve actions only, but no resistances.

The governing limit state condition is:

$$F_d = F_k \cdot \gamma_{dst} \leq G_k \cdot \gamma_{stb} = G_d; \qquad (1.1)$$

i.e. the destabilising actions F_k, multiplied by the partial safety factor $\gamma_{dst} \geq 1{,}0$, may only be as large as the stabilising action G_k, multiplied by the partial safety factor $\gamma_{stb} < 1{,}0$.

(9) The GEO-2 limit state describes the failure of structures and structural elements or the failure of the ground. It includes:

a) Analysis of the bearing capacity of structures and structural elements subjected to loads from the ground or supported by the ground;
b) Analysis of the bearing capacity of the ground to demonstrate that it is not exceeded, e.g. passive earth pressure, bearing resistance, pile resistance or sliding resistance.

The analysis to demonstrate that the bearing capacity of the ground is not exceeded is performed exactly as for any other construction material. The limit state condition is always the governing condition:

$$E_d = E_k \cdot \gamma_F \leq R_k / \gamma_R = R_d; \tag{1.2}$$

i.e. the characteristic effect E_k, multiplied by the partial safety factor γ_F for actions or strains, may only become as large as the characteristic resistance R_k, divided by the partial safety factor γ_R.

(10) The GEO-3 limit state is a peculiarity of earthworks and ground engineering. It describes the loss of overall stability. It includes:

a) Analysis of safety against slope failure;
b) Analysis of safety against global failure.

The limit state condition is always the governing condition:

$$E_d \leq R_d; \tag{1.3}$$

i.e. the design value E_d of the effects may only become as large as the design value of the resistance R_d. The geotechnical actions and resistances are determined using the design values for shear strength:

$$\tan\varphi'_d = \tan\varphi'_k / \gamma_\varphi \quad \text{and} \quad c'_d = c'_k / \gamma_c \quad \text{and, respectively} \tag{1.4a}$$

$$\tan\varphi_{u,d} = \tan\varphi_{u,k} / \gamma_\varphi \quad \text{and} \quad c_{u,d} = c_{u,k} / \gamma_c; \tag{1.4b}$$

i.e. the friction $\tan \varphi$ and the cohesion c are reduced from the outset using the partial safety factors γ_φ and γ_c.

(11) The serviceability limit state (SLS) describes the state of a structure or structural element at which the conditions specified for its use are no longer met, but without loss of its bearing capacity. The analysis is based on the anticipated displacements and deformations being compatible with the purpose of the structure.

1.2.4 Transitional regulations for applying of the Recommendations on Piling in conjunction with the EC 7-1 German Handbook

(1) These Recommendations on Piling are based on the provisions of the EC 7-1 German Handbook [44].

(2) A decisive difference between the provisions of the EC 7-1 German Handbook [44] and DIN 1054:2005-01 is the specification of different partial safety factors γ_P (lower) and the correlation factors ξ (higher) for pile foundations. Overall, however, γ_P and ξ result in a comparable magnitude on the resisting side, like DIN 1054:2005-01, see [59].

(3) The existing limit states to DIN 1054:2005-01 are replaced as follows in the EC 7-1 German Handbook [44]:

a) The previous GZ 1A limit state to DIN 1054:2005-01 corresponds without restrictions to the EQU, UPL and HYD limit states as in the EC 7-1 German Handbook [44].
b) The previous GZ 1B limit state to DIN 1054:2005-01 corresponds without restrictions to the STR limit state as in the EC 7-1 German Handbook [44] as the "internal" pile capacity (material strength). In addition, there is the GEO-2 limit state as in the EC 7-1 German Handbook [44], in conjunction with the "external" design of the foundation elements, e.g. the "external" pile capacity.
c) The previous GZ 1C limit state as in DIN 1054:2005-01 corresponds to the GEO-3 limit state as in the EC 7-1 German Handbook [44], in conjunction with the analysis of the overall stability, i.e. when utilising the shear strength for analysis of safety against slope failure and global failure.

1.3 Planning and Testing Pile Foundations

(1) These Recommendations use the terms "designer" (planner) as the EC 7-1 German Handbook [44] and the EC 7-2 German Handbook [45].

(2) The designer should consult a specialist pile foundation planner if not in possession of the requisite knowledge and experience for pile foundation planning.

(3) A geotechnical design report must be compiled for pile foundations in accordance with the EC 7-1 German Handbook [44], which must make reference to the ground investigation report in accordance with the EC 7-2 German Handbook [45].

(4) Pile foundations are classified as either Geotechnical Category GC 2 or Geotechnical Category GC 3. See 3.2 and the EC 7-1 German Handbook [44] for classification criteria.

(5) Where pile foundations are classified as Geotechnical Category GC 3, a geotechnical expert with requisite experience must be consulted during the

structural engineering appraisal of the ground investigation report and the geotechnical design report. In particular, the geotechnical expert should verify the ground model, the soil parameters and the design approach, and, in agreement with the independent verifier, should carry out comparative analyses.

(6) The execution of pile foundations should be supervised by a specialist planner or a geotechnical expert with the requisite experience and specialist knowledge of pile foundations. Where pile foundations are classified as Geotechnical Category GC 3, the expert mentioned in (5) must also be consulted to check the detailed design and to supervise the pile execution works.

2 Pile Systems

2.1 Overview and Classification into Pile Systems

(1) The available pile systems, highly variable in their structure and their application options, differentiate between three groups in accordance with the respective execution standards, see Figure 2.1:

a) **Bored piles** to DIN EN 1536 and DIN SPEC 18140:
Bored piles are characterised by ground being loosened and transported during installation. The excavated ground volume can correspond to the total or part of the pile volume.

DIN EN 1536 applies to vertical and inclined bored piles with rakes up to 4 : 1 (3 : 1 if the casing remains in place), and diameters between 0,3 m and 3,0 m, and a depth to diameter or width ratio ≥ 5.

The system adopted in the standard differentiates bored piles according to the type of support of the borehole and the methods used for excavation, concreting and reinforcement installation. The combinations are summarised in Table 2.1.

- Unsupported excavation is suitable for installing piles in stable ground (for the duration of pile installation). These piles are generally bored using intermittent excavation methods (Kelly operated augers or buckets or with cable grabs and shells) and concreted in dry conditions. The reinforcement can be placed before or after concreting. The piles can be underreamed to form enlarged bases or, in special cases, be provided with shaft enlargements.
- Cased excavation is suitable for installing piles in unstable ground and for boring below the groundwater table. Excavation is either intermittent as described above, or continuous, e.g. using continuous flight augers or flushing for soil removal (the latter is less common for pile construction in Germany). The piles can be installed with enlarged bases and, for greater depths, be telescoped, i.e. with stepwise reduced diameter at increasing depth. The concrete is placed using a tremie pipe. The concrete may only be placed under the same conditions as for dry concreting if groundwater ingress to the bore can be ruled out. The reinforcement can be placed before or after concreting. The casing is either temporary and is removed during concreting, or can be required permanently, e.g. in free water, where groundwater flow is strong or high groundwater or ground aggressivity to concrete.
- Excavation under a supporting fluid is possible for the same ground conditions as for cased excavation and generally the same excavation tools are employed. In addition to circular piles, rectangular, T, L or cruciform section piles can be constructed. Non-circular piles (barettes) are constructed using diaphragm wall techniques. Stability analysis for

Recommendations on Piling – EA Pfähle.
Editors: German Geotechnical Society (*Deutsche Gesellschaft für Geotechnik e.V.*)

the fluid supported borehole and quality control for the fluid must be performed as for diaphragm wall methods

- Soil supported piles, i.e. continuous flight auger piles with soil-filled flights. Augers with large and small hollow stem diameters are differentiated, also see 2.2.1.4. During boring, the auger helix fills with loosened ground material. This material supports the borehole wall and is transported upwards on the helix flights. If augers with small diameter hollow stems are employed, the concrete is pumped through the hollow stem as the auger is extracted. The reinforcement cage is inserted after concrete placement into the fresh concrete and the auger is completely withdrawn. Excavation material accumulated on the auger flights is removed while the auger is retracted. When using augers with large diameter hollow stem, the reinforcement cage is generally inserted into the hollow stem after completion of excavation and prior to concreting. Piles installed using a large diameter hollow stem are also known as "partial displacement piles", because only part of the volume of the excavated ground is transported to the surface and the remainder is displaced into the surrounding ground. The partial displacement pile subgroup includes those systems with boring tools possessing a continuous auger only along their bottom lengths. When boring into load-bearing ground, excavated material can be at least partially redistributed to the overlying strata. When retracting the auger, the soil material remaining on the helix flights is transported to the surface.

b) **Displacement piles** to DIN EN 12 699 and DIN SPEC 18538:
Displacement piles are characterised by being installed without excavation or removal of soil material – except to limit heave or vibrations, to remove obstacles or as an installation aid.

The soil accumulating at the surface when installing displacement piles is the result of the displacement effect and the lack of soil surcharge close to the surface. It does not represent relevant soil removal.

The minimum pile diameter or corresponding cross-sectional dimensions are currently specified in DIN EN 12699 as 150 mm.

Note: This limit will be dispensed within the current revision of this standard.

The displacement pile group differentiates:

- Prefabricated piles consisting of reinforced concrete, prestressed concrete, steel, cast iron or timber;
- Driven cast-in place concrete piles;
- Grouted displacement piles;
- Screw piles (full displacement piles). These are pile systems in which the drive tube enters the ground through rotation and downward pressure by means of a displacement head located at the lower end, without

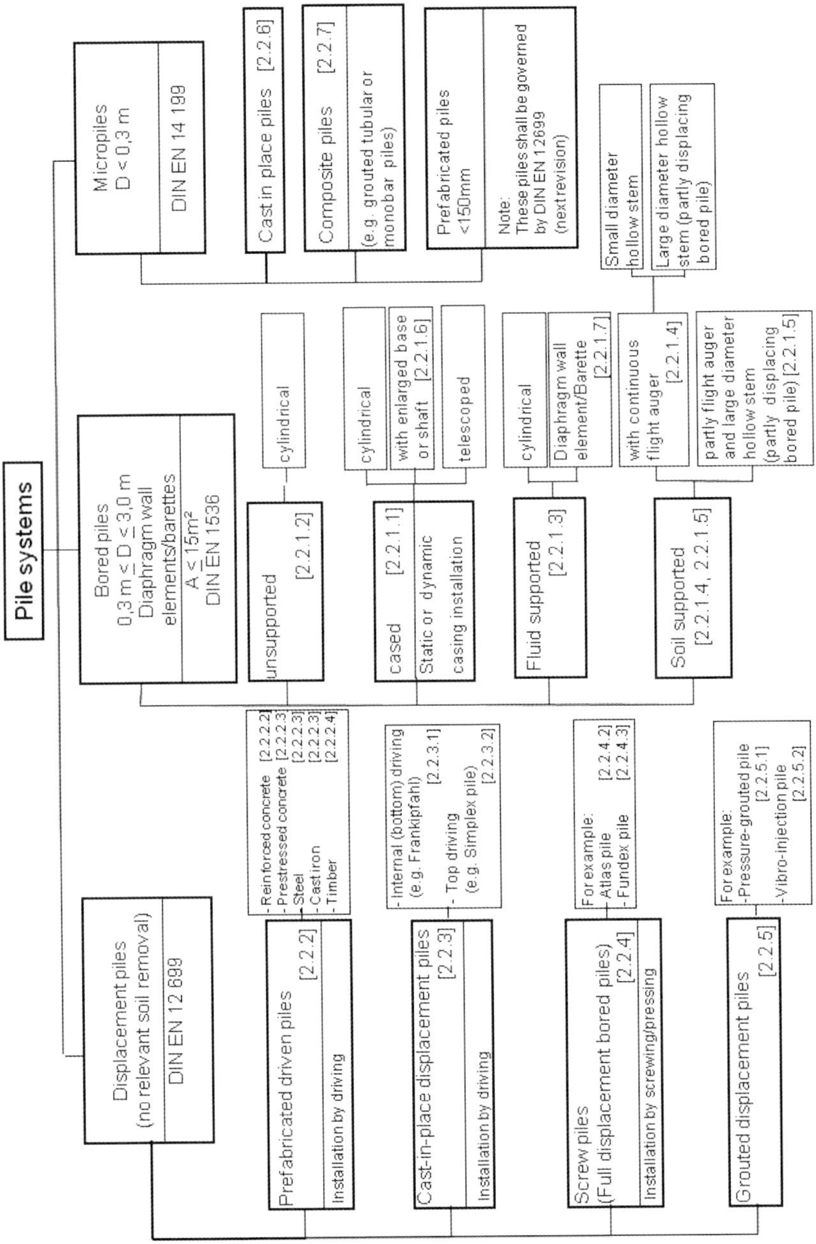

Figure 2.1 Overview of pile systems standardised in the pile execution standards DIN EN 1536, DIN EN 12 699 and DIN EN 14 199

relevant soil transport or redistribution from the embedment zone in the load-bearing stratum into the overlying strata.

Note: During boring, displacement pile systems employing boring tools possessing more than one helix in the base region can redistribute soil out of the load-bearing stratum and into the overlying stratum. When retracting the boring tool, the soil material remaining on the helix is extracted from the bore. These effects must be taken into consideration in the design. They can result in a reduction of bearing capacity compared to “full displacement piles”. No empirical pile resistance data are given in Section 5 for these pile systems.

c) **Micropiles** to DIN EN 14199 and DIN SPEC 18539:
At the current stage of standardisation the diameter range of bored micropiles is less than 0,30 m, and less than 0,15 m for displacement piles. The load transfer from micropiles to the surrounding soil is achieved through skin friction mainly.

This group differentiates:

- Cast-in-situ piles with small reinforcement cages and using fine-grained concrete or grout, for which cased or slurry-supported pile construction methods are generally employed;
- Composite piles, which possess a tendon or tubular bearing element encased in a cement grout sleeve and;
- Precast piles (< 0,15 m) with or without subsequent grouting.

Note During the current revisions of EN 12699 and EN 14199 the diameter restriction shall be abolished for displacement and all displacement piles shall be covered by EN 12699. Consequently, DIN EN 14199 shall in future cover only bored micropiles.

2.2 Pile Construction

2.2.1 Bored piles

2.2.1.1 Cased bored piles

(1) When constructing cased bored piles, the soil is loosened and extracted within a casing. Reinforcement can then be placed and concrete is cast into the temporary, cased void. Installation complies with DIN EN 1536, also see [80] and [136].

(2) The casing can be installed in a number of different ways:

- By rotation either back and forth (oscillation) or in one direction (rotation) together with axial advance using a casing rig or casing rotator;
- Using the rotary head of a rotary rig;
- Using a piling or a vibrating hammer.

(3) For large depths and high skin friction values, a telescopic casing sequence can be employed. Boring is started using the largest diameter. Before reaching the capacity of the casing rig, the first casing is held in position and boring continues using a smaller diameter casing.

(4) In unstable ground the casing shall be installed in advance of the excavation. Only in firm, or at least temporarily stable, ground excavation may advance ahead of the casing. At its bottom, the casing is equipped with a cutting shoe with special, hardened teeth, which overbreak the bore slightly to reduce friction on the casing wall.

(5) The ground can be loosened and extracted in different ways:

- Discontinuously, using rotary excavation methods and rotary rigs with Kelly bar-driven augers, drilling buckets and coring tools;
- Discontinuously using grabbing methods with grabs, gravel pumps or chisels operated with cable cranes;
- Continuously using continuous flight auger (double rotary drilling method);
- Continuously using a drag or roller bit and flush boring.

(6) Using double rotary drilling, the boring tool is advanced into the ground together with the casing whilst being constantly rotated, whereby the soil is loosened by the augering tool and conveyed upwards on the helix flights.

(7) When using boring methods with direct or reverse flushing, the ground is constantly loosened by the roller or drag bit. Material is transported by means of air and/or water flushing. These boring methods are widespread in Germany for installing deep wells or for geothermal energy bores, but rarely for the construction of piles.

(8) The principles described in 11.2.1 shall be followed in order to avoid loosening of the surrounding ground. A reinforcement cage is generally inserted after reaching the final depth, and the pile is concreted using a tremie pipe while simultaneously the casing is withdrawn. However, the reinforcement cage can also be inserted after concreting into the fresh concrete column (e.g. when using the double rotary drilling method).

(9) Special cutting tools are used to produce enlarged pile bases. As they cannot be used inside casings, enlargements can be constructed only in stable ground or ground supported otherwise in the zone of the pile base.

(10) Cased bored piles are constructed with shaft diameters up to 2,5 m, in special cases even larger. The standard range lies between 0,6 m and 1,8 m. Depths up to approximately 60 m can be achieved, depending on the ground conditions and pile diameter. Measures as described in (3) can be necessary for greater depths.

(11) The characteristic pile resistances in the serviceability limit state are in the order of 1 MN to 10 MN, depending on diameter and ground conditions. They can be higher for larger diameters and favourable ground conditions.

(12) Because of their high flexural rigidity, large diameter piles, in particular, can transmit considerable horizontal loads in suitable ground.

(13) Cased bored piles can be installed in all soil types including rock (with limitations). Obstacles can be penetrated using special tools (chisels, rock augers and buckets, core barrels).

2.2.1.2 Unsupported excavations

(1) In sufficiently stable ground bored piles can be constructed without a casing or other bore support.

(2) In such cases, a protective casing is only necessary for stabilisation of the top of the bore and for guidance of the tools.

(3) If unstable strata are penetrated, this length of the bore shall be supported.

(4) This installation method should only be employed for vertical and raked piles up to an inclination of 15 : 1.

2.2.1.3 Fluid-supported excavations

(1) Bored piles constructed under a fluid support can have either circular cross-sections or cross-sections composed of rectangles. The latter are also known as barettes. Bentonite suspensions are generally employed as support fluids to stabilise the bore walls.

(2) When work commences, a lead-in tube (circular) is placed or a guide wall (barette) installed. The soil is then excavated under the support fluid. The fluid level within the bore must at all times remain at a level within the leader pipe or the guide walls, such that sufficient positive pressure is exercised on the bore wall. The stability of the fluid-supported bore shall be analysed compliant with DIN 4126. A reinforcement cage is generally inserted after reaching the final depth, and concrete is subsequently placed by tremie method.

(3) Circular cross-sections are generally excavated using the discontinuous boring tools described above. Diaphragm wall grabs or hydro fraises are generally used to construct non-circular cross-sections.

(4) Circular piles are executed with diameters up to approximately 3 m or as rectangular barettes (see 2.2.1.7) with widths between 0,4 m and 2,0 m, and lengths between 2,4 m and 7 m. Cross-sections composed of rectangles, e.g. T-sections, are also possible.

(5) Due to the large possible cross-sectional dimensions, these piles can support very large vertical and horizontal loads. Pile lengths in excess of 60 m are possible.

(6) The method is also suitable for construction of primary supports (e.g. plunged columns) when using top-down execution methods. These are vertical structural elements which support floor slabs and building loads when struc-

tures or basements are built downwards without excavation pits. The primary supports consist of cast-in-place piles below the basement level and prefabricated steel, reinforced concrete or composite columns within the basement.

2.2.1.4 Soil-supported, continuous flight auger bored piles

(1) Soil-supported, continuous flight auger bored piles are constructed by rotating a continuous flight auger into the ground and pumping in concrete as the auger is retracted. The piles can be installed with or without reinforcement. Installation is vibration-free.

(2) These pile systems are also known as auger piles. Augers with large and small hollow stem diameters are differentiated. Augers with small hollow stems generally have internal vs. external hollow stem diameters of $D_i/D_a < 0,4$ and such with large hollow stems of $D_i/D_a > 0,6$ (partial displacement bored piles).

(3) Common auger diameters range between 0,4 m and 1,2 m.

(4) Continuous flight augers piles with large diameter hollow stems are also known as partial displacement piles, also see 5.4.7 and 11.2.6. These piles laterally displace some of the soil. The extracted soil volume should not exceed approximately 70% of the pile volume.

(5) During boring, the bore wall is supported by the soil being transported upwards on the auger helix. The advance speed and rotation speed of the auger must be adapted to suit the ground conditions in order to limit soil extraction and thus retain the ground support.

(6) During boring, the hollow stem must be sealed, e.g. by an expendable end plate to ensure it remains free of soil material and water.

(7) When using augers with small hollow stem diameter and resultant geometrical constraints, a reinforcement cage can only be inserted into the fresh concrete after the pile is concreted. In contrast, when using augers with a large hollow stem diameter, a reinforcement cage can be installed inside the hollow stem prior to concreting.

(8) To prevent soil transport from the auger downwards into the fresh concrete, the auger should be retracted without rotating, or only rotated slightly in the same direction as used for boring.

(9) In accordance with DIN EN 1536, continuous flight auger bored piles should not be installed in uniform, non-cohesive soils ($d_{60}/d_{10} < 1,5$) below the groundwater table, or in loose, non-cohesive soils with densities $D < 0,3$ or in soft, cohesive soils with undrained shear strengths $c_u < 15$ kN/m², unless the feasibility of the installation method has been demonstrated on test piles or through local experience before commencing works. The standard also points out that uniform, non-cohesive soil where $1,5 < d_{60}/d_{10} < 3,0$ can be sensitive when situated below the groundwater table.

(10) The characteristic pile resistances in the serviceability limit state are in the order of 0,5 MN to 2 MN, depending on the pile diameter and ground conditions.

2.2.1.5 Soil-supported, partial flight auger bored piles

(1) Partial flight augers piles and large diameter hollow stem piles are also known as partial displacement piles, also see 5.4.7.

2.2.1.6 Bored piles with enlarged bases

(1) It can be expedient to enlarge the pile base if the capacity of the pile relies entirely or predominantly on the base resistance, or if pile bases rest on a load-bearing stratum and/or only slightly embedded in the stratum. In addition, an enlarged base allows improved utilisation of concrete strengths, and thus allows material savings in the pile shaft, because the adopted base resistances are generally considerably lower than the concrete strengths.

(2) Ground improvement measures executed around the base region of a pile, such as jet grouting, deep vibration or similar methods are not understood as enlarged bases.

(3) Enlarged bases can only be installed in stable soils or soils that can be suitably stabilised for the purpose, e.g. using bentonite suspension or cement slurry.

(4) Construction of enlarged bases is generally only possible for vertical piles and in cased pile bores, and requires special belling-out tools; these include special, extended belling buckets attached to the Kelly bar and fitted with retractable reaming wings. The reaming wings can generally be extended by the advance of the Kelly bar in conjunction with a scissor mechanism inside the belling bucket. The rate of advance correlates to the extension of the wings and can therefore be controlled. For structural reasons, the reaming wings are arranged slightly above the base of the belling bucket, and this section of the bucket is designed to accept the loosened spoil. The shape of an enlarged base therefore corresponds to a truncated cone with a spherical base and a lower, cylindrical protrusion.

(5) When constructing an enlarged base, boring is carried out first using one of the tools described in 2.2.1.1 (e.g. a standard drilling bucket), such that the reaming wings of the belling-out bucket can later operate from the design bottom level of the base enlargement. The casing is then retracted as far as the top of the reaming wings and, if necessary, the now uncased lower section of the bore stabilised. The Kelly bar with the reaming tools is then inserted into the bore, and the enlarged base (bell) is produced by rotating the bucket and conically milling the uncased bore wall. Depending on the required size of the enlargement and the amount of milled material present in the lower section of the bucket, the reaming tool must be retracted and emptied, and the enlarging process continues repeatedly.

(6) The bore is cleaned after enlarging, the reinforcement inserted and the pile concreted. The enlarged section must be concreted without interruption up to the inside of the casing, as there would be otherwise a danger of entrapment of soil and other materials.

(7) In case of the enlarged base the conical section protruding outside the diameter of the cylindrical pile shaft is left unreinforced. Load distribution from the shaft into the base area is achieved solely by shear strength of the concrete. This load distribution and the stabilisation of the overhanging bell wall are the reasons for the geometric limitations of the bell and the restriction in the base diameter in relation to the shaft diameter.

(8) See 5.5 for the base resistances for piles with enlarged bases.

2.2.1.7 Diaphragm wall elements/barettes

(1) Diaphragm wall elements/barettes are bored piles compliant with DIN EN 1536, if:

- They are concreted in a single process;
- They have square, rectangular, T- or L-shaped, or similar cross-sections;
- The shortest side measures $W_i \geq 0{,}4$ m;
- The ratio of the largest dimension L_i to the smallest dimension W_i of the element is $W_i/L_i \leq 6$;
- The smallest dimension of precast concrete elements installed within diaphragm wall elements measures $D_p \geq 0{,}3$ m or $W_p \geq 0{,}3$ m (D_p being the diameter of a round precast element and W_p the smaller side of a rectangular precast element) and;
- The overall cross-sectional area is smaller than 15 m^2.

2.2.2 Prefabricated driven piles

2.2.2.1 Introduction

(1) Prefabricated driven piles to DIN EN 12699 comprise prefabricated pile elements of reinforced concrete, prestressed concrete, steel or timber, which completely displace the soil when driven or pressed into the ground, thereby generally displacing and compacting it. The piles are manufactured off-site with full length or in sections, transported to the site and driven using suitable equipment. Prefabricated driven piles can be installed vertically or at rake. Depending on the required foundation depth, piles delivered in sections must be extended using couplings or by welding.

(2) As a result of the displacement effect and the method of installation, which acts like dynamic or static pile loading, only minor settlement or heave of the completed pile is required to mobilise the pile's external capacity.

2.2.2.2 Precast driven concrete piles

(1) Precast driven reinforced concrete piles mostly have square cross-sections of 20 × 20 cm to 45 × 45 cm. Less frequently spun concrete piles are employed as round piles of similar dimensions but with a hollow cross-section. The piles are reinforced or prestressed for loads acting during transport and installation, and for the loads of the superstructure (compression, tension, bending). For usual applications, precast concrete piles are produced in lengths between 6 m and 15 m, in exceptional cases up to approximately 25 m. For pile lengths greater than 15 m it is more practical to join sections using tested and approved steel couplings. As result, the piles can be extended almost without limit; precast driven concrete piles up to a length of 80 m have been installed. Special purpose piles can be fitted with embedded grouting tubes for postgrouting (e.g. to increase capacity).

2) Today, precast driven concrete piles are generally percussion driven (hydraulic hammer, diesel hammer, drop hammer). A driving helmet containing a buffer material such as wood, plastic or similar material is located between the driving ram and the pile. Because of the mechanical loads imposed during driving, concrete C 50/60 or higher is used, which is suitable for exposure classes XC 4 to XS 3. The piles are reinforced and contain horizontal links along their whole length to resist the driving forces (compression, tension splitting). About 1 m lengths at both pile or section ends, the reinforcement links are at closer intervals. The piles can be manufactured with a driving shoe, but normally the head and the base are both formed with similarly flat ends. If driving is in hard ground or rock, a steel tip can be cast in the concrete; this is generally necessary for spun concrete piles.

(3) The characteristic pile resistances of precast concrete driven piles in the serviceability limit state are in the order of 0,5 MN to 2 MN, depending on the cross-section and ground conditions.

2.2.2.3 Prefabricated driven steel and cast-iron piles

(1) Steel piles are formed from either H-sections, steel tubes, joined sections or sheet piles with a variety of cross-sections and wall thicknesses. For increase of base resistance, reinforcement of pile bases such as plates or wings are welded on as required.

(2) Driving is in general like that for concrete piles. For hard driving and to prevent bulging and buckling, the pile head is also reinforced. For reasons of economy, steel piles are mainly used where their comparatively high tensile strength and durability can be properly exploited, for example in offshore or harbour facilities, occasionally also where high tensile loads need to be transferred.

(3) The characteristic pile resistances of steel piles in the serviceability limit state are in the order of 0,5 MN to 2 MN, depending on the cross-section and ground conditions.

(4) Prefabricated driven cast-iron piles, also referred to as ductile piles, consist of ductile, spun cast pipes, driven to the load-bearing stratum with the aid of a hydraulic hammer. Because of the low mass of the cast-iron pipes, they can be installed using smaller equipment.

(5) The generally 5 m long pipe segments are fitted with conical sleeves, allowing them to be connected to make up the required pile length during driving. The driving process creates a rigid, frictional bond within the sleeves. Special care must be taken to ensure that the pipe sections are aligned axially to prevent damage during the driving process.

(6) The diameter of the cast-iron pipes is D = 118 mm, with wall thicknesses of 7,5 mm, 9,0 mm and 10,5 mm, and D = 170 mm with wall thicknesses of 9,0 mm and 10,5 mm.

(7) Ductile piles can be constructed with or without grouting.

(8) If executed without grouting, the leader pipe driving shoe is of the same diameter as the cast-iron pipe. This pile type is generally used as an end-bearing pile on very firm bearing strata.

(9) If the pile is executed with shaft grouting, an oversize pile shoe projecting in radius approximately 40 mm is used. During driving, cement grout or fine-grained concrete is permanently pumped to the pile base through the inside of the pipe and fills the annulus formed by the oversize pile shoe. This pile type is generally used in non-cohesive and cohesive ground.

(10) The characteristic pile resistances of ductile piles in the serviceability limit state are order of 0.5 MN to 1,1 MN, depending on the cross-section, kind of construction and ground conditions.

2.2.2.4 Prefabricated driven timber piles

(1) Timber piles only play a subordinate role in today's construction practice. They are predominantly used for temporary construction measures, e.g. as foundations for falsework, etc. Because of the disadvantages presented in terms of material and the comparatively low bearing capacity, they are very rarely used as foundations for structures. Timber piles are also employed for aesthetic reasons, e.g. as retaining walls along river banks or lake shores, for terracing terrain in landscaping, or for light foundations, e.g. in nature conservation areas.

(2) Timber piles are manufactured as round piles with diameters between 15 cm and 35 cm and can be reinforced at the head (driving band) and base (shoe) for driving. Normally, light driving rigs with drop weights up to 1 t maximum are used with timber piles.

(3) The characteristic pile resistances of timber piles in the serviceability limit state range from 100 kN to 600 kN, depending on the cross-section, length and ground conditions.

2.2.3 Cast-in-place concrete piles

2.2.3.1 Cast-in-place concrete piles with internal driving tube (Franki pile)

(1) The Franki pile is a cast-in-place concrete pile with internal tube driving to DIN EN 12699 and is installed with an enlarged pile base.

(2) The drive tube is driven into the ground by means of free fall impact driving within the tube (internal tube driving). Before beginning the driving process, a plug is formed at the base of the driven tube, consisting either of an almost dry concrete (semi-dry concrete), or of a sand or gravel-sand mixture. The plug material is compacted through the tamping of the hammer and blocks the tube. This plug fulfils a dual purpose. First, it serves as a driving cushion onto which the drop hammer impacts and drags the tube into the ground. Second, it seals the bottom of the tube and prevents water and soil ingress. Once the final depth is reached, the drive tube is held in suspension by cables fixed to the leader, and the enlarged base installation begins by driving out the plug material. Depending on the local ground conditions and the planned pile capacity, additional material can be filled in and be driven out and the pile base be enlarged, until the requisite design base volume is achieved.

(3) A reinforcement cage is then generally installed and the tube filled in dry conditions with concrete of high plasticity. The pile shaft is formed by withdrawing the tube and filling-up with concrete.

(4) Pile installation can be modified prior to concreting for the purpose of ground improvement in both the pile shaft and base areas, and for enhancing pile capacity.

(5) For ground compaction in the base area, the tube is driven 1 m to 2 m below the planned pile base level, filled with gravel or crushed stone and withdrawn approximately 2 m to 4 m while tamping and compacting the filled-in material. The tube is then driven back through the compacted material and to the planned final depth. Penetration resistance is increased considerably compared to the first driving, reflecting the ground improvement achieved. If necessary, further gravel compaction can take place until the requisite degree of improvement is achieved. Subsequent pile base and shaft installation follows as described in (2) and (3).

(6) Franki piles are constructed with tube diameters of 42 cm to 71 cm. They can be manufactured either as vertical piles or raked up to 4:1.

(7) The characteristic pile resistances in the serviceability limit state are in the order of 1 MN to 6 MN, depending on ground conditions and pile diameter.

2.2.3.2 Cast-in-place top-driven piles (e.g. Simplex piles)

(1) The cast-in-place top-driven pile is a displacement pile to DIN EN 12699.

(2) Using this method, the drive tube is driven into the ground by means of top-driving by a diesel, hydraulic or vibrating hammer. The tube is sealed at

the base by a sacrificial base plate. Once the final depth is reached, generally a reinforcement cage is placed and the dry bore filled with high plasticity concrete. When the tube is withdrawn, the base plate disengages and the concrete can flow out. The pile is completed by filling-up the tube with concrete while the tube is gradually withdrawn. The base plate remains in the ground and forms the base of the pile.

(3) Ground improvement by gravel compaction can also be achieved using this method. The drive tube with the sacrificial base plate is driven to 1 m to 2 m below the planned pile base, gravel is filled in and the tube is extracted to the working level. The loose gravel remains and fills the exposed void. The tube is then sealed again with a base plate and driven through the gravel column to the required depth. Pile installation then follows as described in (2).

(4) Simplex piles are generally executed in diameters between approximately 40 cm and 60 cm. They can be constructed either as vertical or raking piles up to 4:1.

(5) The characteristic pile resistances in the serviceability limit state are in the order of 0,5 MN to 2,5 MN, depending on ground conditions and pile diameter.

(6) Bottom-driving for formation of enlarged pile bases can also be employed for cast-in-place top-driven piles. Two varieties are differentiated. For the first option, the drive tube is withdrawn for approximately 2 m once the final depth is reached, the reinforcement cage is installed, and the high plasticity concrete is placed. The drive tube is then completely filled with concrete, sealed at the head and driven back to the original depth using top-driving.

(7) In the second option, the tube is top driven to final depth with a diesel or hydraulic hammer and filled with semi-dry concrete. Then a tamping piston is placed on the concrete inside the drive tube, and the piston and the drive tube are top-driven to the original depth, forming the enlarged base of semi dry concrete. Pile construction then continues as described in (3).

(8) The empirical pile base resistance and pile skin friction data given in 5.4.5.2 may not be adopted for the pile systems described in (6) and (7).

2.2.4 Screw piles (full displacement bored piles)

2.2.4.1 Introduction

(1) Screw piles are displacement piles to DIN EN 12699.

(2) The drive tubes of screw piles are screwed into the ground without dynamic impacts. For this reason, and because of the full displacement, the embedment depth of these piles in dense or hard ground layers is limited. They are installed vibration-free.

2.2.4.2 Atlas piles

(1) The Atlas pile is constructed by screwing a drive tube with a displacement body carrying a single-flight auger into the ground. A powerful rotary drilling head and down-thrust is used and the ground is fully displaced. The displacement body has a watertight seal at the base formed by a sacrificial tip.

(2) When the target depth is reached, generally a reinforcement cage is placed, and the tube and its top mounted hopper are filled with high plasticity concrete. By unscrewing and withdrawing the tube, the sacrificial tip is lost and the concrete flows out, filling the void formed by the displacement body. Because of the single-flight displacement body, the finished pile shaft possesses a helical, concrete bulge and the completed pile resembles a screw.

(3) Because the pile is constructed using static forces only, the method is vibration-free.

(4) The diameter of the pile shaft depends on the size of the replaceable displacement body. Common diameter combinations are 31/41 cm and 51/56 cm. The first value gives the minimum diameter of the required concrete cross-section (governs the internal capacity), the second value the external diameter of the helical concrete body (governs the external capacity).

(5) The piles can be constructed vertically or raking up to 4:1.

(6) Common characteristic pile resistances in the serviceability limit state are in the order of 0,5 MN to 1,7 MN, depending on ground conditions and pile diameter.

2.2.4.3 Fundex piles

(1) For construction of Fundex piles, a smooth drive tube with a diameter of 38 cm or 44 cm, sealed by a sacrificial helical tip, is rotated and pressed into the ground. The ground is fully displaced. Common diameter combinations are 38/45 cm and 44/56 cm (the second value indicates the diameter of the tip).

(2) Once the final depth is reached, generally a reinforcement cage is placed in the drive tube and the tube filled with high plasticity concrete. The tube is then withdrawn under oscillation, whereby the tip disengages from the tube. It remains in the ground and forms the pile base.

(3) The piles can be constructed vertically or raking up to 4:1.

(4) The entire pile installation is vibration-free.

(5) Common characteristic pile resistances in the serviceability limit state are in the order of 0,5 MN to 1,5 MN, depending on ground conditions and pile diameter.

2.2.5 Grouted displacement piles

2.2.5.1 Pressure-grouted piles

(1) A pressure-grouted pile is a driven steel pile with a specially formed base, which is driven into the ground while adding grout. This term also covers what were previously referred to as MV- or RV-piles, which were patented driven grouted piles.

(2) A wedge-shaped enlarged pile shoe of rectangular or square cross-section, welded to the pile base, is characteristic for this pile type. During driving, the shoe forms an annulus around the steel shaft, which is filled with cement grout during the installation process.

(3) Contrary to micropiles to DIN EN 14199, the annulus is open to atmosphere, and the grout is applied with hydrostatic pressure only. The grouting process should be controlled to ensure that the piles are completely enclosed in grout.

(4) Pressure grouted piles are predominantly employed as tension piles. Common characteristic pile resistances in the serviceability limit state are in the order of 1 MN to 2,5 MN, depending on ground conditions and pile cross-section.

2.2.5.2 Vibro-injection piles

(1) Vibro-injection piles have a similar construction principle like pressure-grouted piles. By using a slightly enlarged pile base relative to the pile shaft, a void is formed, which is continuously filled with grout under pressure while the pile is driven by vibration.

(2) The base of the vibro-injection piles consists of the steel section with steel plates welded to all sides. During installation, a narrow gap is created between the pile shaft and the ground around the entire circumference, and filled with grout during vibration. In contrast to this, the enlarged base of the pressure-grouted pile (2.2.5.1) displaces a much larger volume of ground.

(3) The vibro-injection pile therefore has a considerably lower penetration resistance, allowing it to be placed by vibration. The cement grout is pumped in through injection tubes, and thus guarantees intimate contact with the ground.

(4) Thanks to the vibration method, the vibro-injection pile produces far lower noise emissions than driven piles. The effects of vibration on the surroundings and neighbouring structures must, however, be taken into due consideration.

(5) Limitations are installation-related and exist for very dense sands and gravels. Installation aids (low-/high-pressure flushing) may be expedient. Limitations exist also in cohesive soils, depending on the consistency.

(6) Vibro-injection piles are predominantly employed as tension piles. Common characteristic pile resistances in the serviceability limit state are in the

order of 0,5 MN to 1,5 MN, depending on ground conditions and pile cross-section.

2.2.6 Micropiles

(1) Micropiles are defined as piles with diameters smaller than 0,3 m and are controlled by DIN EN 14199.

(2) Among others, they include the previously common mini-piles and the monobar or tubular grouted piles predominantly employed in the last 20 years.

(3) Their advantages are that they can be installed using compact rigs even under limited conditions, and installation is generally low-noise and low-vibration.

(4) Transfer of forces to the surrounding ground is achieved by grouting with concrete or cement grout. The following are differentiated:

a) Cast-in-place pile containing continuous, longitudinal, steel reinforcement. They can be constructed using concrete or cement grout. The minimum required shaft diameter is 150 mm.
b) Composite pile, characterised by a steel bearing element with a hardened cement grout cover. The bearing element is inserted in a bore.

(5) Percussion and rotary boring methods using internal and external flushing, driving and vibratory methods are suitable for producing the bore. Loosening the ground using flushing methods alone is not permitted. When boring below the groundwater table, positive pressure in the flushing or support fluid must be used to prevent soil entering the bore. The bore must be cleaned of boring residues.

(6) Grouting in this context is defined as placing the grout under a pressure greater than the hydrostatic pressure at the discharge point. For post-grouting, one or more grouting stages are carried out after the original grout has set or hardened. Grout composition, pressures and quantities shall be adapted to suit the ground and local conditions. The grout must be composed such that the fractures in the set or hardened material of the original grout can be filled. Piles already under load may not be post-grouted. Post-grouting is generally performed using thin grouting tubes and at pressures between 10 bar and a maximum of 60 bar.

(7) Common characteristic pile resistances in the serviceability limit state are in the order of up to 1,0 MN, depending on ground conditions and pile cross-section.

2.2.7 Tubular grouted piles

(1) The tubular grouted pile is a micropile to DIN EN 14199, provided its diameter is less than 30 cm. It is classified as a composite pile. For diameters

greater than 30 cm it can be regarded as bored pile with a steel tube as special reinforcement to DIN EN 1536.

(2) If the bond between the external grout and the steel tube is adopted for pile capacity and as corrosion protection, technical approval can be required.

(3) The tubular grouted pile is a bored, thick-walled, tubular steel pile with a cement grout cover grouted along the shaft and at the base.

(4) The pile is suitable for transferring high tensile loads and is installed with low vibration and noise. In immission-sensitive areas and in cases requiring monitoring of working loads, it is therefore a useful alternative to driven anchoring elements being structurally connected to a tension pile.

(5) This pile type is unsuitable for ground conditions where large flushing losses are anticipated unless specific additional measures are taken.

(6) The tubular grouted pile consists of approximately 3 m long steel pipe sections, coupled during installation by screwing or welding.

(7) The tube can be provided with a rolled-on external thread to improve bonding of grout.

(8) The piles are bored in sections using rotary methods and external flushing. A 30 cm to 50 cm long boring shoe at the pile tip centres the tube for installation and creates an annulus around the pile shaft corresponding to its greater outside diameter.

(9) The flushing fluid, consisting of cement grout, exits at the boring tip and transports the loosened soil to the surface via the annulus between the tube and the bore wall. Depending on the ground conditions and the applied flushing pressure, diameters up to 70 cm can be achieved by the jet emanating radially from the boring tip.

(10) These pile base enlargements are subsequently grouted at high pressure using a cement grout. The annulus between the steel tube and the bore wall remains filled with hardening cement grout up to the pile head.

(11) Depending on ground conditions, cross-section and the length of the grout body, common characteristic pile resistances in the serviceability limit state are in the order of 0,75 MN to 1,5 MN, for tube dimensions of 73 mm to 102 mm outside diameter, and 12,5 mm to 20 mm wall thickness, or up to 2,5 MN for an outside diameter of 114 mm and 28 mm wall thickness.

2.3 Foundation elements similar to piles

(1) Apart from pile systems described in 2.1 and 2.2, which are associated with the pile execution standards DIN EN 1536, DIN EN 12699 and DIN EN 14199, further deep foundation elements, referred to as "elements similar to piles", are utilised in construction practice.

(2) Elements similar to piles can be, for example:

- Vibrated concrete columns (DIN EN 14731 – ground treatment by deep vibration);
- Grouted vibro columns (DIN EN 14731 – ground treatment by deep vibration);
- Prefabricated tamped mortar columns, mixed-in-place columns (MIP) to DIN EN 14679 – deep mixing or;
- Jet-grouted columns (DIN EN 12716 – jet-grouting).

Such elements can require a national technical approval.

(3) These Recommendations do not apply directly to the above or similar elements. In particular, the empirical pile resistance data (5.4) shall not be adopted for these elements. Irrespective of this, however, the EC 7-1 Handbook [44] analyses, specified for pile foundations and adopted in these Recommendations, can be adopted for these elements, unless different analyses are required in technical approvals.

(4) Well foundations and caissons are generally not regarded as pile-similar foundations, but instead are viewed as sunken shallow foundations.

3 Pile Foundation Design and Analysis Principles

3.1 Pile Foundation Systems

3.1.1 Single pile solutions

(1) Single piles are understood as piles that do not interact with other piles, either through the ground or through the superstructure, or where individual piles only interact to a negligible degree.

(2) See 1.1 for details of the relevant German standards in terms of pile installation, analysis and design.

(3) "Internal" and "external" pile capacities are differentiated when designing piles. The internal capacity comprises the analysis of the safety against failure of the pile material, generally concrete, reinforced concrete, steel or timber. The external capacity comprises the analysis of the safety against failure of the ground around the pile.

(4) The ultimate limit state (ULS) and the serviceability limit state (SLS) must be analysed for both the internal and the external pile analysis, as stipulated in the EC 7-1 Handbook [44]. The ultimate limit state of pile foundations is normally governed by the failure of the structure and structural elements (in accordance with the EC 7-1 Handbook [44]: STR – structural failure for internal capacity, GEO-2 – geotechnical failure for external capacity). Tension piles require also the analysis of the limit state of loss of static equilibrium by buoyancy or uplift (UPL – uplift). The limit states are also dealt with in 1.2.3 and 6.

(5) Foundation piles are predominantly axially loaded. The external resistance in axial direction comprises the components base resistance R_b and shaft resistance R_s. The pile resistance R is a function of the respective settlement or uplift and can be described by a resistance-settlement (or uplift) curve for the entire resistance zone (RSC or RUC, respectively). These curves should be determined from pile test results, see 5.2, 5.3, and 9 and 10. However, they can also be determined on the basis of empirical data, see 5.4.

(6) Figure 3.1 shows the resistance-settlement curve (RSC) for the overall resistance and, separately, for the pile shaft and the pile base resistances. The variables with the indices g and ult apply to the respective limit settlement $s_g = s_{ult}$, which refer to the ultimate limit state resistance $R = R_{ult} = R(ULS)$. Pile resistances in the serviceability limit state are abbreviated below by $R = R(SLS)$.

Note: EC 7-1 designates the compression pile resistance in the ultimate limit state R_c (ULS), shortened to R_c and the tension pile resistance R_t (ULS), shortened to R_t.

Recommendations on Piling – EA Pfähle.
Editors: German Geotechnical Society (*Deutsche Gesellschaft für Geotechnik e.V.*)

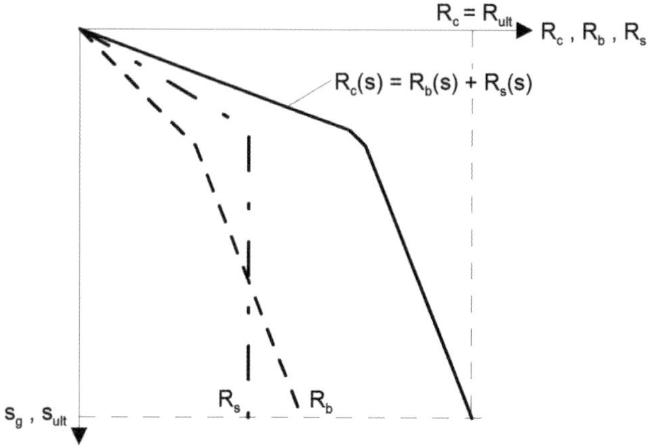

Figure 3.1 Resistance-settlement curve for a pile under compressive load

(7) The limit value of skin friction q_s is normally achieved at relatively small pile displacements. This leads to a pronounced curvature of the load displacement line for a pile predominantly subjected to shaft resistance "friction pile". Normally, only the base resistance increases notably with larger displacements, see [59].

(8) A less pronounced curvature displays the resistance-settlement curve for end-bearing piles. Load transfer of these piles is predominantly via the base resistance, which increases up to very large settlements, see [59].

(9) For laterally loaded piles, two limit cases are differentiated:

- short, almost rigid piles, with a restraint effect in the ground that can be derived from a force couple of three-dimensional passive earth pressures and;
- slender, flexible piles, generally analysed using the modulus of subgrade reaction method.

(10) Normally, the mutual influence of the external load-bearing behaviour resulting from simultaneous axial and lateral actions is insignificant. Both actions can be considered separately. However, it should always be considered that piles are subject to both vertical and horizontal actions and resulting bending moments, e.g. from horizontal forces. One component often substantially dominates over the others.

3.1.2 Pile grillages

(1) Pile grillages consist of single piles connected by a superstructure and spaced so far apart that no substantial interactions between neighbouring piles occur in terms of pile load-bearing behaviour.

(2) See 3.1.3 and Section 8 for details of interactions and pile distances.

(3) See Section 7 for details of grillage analyses.

3.1.3 Pile groups

(1) Several piles form a group if they are incorporated in a common pile cap and their load-bearing behaviour is mutually influenced. The mutual influence of the piles is called group effect or pile-pile interaction.

(2) The group effect of axially loaded piles can refer to both the settlement and the resistances. The settlement-related group effect G_S is defined by the factor

$$G_s = \frac{s_G}{s_E}, \qquad (3.1)$$

the resistance-related group effect G_R by the factor

$$G_R = \frac{R_G}{n_G \cdot R_E}; \qquad (3.2)$$

with:

s_G the mean settlement in a pile group;
s_E the settlement of an equivalent single pile at the mean pile load of the group piles;
R_G the overall resistance of the pile group;
R_E the resistance of the single pile at the mean settlement of the pile group and;
n_G the number of piles in the group.

(3) The limit distance, at which the interaction effects between two neighbouring piles become negligibly small, is often assumed to be 6 to 8 times the pile diameter. However, the limit distance is also a function of the length of the pile in the load-bearing ground. The limit distance increases with increasing embedment depth d, also see Section 8.

(4) For small and given the same settlements group piles normally exhibit smaller resistances than single piles. For large settlements the resistance of the group pile can exceed the resistance of the single pile.

(5) The load-bearing behaviour of the axially loaded group piles differs with the respective pile position. The piles shown in Figure 3.2, for example, are differentiated into corner, edge and inner piles. In low-settlement bored pile groups the corner piles display the greatest pile resistances and the central piles the least. Interlocking effects can occur at larger settlements, leading to a reversal of this distribution.

(6) Paragraphs (4) and (5) predominantly refer to bored piles in pile groups without, or with only minor, ground displacement effects. In pile systems with

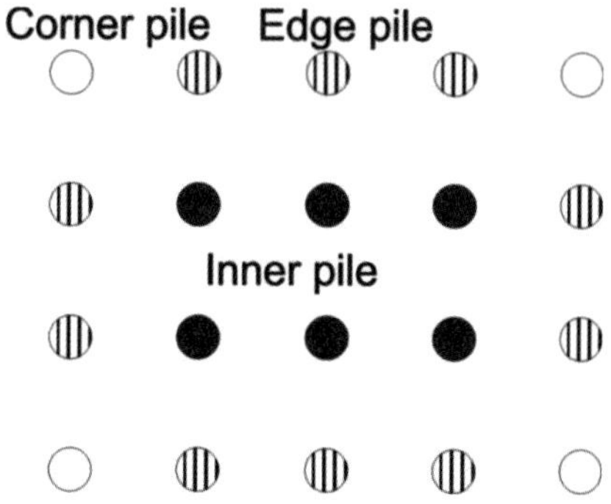

Figure 3.2 Pile designations within a group

greater ground displacements, normally different behaviour can be expected, also see 8.2.1.4.

(7) In horizontally loaded pile groups, in which all piles display practically equal horizontal head displacements, the individual piles are subject to differing shares of the actions imposed on the pile group. The distribution of pile resistances within the group is dealt with in 8.2.3.

(8) When adopting the governing actions it should be noted that the various actions are, to a degree, mutually dependent. For example, horizontal forces not only generate a moment action on the foundation, but, due to the location of the moment transition point below the pile head slab, additional vertical pile loads are also generated.

(9) See Section 8 for details of numerical estimation of group effects.

3.1.4 Piled raft foundations

(1) According to the EC 7-1 Handbook [44], piled raft foundations (PR) are composite geotechnical structures for transferring structural loads to the ground with foundation slabs and piles exercising a mutual load-bearing effect. The interactions shown in Figure 3.3 must be considered simultaneously.

(2) The load-bearing effect of a piled raft is described by the piled raft coefficient α_{PR}. This coefficient indicates the proportion of the total action transmitted by the piles. The remaining proportion of the action is transferred to the ground via the bearing pressure of the foundation slab. A piled raft coefficient of $\alpha_{PR} = 1{,}0$ corresponds to a pure pile foundation and $\alpha_{PR} = 0$ represents a pure shallow foundation, see Figure 3.4.

(3) Analysis of piled raft foundations follows the German guidelines for the design, dimensioning and construction of piled raft foundations (Richtlinie für

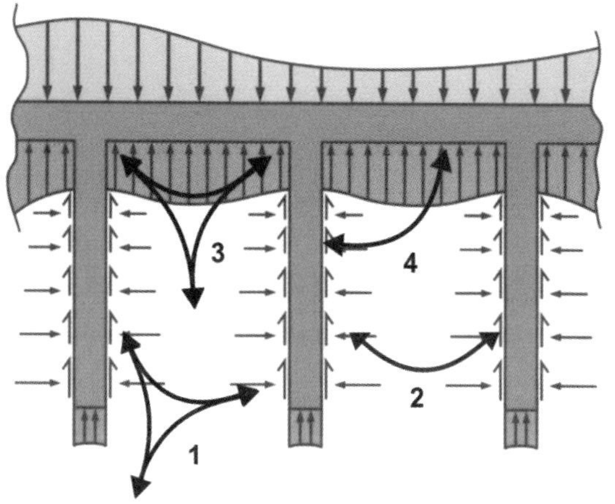

Figure 3.3 Soil-structure interaction, after [46]

den Entwurf, die Bemessung und den Bau von Kombinierten Pfahl-Plattengründungen (KPP-Richtlinie) [46]. Additional guidance can be found in the EC 7-1 Handbook [44].

(4) The guidelines apply predominantly to vertically loaded, piled raft foundations. They do not apply in cases where soil strata of relatively low-stiffness (e.g. soft, cohesive or organic soils, collapsible fill) are present below the foundation slab or for stratified ground with a stiffness ratio for the upper and the lower strata of $E_{S,upper}/E_{S,lower} \leq 1/10$, nor in other cases with a piled raft coefficient $\alpha_{PR} > 0{,}9$.

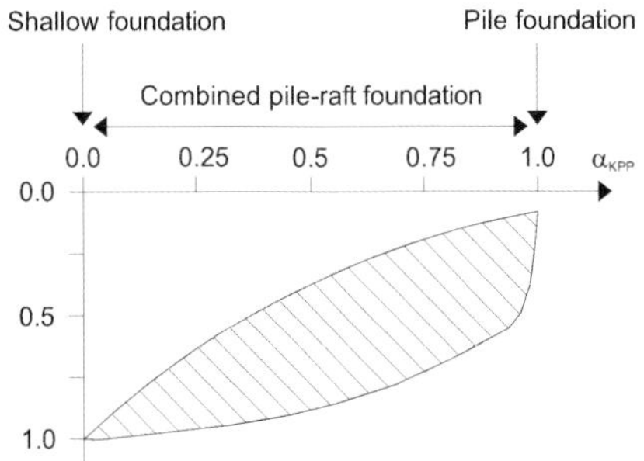

Figure 3.4 Possible settlement reduction for a combined piled raft depending on the piled raft coefficient α_{PR}, qualitative example, after [46]

Note: The analysis of piled raft foundations shall in future be covered also in DIN 4018.

(5) Piled raft foundations shall always be assigned to Geotechnical Category GC 3 in accordance with the EC 7-1 Handbook [44].

3.2 Geotechnical Investigations for Pile Foundations

(1) The structure and properties of the soil and rock and the groundwater conditions must be known in sufficient detail for any construction project. This is necessary to reliably assess the stability and serviceability of the pile foundations and of the overall structure as required by EC 7 and DIN 1054 and to assess the effects of pile foundations on their surroundings.

(2) Moreover, the information above (1) must also be sufficient to allow technically the competent pile installation or construction, e.g. based on DIN EN 1536, DIN EN 12699 and DIN EN 14199, taking DIN 18301 into consideration.

(3) To this end, project-specific geotechnical investigations shall be carried out in accordance with the EC 7-2 Handbook [45].

(4) The results shall be summarised in the Geotechnical Investigation Report and be evaluated in the Geotechnical Design Report regarding the technical consequences for the construction.

(5) The EC 7-2 Handbook [45] stipulates that the type and scope of geotechnical investigations depend on the geotechnical categories and shall be specified in detail by the geotechnical expert.

(6) The EC 7-1 Handbook [44] classifies pile foundations into the following geotechnical categories:

- Geotechnical Category GC 1:
  pile foundations shall not normally be assigned to the Geotechnical Category GC 1.
- Geotechnical Category GC 2:
  a) pile foundations for which the pile resistances are determined on the basis of empirical data, e.g. as described in 5.4, in cases where simple ground conditions exist;
  b) common cyclic, dynamic and impact actions after 4.1 (1);
  c) piles subjected actively to lateral actions with respect to the pile axis, e.g. resulting from structural loads;
  d) piles with negative skin friction.
- Geotechnical Category GC 3:
  a) substantial cyclic, dynamic and impact actions after 4.1 (1), and seismic actions;
  b) raked tension piles with inclinations less than 45° to the horizontal;

c) tension pile groups;
d) grouted pile systems (micropiles to DIN EN 14199 and grouted displacement piles to DIN EN 12699) as anchorage elements;
e) determination of tensile pile resistances;
f) loading lateral to the pile axis or bending resulting from lateral ground pressure or settlements;
g) highly utilised piles in conjunction with special serviceability requirements;
h) piles with shaft and/or base grouting;
i) combined piled raft foundations.

(7) The geotechnical investigations must extend to sufficient depth to record all ground formations and strata influencing the structure and its execution, and to identify the load-bearing and deformation properties of the ground, also see the EC 7-2 Handbook [45] and Figure 3.5. In addition to the stipulations in Figure 3.5, the ground investigations should extend to a depth of at least $z_a \geq 4\ D_b$ below the pile base, if the pile resistances are determined based on empirical data according to 5.4.

(8) If there are any doubts about the safe execution of piling works, trial piles should be installed to examine the execution and load-bearing aspects.

(9) When specifying the extent of the ground investigations important findings gathered on comparable foundation projects under similar conditions and/or in the vicinity of the current site should be taken into due consideration.

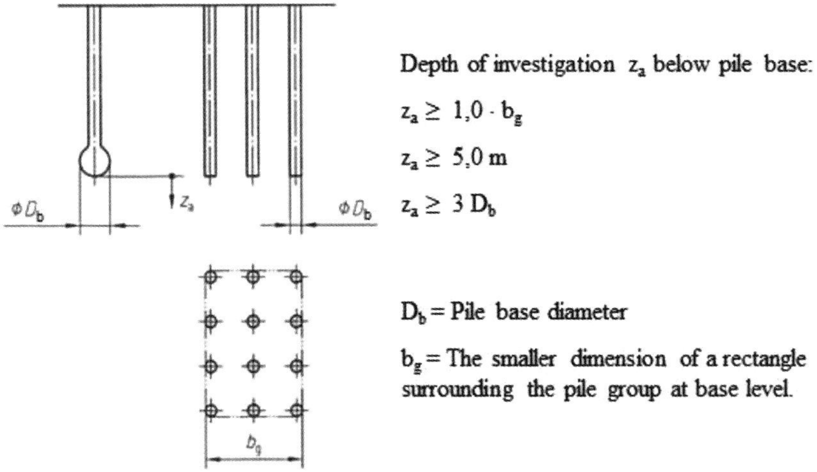

Figure 3.5 Minimum ground investigation depths for pile foundations, from EC 7-2 Handbook [45]; Note: if the pile resistances of compression pile foundations are determined based on data from proven experience to 5.4, the ground investigations should extend to a depth below the pile base of at least $z_a \geq 4D_b$

(10) It is permissible to correlate empirical data, if similarity can be demonstrated by means of suitable investigations, e.g. penetration tests, vane tests, pressiometer and similar tests.

(11) The Geotechnical Investigation Report and the Geotechnical Design Report, respectively, must contain all relevant data that can affect pile capacity, choice of the execution method and the pile installation. Characteristic soil parameters shall be derived for the specific pile foundation. The ground information and the parameters relate not only to pile capacity, but also to drillability, driveability, etc. If this information can not be given in general, different sets of parameters shall be given for each respective application.

(12) In addition to the general information according to the EC 7-2 Handbook [45], the Geotechnical Investigation Report shall contain the following information, which is particularly relevant to pile foundations:

a) the ground elevations of all points at which investigations or tests were carried out, relative to the principle datum or to another established reference;
b) occurrence and properties of loose or soft soil, or of ground that can alter unfavourably during the excavation or installation of piles (softening, instabilities, etc.);
c) occurrence of soil or rock strata with tendency to swelling;
d) occurrence of coarse-grained soils with an open structure (high permeability, voids), which can cause a sudden loss of support fluids or sudden drop of the level of concrete during placing;
e) occurrence of stones or blocks, or other natural or artificial obstructions, which can present difficulties during either excavation or installation, or which require special methods or tools for penetration or removal;
f) thickness, depth and geotechnical parameters of all strata, in particular the bearing stratum the piles are embedded in;
g) occurrence, extent and thickness of all strata which can react sensitively to water or loads during pile installation (e.g. impact, vibration or oscillation);
h) groundwater levels and their fluctuation ranges, including information on confined groundwater conditions;
i) strata in which high groundwater flow velocities are prevalent;
j) aggressivity of groundwater and/or ground, which can impair either the durability or the properties of the pile materials, e.g. of the support fluid, the fresh or hardened concrete, etc., also see EN 206 and DIN 4030;
k) depth, strike and fall of all rock formations;
l) thickness and extent of weathered rock;
m) occurrence, type and quality of rock, in particular strength, filled or open fractures/joints, weak zones, voids, erodibility;
n) extent, thickness and type of contaminated ground or waste which can influence the pile properties, the handling and disposal of spoil, or lead to contamination of the underlying strata;

o) mining and any possible mining impacts;
p) stability problems in the region of the site.

(13) Moreover, the ground investigations shall demonstrate that no soft stratum directly underlies a stratum classified as load-bearing, if a punching failure is possible.

(14) In addition, the specific investigations required as a function of the pile type to DIN EN 1536 for bored piles, DIN EN 12699 for displacement piles and DIN EN 14199 for micropiles, shall be observed.

(15) If no pile load tests have been carried out or comparable pile test data are not available, the characteristic axial pile resistance of a single pile can be determined based on empirical data given in Section 5.4. In this case, reliable data on the ground composition, stratification, groundwater conditions and strength properties shall be available. Empirical data on the characteristic axial pile resistances of bored piles, displacement piles and grouted micropiles can be taken from the data in Section 5.4 for non-cohesive and cohesive soils, and for rock, if the relationships given in (16) are adopted. Irrespective of this, the provisions of the EC 7-2 Handbook [45], and DIN EN ISO 22476-1, -2, -3 and -9 should also be considered. In addition see DIN 1055-2 and DIN EN ISO 14688-1.

The relationships, described in (16), between penetration test resistances and the density of non-cohesive soils on one side, and the consistency or undrained shear strength of cohesive soils on the other, as well as rock strength designations and the associated unconfined compressive strengths, display mean values that can deviate from the data given in the EC 7-2 Handbook [45], DIN EN ISO 22476-2 and DIN 1055-2. The relationships mentioned above shall always be confirmed by ground investigations. In this context reference is made also to literature dealing with this topic, e.g. [27], [87], [96], [97], [110], [123], [132].

(16) The strength of non-cohesive soils should be verified by the cone resistance q_c of the cone penetrometer. Alternatively, the orientation values given in Tables 3.1 to 3.5 may be adopted for correlations between the different types of investigation for pile foundations. The applicability of the tabled data for the respective, specific application must be confirmed by the geotechnical expert.

Note 1: The data in Table 3.1 may be adopted in approximation for converting to q_c values. The relative densities below the water table tend to be slightly higher for the same penetration resistances or blow counts.

Note 2: The density data refer to non-cohesive soils with values of $U \leq 3$. For $U > 3$ other limits also apply, see the EC 7-1 Handbook [44], for example.

Table 3.1 Orientation values for relationships between relative densities and penetration resistances in non-cohesive soils above the groundwater for use with pile foundations

Relative Density D	Density Index I_D	Description	Penetration resistances		
			q_c [MN/m²] CPT	N_{30} BDP	N_{10} DPH
< 0,15	< 0,15	Very loose	< 5,0	< 7	< 4
0,15 ... 0,30	0,15 ... 0,35	Loose	5,0 ... 7,5	7 ... 15	4 ... 9
0,30 ... 0,50	0,35 ... 0,65	Medium-dense	7,5 ... 15,0	14 ... 30	8 ... 18
0,50 ... 0,70	0,65 ... 0,85	Dense	15,0 ... 25,0	23 ... 50	14 ... 30
> 0,70	> 0,85	Very dense	> 25	> 50	> 25

In Table 3.1 are:

q_c cone resistance of the cone penetrometer (CPT);
N_{30} blow count per 30 cm penetration of the borehole dynamic probing (BDP);
N_{10} blow count per 10 cm penetration of the heavy dynamic probing (DPH).

The relationship between CPT cone resistance q_c and the blow count per 10 cm DPH penetration N_{10} can generally be described as follows:

$$q_c\ (\text{CPT}) = \kappa \cdot N_{10}\ (\text{DPH}). \qquad (3.3\text{ a})$$

The value of κ shall be specified by the geotechnical expert on the basis of local experience. For non-cohesive soils the value of κ normally varies between 0,75 and 1,25. Using the average value κ = 1,0 formula (3.3 a) can be simplified to:

$$q_c\ (\text{CPT}) \approx N_{10}\ (\text{DPH}). \qquad (3.3\text{ b})$$

This relation and Table 3.2 have proven reliable for pile foundations in non-cohesive soils; as no incidents are known to have resulted from this.

Table 3.2 Orientation values for conversion from CPT cone resistances q_c in MN/m² and blow count N_{30} of borehole dynamic probing (BDP)

Soil type	q_c/N_{30} [MN/m²]
Fine to medium sands or slightly silty sand	0,3 to 0,4
Sand, or sand with some gravel	0,5 to 0,6
Widely-graded sand	0,5 to 1,0
Sandy gravel or gravel	0,8 to 1,0

Table 3.3 Orientation values for relationships between consistency and shear strength of an undrained, cohesive soil

Consistency index I_c	Consistency	Shear strength of the undrained soil c_u [kN/m²]
0,5 ... 0,75	Soft	15 ... 50
0,75 ... 1,00	Stiff	50 ... 100
> 1,00	Solid	> 100

Correlations between penetration resistances and shear strength of undrained soils c_u can be found in [55] and [87], for example.

It is expressly pointed out that the relationships, in particular in Tables 3.1 to 3.3, can have very large regional uncertainties and that their validity for the specific application shall be confirmed by the geotechnical expert.

It is recommended to preferably determine the shear strength of the undrained soil c_u for analysis of pile capacities on the basis of unconfined compression tests or UU triaxial tests.

For soft soils, the shear strength c_u of the undrained soil can also be determined with the aid of vane tests to DIN 4094-4 or pr DIN EN ISO 22476-9. The maximum measured shear resistance of the soil $c_{f,max}$ on initial shearing shall be reduced using the correction factors given in Table 3.4 below in order to obtain the undrained shear strength of the soft, normally consolidated soil:

$$c_u = \mu \cdot c_{f,max} \,. \tag{3.4}$$

Table 3.4 Correction factors μ for soft, normally consolidated soils

Plasticity index I_P [%]	0	30	60	90	120
Correction factors μ [–]	1,0	0,8	0,68	0,60	0,55

Note: Also see [26].

For pile foundations in rock, the usual rock classification methods shall be applied, including features such as compressive rock strength, discontinuities, degree of weathering, degradation and decomposition status, as well as mineralogy and texture. Table 3.5 provides rough orientation values based on DIN 1054:2005-01, 7.7.4 (3) and EN ISO 14689-1 for the relationship between common rock strength designations and compressive rock strength.

It is generally necessary to carry out unconfined compressive strength tests for pile foundations in rock. In addition – and in exceptional cases also alternatively – point load tests can be performed to estimate the unconfined compressive strength.

Table 3.5 Rough orientation values for the relationship between strength designation and unconfined compressive strength

Strength designation	Field test, description to EN ISO 14689-1	Unconfined compressive strength q_u [MN/m²]
Extremely low[1)]	Can be scratched by finger nail	Less than 1
Very low	Can be scratched by knife, crumbles under firm blows with the point of the geologist's hammer	1 to 5
Low	Can only be scratched with difficulty by knife, can be slightly indented under firm blows with the point of the geologist's hammer	5 to 25
Moderately high	Can no longer be scratched by knife, breaks under a single firm blow with the geologist's hammer	25 to 50
High	Can only be broken by several blows with the geologist's hammer	50 to 100
Very high	Can only be broken by a very large number of blows with the geologist's hammer	100 to 250
Extremely high	Only chips are loosened under blows with the geologist's hammer	More than 250

1) Some very soft rocks behave similar to soil and should therefore be described as such.

Table 3.6 Degrees of rock weathering (after [33])

Designation	Rock property	Rock mass property
Unaltered (*VU*)	Unaltered, fresh, no influence of weathering recognisable	No loosening at discontinuities related to weathering
Partially weathered (*VA*)	Weathering of individual mineral constituents recognisable on a fresh fracture, early mineral alteration and discolouring	Some loosening at discontinuities
Disintegrated (*VE*)	Mineral structure loosened by weathering processes, but still forming a mass, mostly in conjunction with mineral alteration at discontinuities	Complete loosening at discontinuities
Decomposed (*VZ*)	Texture of the rock mass preserved, altered by the formation of new minerals, with no rock properties	Jointed rock body (fault) without rock properties

The weathering status is particularly relevant to pile foundations in soft rock of varying strength. Weathering describes the decomposition and fracturing of rock as a result of exogenic influences, predominantly physical and chemical processes. The state of weathering is described using features, which can visibly be recognized. For construction purposes classification according to weathering intensity has proved useful. The German guidelines for rock descriptions for construction purposes in highway engineering (*Merkblatt über Felsgruppenbeschreibung für bautechnische Zwecke im Straßenbau* [33]) differentiate four homogeneous classes (Table 3.6).

3.3 Classification of Soils for Pile Foundations

(1) Soils meeting the following criteria are regarded as non-cohesive by the EC 7-1 Handbook [44]:

a) Soils such as sand, gravel, stones and their mixtures, if the mass proportion of components with grain sizes < 0,06 mm is less than 5%. This corresponds to the coarse-grained soils of soil groups GE, GW, GI, SE, SW and SI to DIN 18196.
b) Mixed-grained soils with a mass proportion of components with grain sizes < 0,06 mm between 5% and 15%. This generally corresponds to soils in soil groups GU, GT, SU and ST to DIN 18196, and in some cases GU*, GT*, SU* and ST*, if the fine-grained proportion does not govern the behaviour of the soil in terms of pile bearing capacity. If in doubt, see (4).

(2) Soils meeting the following criteria are regarded as cohesive by the EC 7-1 Handbook [44]:

a) In this standard, clays, clayey silts, silts and their mixtures with non-cohesive soils are regarded as cohesive if the mass proportion of components with grain sizes < 0,06 mm is greater than 40%. This corresponds to the fine-grained soils of soil groups UL, UM, and UA, as well as TL, TM and TA to DIN 18196.
b) Mixed-grained soils with a mass proportion of components with grain sizes < 0,06 mm between 15% and 40% are also generally classified as cohesive soils. This corresponds to soils in soil groups GU*, GT*, SU* and ST* to DIN 18196, and in some cases GU, GT, SU and ST, if the fine-grained proportion governs the behaviour of the soil. See (4).

(3) The designation of DIN 18196 of soil groups to the terms “non-cohesive” or “cohesive”, which are also relevant to pile capacities to 5.4, is shown in Figure 3.6 for normal cases.

(4) When assigning mixed-grained soils to either non-cohesive or cohesive soils, not only the grain-size distribution and plastic properties but also the regional geological formation history, e.g. pre-consolidation, must be taken into consideration. Moreover, impacts on the bearing capacity of these soils arising from pile installation, e.g. pile type, dynamic actions, water, etc. must

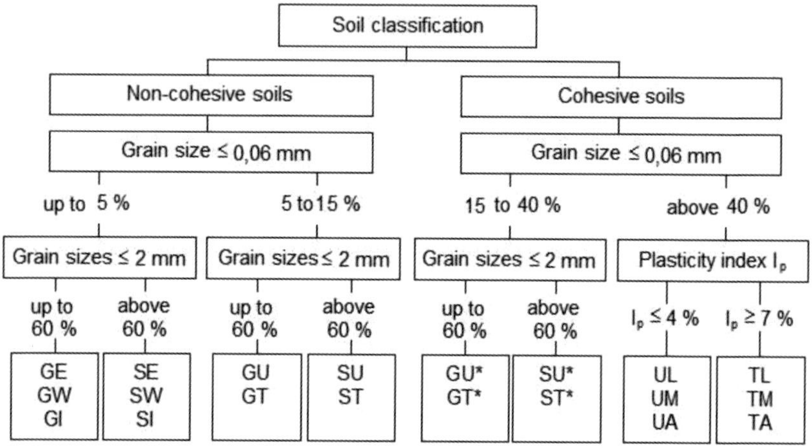

Figure 3.6 Classification of soil groups to DIN 18196 to non-cohesive and cohesive soils, for normal cases

be taken into consideration. For example, proven, glacially pre-consolidated boulder clay may, in individual cases, be classified as a non-cohesive soil, despite its cohesive character as stated in the EC 7-1 Handbook [44] and DIN 18196. The prerequisite for this is the reliable, regional geological assessment by the geotechnical expert. Additional information can also be found in [39].

3.4 Pile Systems for the Execution of Excavations and for Retaining Structures

3.4.1 General

(1) Piles are used to construct walls for excavation pits and for retaining walls. After [26] differentiation is made between:

a) H-piles as soldier piles in soldier pile walls, which are driven, vibrated or set in bores and;
b) cast-in place concrete walls with bored piles.

(2) In terms of flexural stiffness, soldier pile walls can generally be regarded as flexible (according to [26]) and bored pile walls as stiff.

(3) Bored pile walls consist of adjacent bored piles. In addition to structural loads these bored pile walls can carry both earth and water pressures.

(4) Bored pile walls can be used for temporary and permanent purposes.

3.4.2 Pile configurations

(1) If bored pile walls are required to accept water pressures, the individual piles must intersect each other ("secant pile wall"). The centre to centre distances at working level shall be such that with respect to anticipated unavoidable bore deviations sufficient pile intersection is ensured for the entire depth.

(2) "Contiguous bored pile walls" are used if only earth pressure and structural loads need to be transmitted. Local water seepage is discharged under gravity. Additional drainage measures can be required.

(3) "Widely spaced bored pile walls" are suitable for (temporarily) stable ground. The piles are arranged with clear spaces between them. The space between the piles is stabilised by shotcrete or in-situ concrete. Local water seepage is discharged under gravity, see (2).

3.4.3 Pile systems and special execution requirements

(1) All bored pile systems to 2.2.1 involving soil excavation, with the exception of partial displacement piles, are suitable for secant and contiguous bored pile walls to 3.4.2 (1) and (2). For widely spaced bored pile walls partial displacement piles to 2.2.1.4 and full displacement bored piles to 2.2.4 are also feasible, assuming the distance between the piles is large enough to prevent horizontal ground movement that could damage already installed piles.

(2) Drilling templates for guiding the casing or drilling tools shall be used to maintain the drilling tolerances at the working level.

(3) Bored pile walls are mostly installed vertically. Raked pile walls can also be installed in exceptional cases and require special measures to align the drilling rig and the drilling tools.

(4) Pile alignment shall be monitored in case of high requirements on boring precision. It is normally sufficient to survey the respective pile head and base positions. Special devices such as rope inclinometers or ultrasonic-echo devices are available for this purpose. Boring methods using continuous drill strings often employ inclinometers integrated in the tools for monitoring and documenting of the bore run during the boring process.

(5) If the deviations exceed an individually specified value, additional measures or corrections of the boring method shall be initiated. If the deviations of walls subject to water pressure lead to gaps between the piles, they can be sealed by grout injections, by installing additional piles or by jet grouting of the gaps between the piles. If the deviations lead to the effect that piles project into the final structure, or that load transfer is no longer assured, corrections or changes of the boring method shall be considered.

(6) See [26] for more details on installing pile walls in sensitive soft soils.

3.4.4 Design

(1) See [26] for the design of pile walls.

3.4.5 Reinforcement

(1) Piles are usually reinforced using reinforcing cages consisting of longitudinal bars, horizontal links (spirals), stiffening rings and spacers. The reinforcing elements should be robustly welded to each other. If the reinforcement is installed after concreting, the cage can have a conical taper at the bottom.

(2) In special cases rolled steel sections are used.

(3) All piles of contiguous and widely spaced bored pile walls are reinforced.

(4) Of secant pile walls only the secondary piles are reinforced. Either every second pile (1-1 system) or every fourth pile (1-3-1 system) is reinforced. The unreinforced (primary-) piles act as fill piles or lagging. They form a horizontal vault between the reinforced piles. Bending moments and shear forces are transferred through the reinforced piles.

(5) In order to meet the reinforcement requirements stipulated in DIN EN 1536 large bar diameters should be preferred even for heavily reinforced piles.

3.4.6 Concrete

(1) The concrete used for the reinforced piles (secondary piles) shall meet the criteria stipulated in DIN EN 1536. For the unreinforced primary piles special concrete mixes with low strength at the time of intersection are recommended. For large pile diameters in particular (> 0,80 m) the casing rig must exert extremely large driving forces if the concrete's compressive strength exceeds 10 N/mm^2.

(2) Smaller coarse aggregate sizes should be selected at the planning stage for highly reinforced piles.

(3) If the reinforcement is installed subsequent to concrete placement, the stability of the fresh concrete shall be assured, in particular, to prevent filtering (bleeding) effects and the associated stiffening of the concrete prior to cage installation.

3.4.7 Impermeability of bored pile walls

(1) During planning, the fact that bored pile walls are not technically impermeable shall be taken into consideration. The cutting shoe on the bottom casing has a slightly larger diameter than the casing (10 to 20 mm overbreak), to prevent excessive friction between the casing and the intersected pile. When intersecting the primary pile some debris is always transported into the joint area and ground up. When concreting and simultaneously withdrawing the

casing, some debris remains on the intersected concrete surface of the primary pile, forming as a separating layer between the hardened and the fresh concrete. This separating layer can be hardened and partially sealed by the penetration of filtered cement paste. However, even small relative movements between individual piles, common when tensioning anchors or prestressing struts, or during excavation at the latest, will necessarily lead to local leaks.

(2) Even if running water is stopped by subsequent sealing works, bored pile walls subjected to water pressure will normally remain wet. Assuming sufficient ventilation and air exchange, this moisture penetration can be tolerated even for permanent structures, without impairing their serviceability. If ventilation cannot be assured, visible wet patches must be anticipated. Very heavy mould formation has been observed in most unfavourable cases.

(3) The "Bored Pile Guidelines" (*Richtlinie Bohrpfähle* [105]), published by the Austrian Association for Concrete and Construction Technology (*Österreichische Vereinigung für Beton und Bautechnik*), establish requirement classes for watertight pile walls in terms of the achievable impermeabilities, as well as a number of execution rules (construction classes). They are formulated on the basis of the "Impermeable Diaphragm Wall Guidelines" (*Richtlinie Dichte Schlitzwände*), by the same publisher, and can serve as a guide for the achievable impermeability. They state that a single-shell construction can be used for up to maximum 5 m of water head, for example for underground car parks, where wet patches and water streaks are acceptable. Wet patches, and running water discharges and streaks, must be anticipated for higher water heads. In these cases twin-shell construction methods should be planned for the final structure. Twin-shell methods are required in conjunction with bored pile walls for higher purpose uses of a structure in its lower floors.

3.5 Piles for the Stabilisation of Slopes

(1) Installing rigid elements as dowels into a stable lower stratum of sliding or creeping slopes and embankments, is one option for maintaining unstable soil masses in position. As an additional remedial measure, or in combination with such measures, for example counteracting surcharges, anchorages, nailing, grouting and dewatering, steel sections, diaphragm wall elements or bored piles can be employed as dowels.

(2) For the arrangement of such dowels the factors causing the sliding should be considered and normally an open pattern should be chosen, i.e. single or double rows downslope of the sliding mass (Figure 3.7). Pile rows in the direction of the sliding movement can also be useful given the specific conditions.

(3) The loads imposed on the dowels as approximately rigid elements result from the movement of the sliding mass above the stable lower stratum.

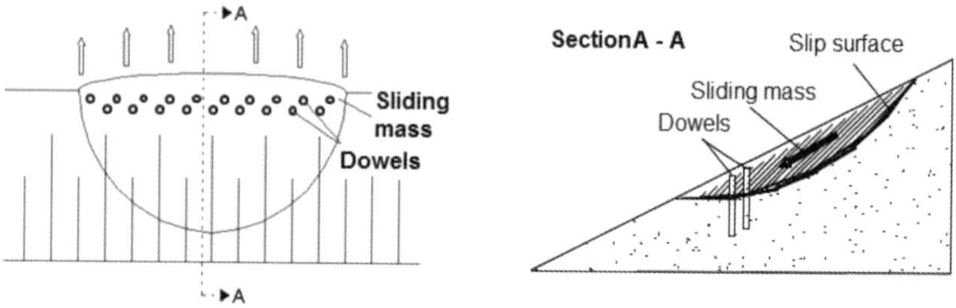

Figure 3.7 Example of dowel pile configuration within a sliding mass

Because the axial distances of the rigid, individual pile elements are mostly greater than twice their diameter, no interlocking effects within the ground will develop which would allow a design approach based on earth pressure theories. Accordingly, the flow pressure resulting from the movement of the sliding mass acts as the maximum possible load on the dowels. This load is transferred by a rigid dowel via a movement (rotation) and bending through its restraint into the stable stratum, see Figure 3.8.

(4) Analysis approaches for slope stabilisation and dowelling measures can be found in [9], [21], [22], [43], [81] and [134], for example.

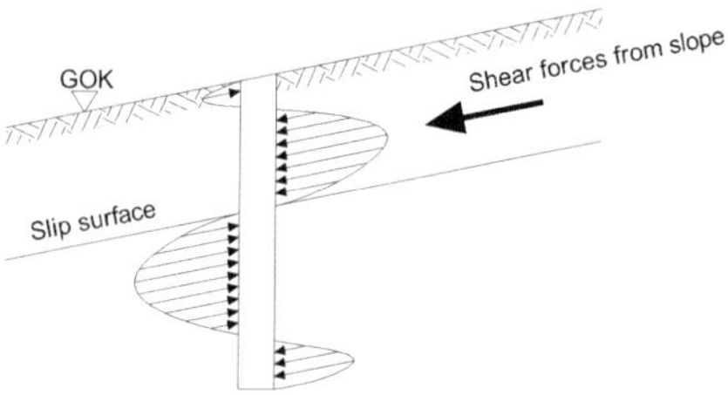

Figure 3.8 Dowel loads from downhill forces

3.6 Use of sacrificial Linings

(1) Regulations on the necessity of sacrificial linings in aggressive ground or ground subjected to strong groundwater flow are included in Sections 7.1.3 and 7.1.4 of DIN EN 1536:2010-12. Section 7.7.4 deals with the concrete cover inside linings. The necessity for, and configuration of, linings in soft strata, in which soil can accidentally enter the bore or concrete be discharged from the bore into the soft strata, are dealt with in Sections 8.4.1.14 and 8.4.5. In addition, linings can also be employed to reduce either negative or positive

skin friction transfer in certain soil zones or depths, or to limit the impacts on existing subterranean structures, for example (see Figure 3.9).

(2) Reduction or elimination of the transfer of skin friction is a complex task and mostly only possible to a limited extent. Preventing the uncontrolled penetration of soft ground into the fresh concrete column of the pile, or the discharge of concrete into soft strata on the other hand, can be achieved reliably. Mostly, single or double linings, with or without anti-friction layers or empty annuli, are employed for this purpose (Figure 3.10).

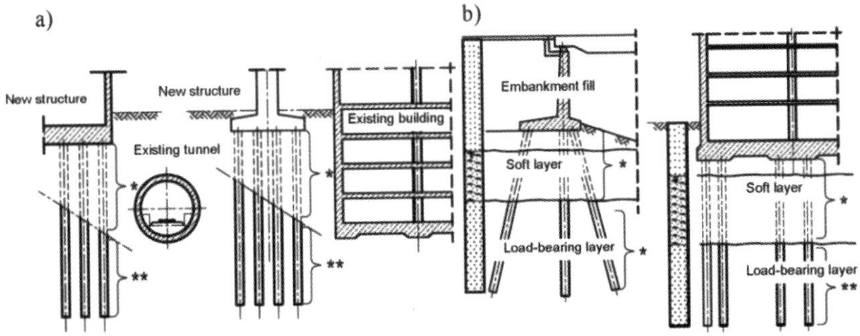

Figure 3.9 Examples for installation of sacrificial linings [79] a) Reduction of positive load transfer to existing underground structures; b) Reduction of negative skin friction or prevention of uncontrolled penetration of soft ground into the fresh concrete column of the pile, or discharge of concrete into soft strata

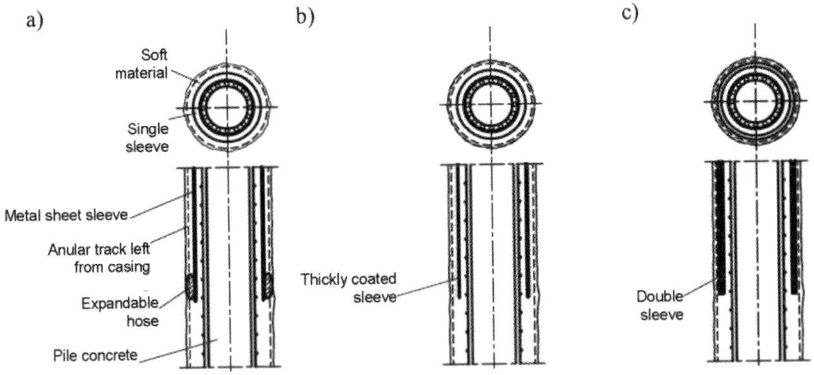

Figure 3.10 Sacrificial lining systems for bored piles (examples, after [79]: a) Single lining with sealed annulus; b) Single lining with coating, without sealed annulus; c) Double lining without sealed annulus

(3) Where bored piles are constructed with the aid of a support fluid, the installation of sacrificial linings, the sealing of the annulus between the wall of the bore and the lining against rising concrete, and filling with soft material, can be reliably achieved. It can be difficult for piles constructed with casings, and is impossible for CFA piles.

(4) If it is necessary to reduce skin friction, the filling of the annulus with soft material can also be problematic, because even soft material is capable of transmitting large friction forces because of its bond to the lining and the neighbouring ground. In such cases it is more favourable to install an open annulus around the lining or to employ double sleeves. See [79] for more detailed information and the results of comparative investigations.

4 Actions and Effects

4.1 Introduction

(1) The actions on pile foundations are differentiated into:

- foundation loads, e.g. imposed by the structure, see 4.2;
- geotechnical actions; in particular actions imposed by the ground, e.g. negative skin friction (see 4.4), lateral pressure (see 4.5) and bending caused by settlement (see 4.6);
- common cyclic, dynamic and impact actions in the sense of DIN 1054: 2010-12, A2.4.2.1, A(8a), e.g. caused by standard loading on traffic areas or from site operations, and taken into consideration as variable static actions, see 4.2;
- considerable cyclic, dynamic and impact actions in the sense of DIN 1054: 2010-12, A2.4.2.1, A(8b), e.g. resulting from jolts due to collisions, concussions, air- or water-borne pressure waves or machinery vibrations, see 4.2, 5.9 and 13.2.

(2) All characteristic actions mentioned in (1) lead to effects which must be accepted by the piles via the external and internal pile capacity, see Section 6.

(3) For the analysis of pile loads and internal forces the values of the determined effects $E_{G,k}$, $E_{Q,k}$ and $E_{Q,rep}$ on the piles are introduced as the actions G_k and Q_{rep}.

(4) With reference to the EC 7-1 Handbook [44], the effects resulting from variable actions are designated by $E_{Q,rep}$, regardless of whether a combination factor has been adopted or not. However, the designation $E_{Q,k}$ without the index "rep" is also permissible if no combination factors have been adopted.

(5) For structural analysis, for example when determining the effects in the pile capping slab or in the framework and slab structures of the superstructure, the piles shall be considered as displacement-dependent resistances. This displacement dependent resistance is often modelled by adopting equivalent springs at the pile locations. The (equivalent) spring stiffness of the respective pile can be determined from the resistance-settlement curve for the governing displacement or loading condition, see 6.4.1 (3). In terms of the computed pile displacement and loading conditions, the results of the structural analysis shall be examined for consistency with the equivalent spring stiffness assumptions.

(6) The resistance-settlement behaviour of pile groups can be influenced by pile group effects. Accordingly, pile location-dependent equivalent spring stiffnesses should be adopted as a function of the three-dimensional orientation of the piles in the group (Section 8), if a governing influence is to be anticipated. It can also be necessary to adopt different spring stiffnesses within a pile group in heterogeneous ground conditions.

Recommendations on Piling – EA Pfähle.
Editors: German Geotechnical Society (*Deutsche Gesellschaft für Geotechnik e.V.*)

(7) It is normally understood that pile settlements resulting from pile self-weight have either ceased or become negligible once the superstructure is erected. The self-weight of tension piles may be taken into account as outlined in 8.1.2 (2).

(8) Also see [26] for pile structures used to support excavations (3.4).

4.2 Pile Foundation Loads Imposed by the Structure

(1) The EC 7-1 Handbook [44] states that actions and effects imposed on the piles by structures (foundation loads) are derived from the analysis of the supported structures carried out with the applicable regulations and standards. For economical geotechnical design, actions and effects shall be provided by the structural engineer as characteristic or representative actions and effects at the top of the foundation structure for each decisive combination of actions in the governing design situations:

- for the ultimate limit state (**ULS**) and;
- for the serviceability limit state (**SLS**).

(2) For pile foundations, it is also permissible to directly adopt the design value of the total effect E_d when analysing the piles in the ultimate limit state.

(3) The foundation loads and the geotechnical actions can lead to the following characteristic or representative actions on the piles:

- $F_{G,k}$ as permanent, axial actions;
- $F_{Q,rep}$ as representative, axial actions;
- $H_{G,k}$ as permanent actions lateral to the pile axis;
- $H_{Q,rep}$ as representative actions lateral to the pile axis;
- $M_{G,k}$ as moments resulting from permanent actions;
- $M_{Q,rep}$ as moments resulting from representative actions.

(4) DIN 1054:2010-12, A2.4.2.1 differentiates between common cyclic, dynamic and impact actions, normally associated with the characteristic, variable, static actions $F_{Q,rep}$, $H_{Q,rep}$ and $M_{Q,rep}$ as mentioned in (3), and considerable cyclic, dynamic and impact actions, which must generally be considered separately as outlined in 5.9 and 13, also see 4.1 (1).

4.3 Installation Effects on Piles

(1) Effects of varying magnitudes on piles are to be anticipated as a result of the installation process. Transport and storage of prefabricated driven piles impose additional effects.

(2) Information on installation effects can be found in DIN EN 1536, DIN EN 12699 and 11.2 and 11.3.

(3) The existence of governing installation effects, which must be taken into account in pile design, shall be considered on a case basis.

4.4 Negative Skin Friction

4.4.1 Introduction

(1) Negative skin friction in piles is to be regarded as a permanent action F_n, originating from relative axial movement between the ground and the pile, when the ground settles more than the pile. This relative movement is generally initiated by settlement of a soft stratum, which can be the result e.g. of surcharges, consolidation processes or groundwater table fluctuations. The self-weight of the settling stratum and of the overlying strata act on the pile via skin friction. This skin friction acts counter to the skin friction from the pile settlement and is therefore known as negative skin friction.

(2) The pile continues to settle until the actions from negative skin friction τ_n, together with the actions imposed on the pile by the superstructure, and the pile resistances resulting from the pile end bearing capacity and supporting skin friction q_s, are in equilibrium. Figure 4.1 shows these relationships for two cases:

- For small effects F_a resulting from structural loads and thus small pile settlement s_a, and large negative skin friction effects F_n, the depth influenced by τ_n is great.
- Reciprocally, a large effect F_b leads to large pile settlements and thus to activation of positive skin friction q_s as a result of relative movement between the ground and the pile.

The following details are taken primarily from [54].

(3) The boundary between numerically positive and negative skin friction is known as the neutral point, see [37] and Figures 4.1 and 4.2.

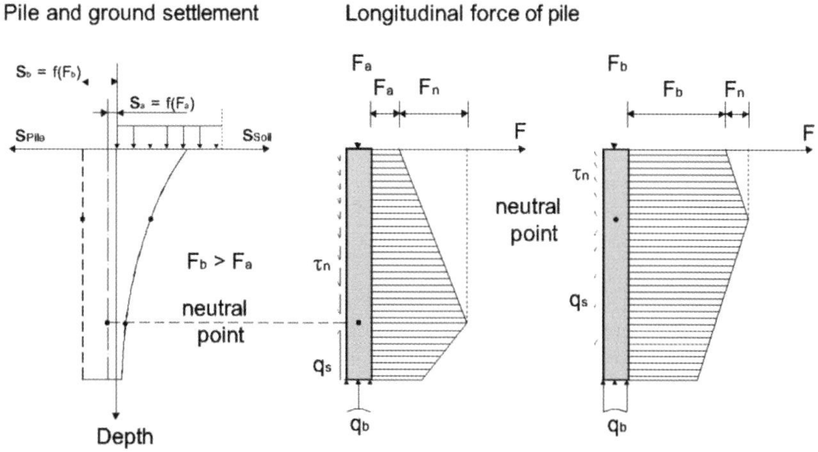

Figure 4.1 Qualitative relationships between pile resistances and effects from structural loads, and negative skin friction in homogeneous ground, and the definition of the neutral point, from [59]. Note: shown is the change of the axial load of the pile.

(4) In the partial safety factor approach the negative skin friction is generally defined as a permanent action on pile foundations, leading to an additional effect on the piles.

4.4.2 Determination of the characteristic action from negative skin friction

(1) An appropriate estimate of the pile's negative skin friction $\tau_{n,k}$ requires information on:

- pile settlements with depth;
- soil strata settlements with depth;
- the resulting relative movements and
- any mobilisation functions of $\tau_{n,k}$ and $q_{s,k}$.

(2) Two principle approaches for deriving the characteristic negative skin friction $\tau_{n,k}$ are given in the literature dealing with negative skin friction:

- Using total stresses for cohesive soils

$$\tau_{n,k} = \alpha \cdot c_{u,k} \tag{4.1}$$

where:

α factor for specifying the value of the characteristic negative skin friction for cohesive soils;
$c_{u,k}$ characteristic value of the shear strength of the undrained soil.

Depending on the soil type and pile type the factor α generally ranges between 0,15 and 1,60, whereby $\alpha = 1$ is often adopted in approximation, which is generally recommended for cohesive soils.

More detailed information on the value of α can be taken from [13], [31], [36], [59], [83] and [145].

- Using effective stresses for non-cohesive and cohesive soils:

$$\tau_{n,k} = K_0 \cdot \tan \varphi'_k \cdot \sigma'_v = \beta \cdot \sigma'_v \tag{4.2}$$

where:

σ'_v effective vertical stress;
K_0 coefficient of at-rest earth pressure;
φ'_k characteristic value of the friction angle;
β factor for specifying the value of the characteristic negative skin friction for non-cohesive and cohesive soils.

According to the literature the factor β generally ranges between 0,1 and 1,0, depending on soil type, e.g. see [13], [14], [31], [36], [59] and [83]. For non-cohesive soils $\beta = 0{,}25$ to 0,30 is often used.

(3) There is no reliable information on whether $\tau_{n,k}$ changes with time depending on the consolidation status of the soft stratum in the around the pile.

(4) A non-cohesive fill overlying a soft stratum can, by calculation, lead to extremely high effects on the pile resulting from negative skin friction. The resulting characteristic effect should therefore not be adopted at greater values than the weight of this stratum within the piles' zone of influence. However, this rule only makes sense when applied to a group of closely spaced piles.

(5) DIN 1054 mentions that the negative skin friction $\tau_{n,k}$ is not to be expected to be greater than any positive skin friction $q_{s,k}$ in similar ground conditions. This assertion in the standard should be expressly confirmed or be modified for the specific project by the geotechnical expert or the geotechnical designer in the Geotechnical Investigation and Design Reports, because the positive skin friction normally represents lower, conservative values and thus $\tau_{n,k} > q_{s,k}$ is also possible as an action.

(6) Negative skin friction affects the pile down to the neutral point. In reality, a transition zone exists between negative and positive skin friction, whereby the transition is assumed to be linear. It is referred to in [37] as the neutral zone.

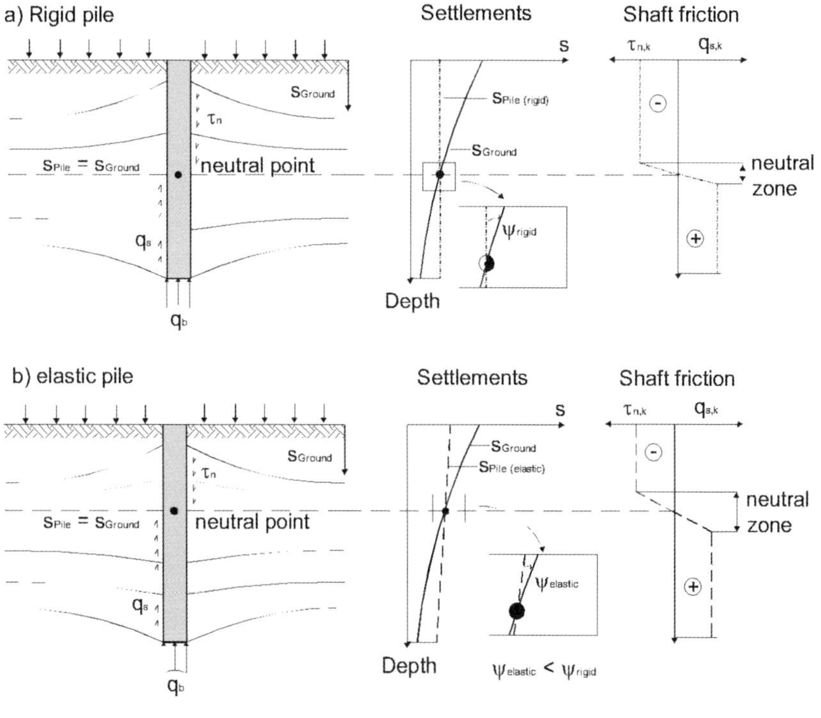

Figure 4.2 Model of negative skin friction and activation of pile skin friction as a function of the angle of intersection ψ of the settlement curves for a) rigid pile and b) elastic pile, after [37], from [54]

Accordingly, skin friction is not fully mobilised within this transition zone. The length of the neutral zone as shown in Figure 4.2 depends on the relative movement between the pile and the ground. The smaller the angle ψ between the intersecting settlement curves, the longer is the transition zone from negative to positive skin friction.

(7) For practical analysis the neutral point may be adopted in approximation between τ_n and q_s, without a transition zone. The largest axial strain of the pile occurs at this point, because the total load is increased by the downward actions resulting from negative skin friction. No load is transferred to the ground until this point, because no pile resistance is mobilised, see Figure 4.1. Moreover, the pile settlements at the neutral point coincide with the settlements in the surrounding ground.

(8) For end bearing piles the neutral point is near the pile base. The neutral point of a long friction pile, in contrast, is often above the middle of the pile.

(9) The actions on the pile resulting from negative skin friction can form a capacity reserve for the external pile resistance. An additional action resulting from structural loads can thus reduce the negative skin friction, instead of increasing utilisation of the pile resistance. On the other hand this means that, due to this reduction, the negative skin friction can represent an external pile resistance reserve, similar to a prestressing of the ground.

(10) To determine the depth of the neutral point, and thus the value of the characteristic action $F_{n,k}$(SLS) in the serviceability limit state it is recommended to normally determine the deformations s_{Ground} of the ground surrounding the pile for the final situation and using characteristic values, i.e. taking consolidation and creep deformations into consideration.

Comparing the deformations resulting from pile settlement s_{Pile} and the settlement of the surrounding soft stratum s_{Ground} gives the location of the neutral point.

(11) In order to determine the neutral point, and thus the characteristic action $F_{n,k}$(ULS) in the ultimate limit state ("external" pile capacity), it is recommended to specify the pile settlement $s_g = s_{ult}$ in the ultimate limit state in accordance with Section 5, depending on the method selected to determine pile capacity.

Comparing the deformations for $s_g = s_{ult}$ with the settlement of the surrounding soft strata s_{Ground} gives the location of the neutral point for the ultimate limit state, which can be located at a different depth than in the serviceability limit state.

(12) In reality, the estimated pile settlements in the ultimate limit state do not occur for the actually acting loads (characteristic actions). Ultimate limit state analyses are therefore performed on the basis of an imaginary deformation state.

4.4.3 Determination of the design values of actions or effects and method of verification

(1) When allocating the action of negative skin friction to a load case it is recommended to allocate it to the persistent design situation DS-P, if the negative skin friction continues to exist throughout the pile's functional lifetime and the deformed soft stratum remains as a permanent action around the pile, even after settlements of the soft stratum have ceased.

(2) When analysing the ultimate and serviceability limit states, the cases given below in (3) and (4) have to be differentiated in conjunction with the pile effects resulting from negative skin friction.

(3) "External" pile capacity:

a) Serviceability limit state (SLS): the characteristic action $F_{n,k}(SLS)$ and the location of the neutral point have to be calculated by the deformation behaviour associated with the pile settlement s_{Pile} and the settlements in the soft stratum s_{Ground}. The design value of the effects is:

$$F_d = F_k = F_{G,k} + F_{n,k}(SLS) + F_{Q,rep} \qquad (4.3)$$

b) Ultimate limit state (ULS): the characteristic action $F_{n,k}(ULS)$ and the location of the neutral point have to be calculated by comparing the deformations associated with the pile settlement $s_g = s_{ult}$ and the settlements in the soft stratum s_{Ground}. The location of the neutral point is normally higher than in the serviceability limit state, because the imaginary pile settlement s_{ult} is greater than s(SLS) (except for piles on rock, for example). The design value of the effects is:

$$F_d = (F_{G,k} + F_{n,k}(ULS)) \cdot \gamma_G + F_{Q,rep} \cdot \gamma_Q \qquad (4.4)$$

(4) "Internal" pile capacity (structural analysis):

Analysis is usually done for the ultimate limit state (ULS) adopting the actions resulting from negative skin friction in the serviceability limit state $F_{n,k}(SLS)$ for pile settlement s(SLS) after Paragraph (3) a).

(5) Appendix B8 contains an example for consideration of negative skin friction for piles and the course of calculations.

4.4.4 Skin friction as a result of heave in the vicinity of the pile

(1) Piles can also be subjected to additional effects as a result of heave in the vicinity of the pile. For example, such effects can occur if a structure to be founded on piles is built within an excavation and the piles are installed prior to final excavation. During the subsequent excavation process the piles are additionally subjected to tension as a result of discharge caused by the excavation and the associated ground heave, see Figure 4.3.

(2) The zone of greatest tension or strain (neutral axis) is generally located in the middle third of the pile length.

(3) The value of the strains depends on the heave, i.e. excavation discharge and ground stiffness, as well as pile length, pile distances and arrangement, and can be estimated e.g. using numerical analyses.

Monitoring of bored piles showed that strains can occur up to around 0,3‰, see [99], which could exceed the numerical yield limit of the pile concrete.

(4) It must be taken into consideration that the tensile pile stresses in the situation shown in Figure 4.3 are normally compensated again by gradually developing compression by subsequently imposed structural loads.

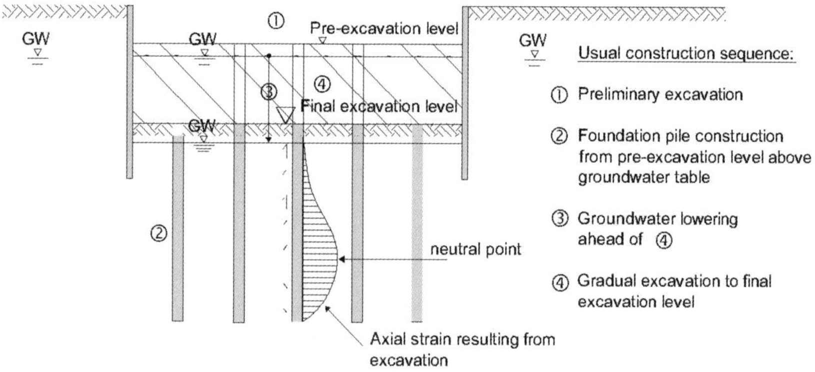

Figure 4.3 Example of effects on piles imposed by heave in the vicinity of the pile; here: as a result of subsequent excavation within the construction pit

4.5 Lateral Pressure

4.5.1 Introduction

(1) Ground movements in soft, cohesive strata impose actions lateral to the pile axis and resulting bending effects. Examples of this type of action are shown in Figure 4.4. Lateral pressure on piles [34] often occurs in the backfill of abutments with pile foundations. Among others, here the size of the action depends on the magnitude and speed of the ground movement, on the pile stiffness and on the geometrical boundary conditions.

(2) Lateral pressure effects on piles can be present, regardless of ground stratification, if the:

- piles are arranged for stabilisation within a slope or
- piles are foundation elements supporting a retaining structure.

See 3.5 for details of pile foundations used for slope stabilisation (e.g. dowels).

See 4.7 for details of effects on foundation piles of a retaining structure (e.g. abutment foundations.

For the following considerations it is understood that lateral pressure on piles is primarily caused by the flow of generally soft, cohesive soils around and lateral to the pile axis. This leads to lateral loads on the pile as an action imposed by the ground, see 4.1.

(3) See 4.6 for details of actions on raked piles (settlement bending).

(4) Appendix B10 contains an example of piles subjected to lateral pressure.

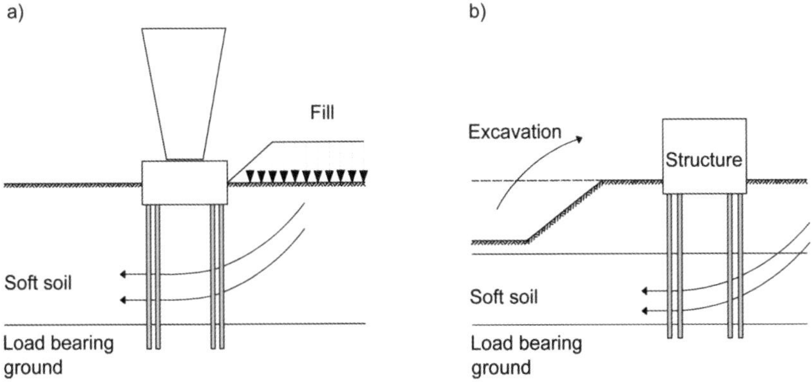

Figure 4.4 Examples of the causes of lateral pressure on piles; a) resulting from fills, b) resulting from excavations, from [59]

4.5.2 Necessity for design of piles for lateral pressure

(1) If cohesive soils are present, in particular normally or slightly overconsolidated and of soft or even more unfavourable consistency, and lateral pressure on the piles as a result of geometrical or load-related boundary conditions cannot be ruled out, the additional actions resulting from lateral pressure must be analysed as outlined in 4.5.3 and 4.5.4.

(2) The necessity of designing piles for lateral pressure can be estimated with the aid of a global failure analysis according to DIN 4084, adopting the partial safety factors from the EC 7-1 Handbook [44] for design approach GEO-3. The procedure is as follows:

- Global failure analysis on the "decoupled system" see Figure 4.5b.
- The design earth pressure E_d imposed on the support structure is adopted as a system support as shown in Figure 4.5, whereby any variable action is disregarded, because it would increase the support force.

– The analyses are generally performed adopting the shear strength $c_{u,d}$ of the undrained soil of the soft strata.

Depending on the utilisation factor μ of the design resistance, an estimate of whether the deformations in the soft strata can lead to lateral pressure effects on the piles can be made, see (3).

(3) If the limit values compiled in Table 4.1 for the utilisation factor μ are adhered to, pile design for lateral pressure may be dispensed with. Otherwise design is to follow (4) and (5).

(4) If the limit values for the utilisation factor μ given in Table 4.1 are exceeded, or if conditions are as outlined in Paragraph (1), the action resulting

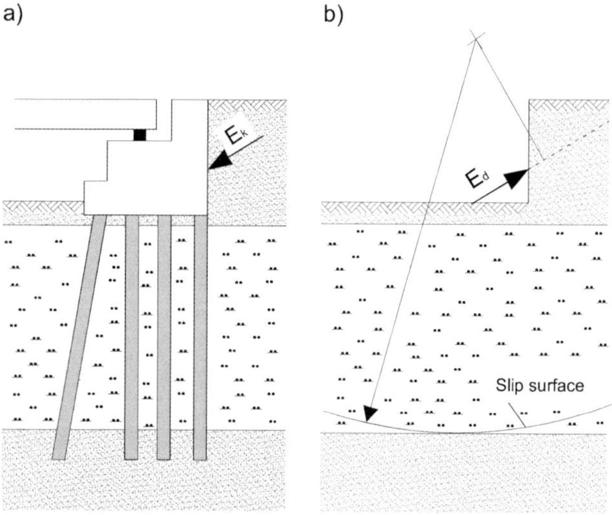

Figure 4.5 Investigation of the necessity for pile design for lateral pressure; adoption of support force when analysing general stability to DIN 4084 a) System; b) decoupled system

Table 4.1 Limit values for the design resistance utilisation factor μ for general stability according to DIN 4084 on the decoupled system

μ	soft strata, imposing a lateral pressure on the piles
0,80	cohesive soils to Para. (1)
0,75	soils with high organic content, where V_{gl} > 15% and w > 75%, e.g. tidal mud, peat, etc.

from ground movements lateral to the pile axis has to determined by the following both approaches

- the characteristic flow pressure $p_{f,k}$ and
- the characteristic resulting earth pressure Δe_k.

(5) The governing factor is the respective smaller total force resulting from lateral pressure on the piles, whereby the effect resulting from flow pressure $P_{f,k}$ and the resulting earth pressure ΔE_k, respectively, are determined over the entire height of the action, see Figure 4.6. The minimum total effect always represents the governing action for each pile, even if one of the lateral pressures mentioned in (4) is lower in some pile sections than in others.

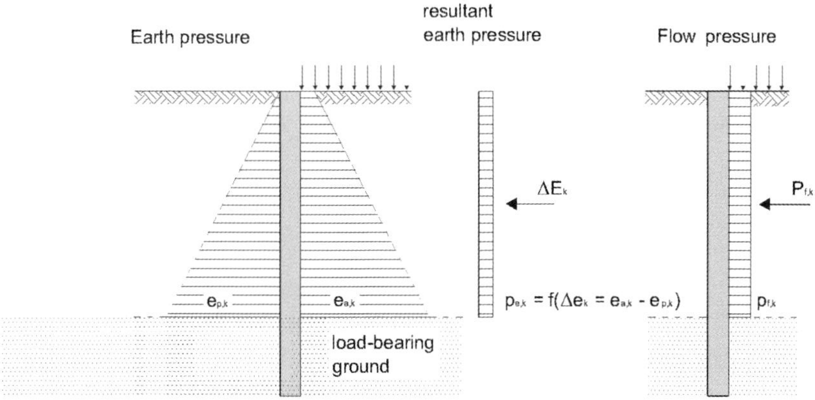

Figure 4.6 Governing total effect from resulting earth pressure or flow pressure (example)

4.5.3 Determination of the characteristic action from flow pressure

(1) It is assumed that the shear strength of the ground is fully used and the plasticised ground flows around the pile.

(2) Based on [157], the value of the characteristic action resulting from flow pressure as a linear load lateral to the pile axis is:

$$p_{f,k} = 7 \cdot \eta_a \cdot c_{u,k} \cdot a_s \text{ bzw. } 7 \cdot \eta_a \cdot c_{u,k} \cdot D_s \qquad (4.5)$$

where

a_s pile width normal to the direction of flow for a square cross-section or
D_s pile diameter for a round cross-section;
η_a calibration factor for the pile arrangement as shown in Figure 4.7.

The increase in the flow pressure in pile groups should be taken into consideration as dictated by the arrangement of the pile group after [155], see

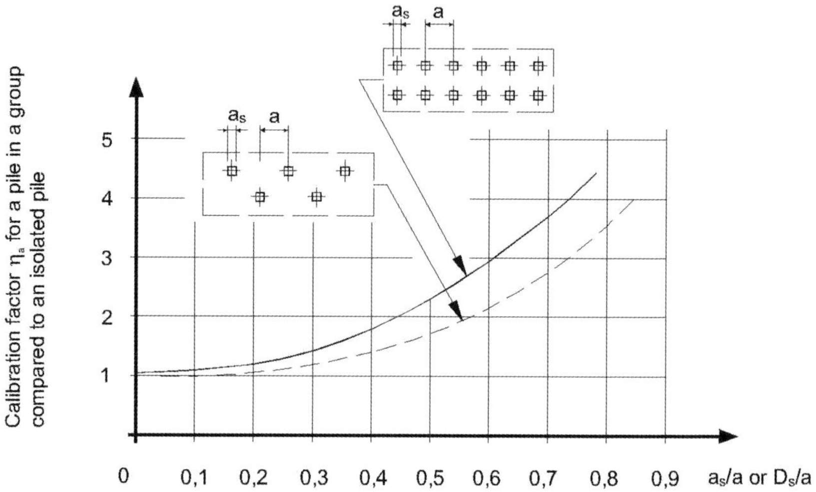

Figure 4.7 Calibration factor η_a as function of the pile arrangement, after [155]

Figure 4.7. The possible flow directions are estimated on the basis of the geometrical situation.

In pile groups the full flow pressure shall be adopted on each group pile, unless the piles are placed unusually close in the direction of the force.

4.5.4 Determination of the characteristic action from the resulting earth pressure

(1) An additional limit value for the lateral pressure effect is estimated using the following earth pressure approach [34], which is however not based on a real deformation state (Figure 4.8). Earth pressure and passive earth pressure are determined for an imaginary, vertical wall in front of and behind the pile group. The structure-ground interface friction angle δ is adopted in approximation as zero. A reduction in the earth pressure effect by shielding, e.g. by rearward horizontal spurs or projecting pile capping slabs, may not be taken into consideration. The earth pressure components are determined as planar earth pressures.

(2) The characteristic, resulting earth pressure Δe_k is calculated as the difference between the active earth pressure $e_{a,k}$ and the passive earth pressure $e_{p,k}$ on the imaginary, vertical wall.

$$\Delta e_k = e_{a,k} - e_{p,k} \tag{4.6}$$

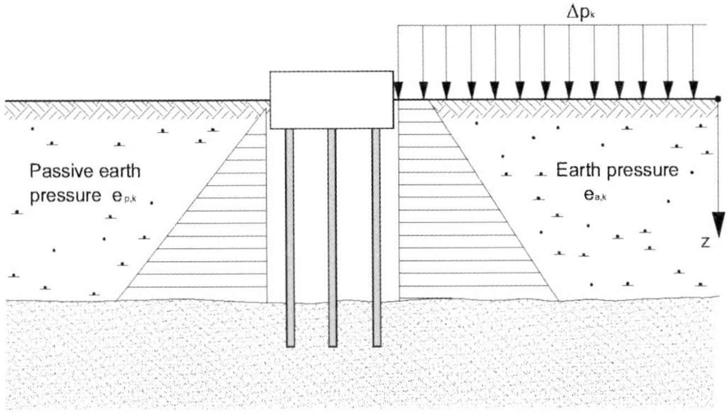

Figure 4.8 System and earth pressure approach

(3) The characteristic active earth pressure is calculated using either:

– the shear parameters of the undrained soil for the initial state;

$$e_{a,k} = \gamma \cdot z + \Delta p_k - 2 \cdot c_{u,k} \tag{4.7}$$

– the effective shear parameters for the final state;

$$e_{a,k} = (\gamma \cdot z + \Delta p_k) \cdot K_{agh} - 2 \cdot c'_k \cdot \sqrt{K_{agh}} \tag{4.8}$$

– or, for partially consolidated states, using:

$$e_{a,k} = (\gamma \cdot z + U_c \cdot \Delta p_k) \cdot K_{agh} + (1 - U_c) \cdot \Delta p_k - 2 \cdot c'_k \cdot \sqrt{K_{agh}} \tag{4.9}$$

where:

γ unit weight of soft stratum, whereby in place of γ for saturated soils
γ_r is adopted above the groundwater table and
γ' below the groundwater table;
Δp_k stresses resulting from a surcharge or other actions generating flow forces;
U_c degree of consolidation in the soft strata as a result of Δp_k.

(4) The characteristic surcharge Δp_k can comprise self-weight components, e.g. resulting from an abutment backfill, live loads or other actions generating flow forces; Δp_k may be regarded in approximation as a permanent action.

(5) The characteristic passive earth pressure is determined for all consolidation states after (3) approximately as:

$$e_{p,k} = \gamma \cdot z \cdot K_{pgh} \tag{4.10}$$

adopting $K_{pgh} = 1{,}0$, in order to ensure the compatibility of the deformations. If the ground level is sloping down on the passive earth pressure side, an appropriately idealised, horizontal ground level can be adopted as the reference level for the z ordinate for the calculation of the passive earth pressure. In saturated soils γ is replaced by:

γ_r above the groundwater table and
γ' below the groundwater table.

(6) The magnitude of the characteristic action resulting from pressure lateral to the pile axis and acting as a linear load on the pile is given by the characteristic resulting earth pressure Δe_k and the width of influence b:

$$p_{e,k} = b \cdot \Delta e_k \quad [\text{kN/m}]. \tag{4.11}$$

The width of influence b of Δe_k on the individual pile is selected as the minimum of one of the following conditions a) to d):

a) the mean pile distance square to the direction of the force as shown in Figure 4.7;
b) three times the pile width a_s or three times the pile diameter D_s;
c) the thickness of the stratum generating the lateral pressure;
d) the overall width of the pile group divided by the total number of piles.

(7) Moreover, it should be checked with Eq. (4.12) whether the effect on the individual pile in a pile group containing n_G piles, with pile centre to centre distances $< 4 \cdot a_s$ or $< 4 \cdot D_s$, becomes greater than calculated with Eq. (4.11), which would then be governing, (see [50]):

$$p_{e,k} = [(B' + 3 \cdot a_s) \cdot k \cdot \Delta e_k] / n_G \quad \text{or}$$
$$[(B' + 3 \cdot D_s) \cdot k \cdot \Delta e_k] / n_G \quad [\text{kN/m}] \tag{4.12}$$

The designations used in Eq. (4.12) are taken from Figure 4.9.

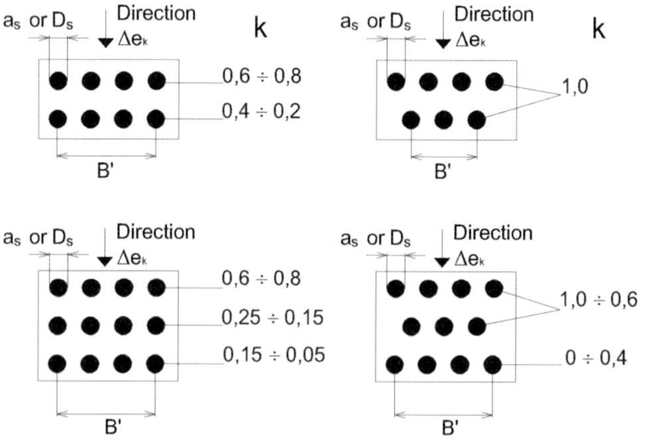

Figure 4.9 Coefficients k for distributing the lateral pressure among the individual piles of a pile group, after [50]

(8) If the minimum action d), given in Paragraph (6), is governing, of extensive pile groups only those piles shall be taken into consideration which are within a length in the direction of the resulting earth pressure corresponding to 1,5 times the height of the stratum generating the lateral pressure.

(9) The distance between the pile foundation and the action causing the lateral pressure can be considered in accordance with 4.5.5.

4.5.5 Influences of distance and minimum moments

(1) If piles or a pile group are at a greater distance l from an action generating a lateral pressure, e.g. a fill as shown in Figure 4.10, an approximate lateral pressure effect after [50] must be taken into consideration in accordance with (2).

(2) For more distant piles the characteristic lateral pressure effect after 4.5.3 and 4.5.4 may be reduced in accordance with Table 4.2 and the boundary conditions shown in Figure 4.10.

Table 4.2 Influence of distance on the magnitude of the characteristic lateral pressure effect, after [50]

Distance l [m]	10 to 15		25 to 40	
Thickness of soft soil layer h_w [m]	15–30	5–15	15–30	5–15
Reduction of resulting earth pressure to %	10–20	5–15	5–15	5

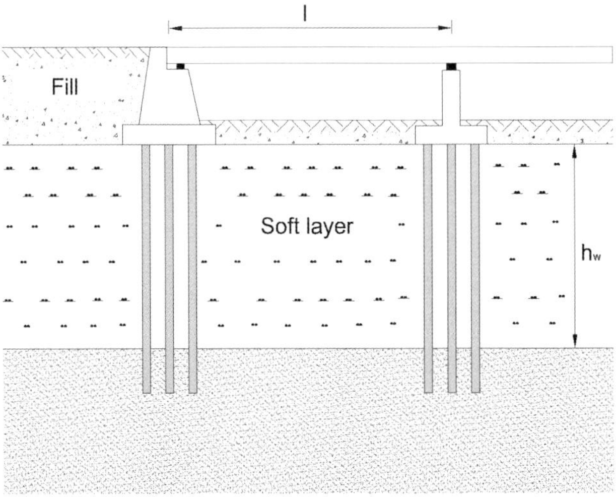

Figure 4.10 System data for distance influence

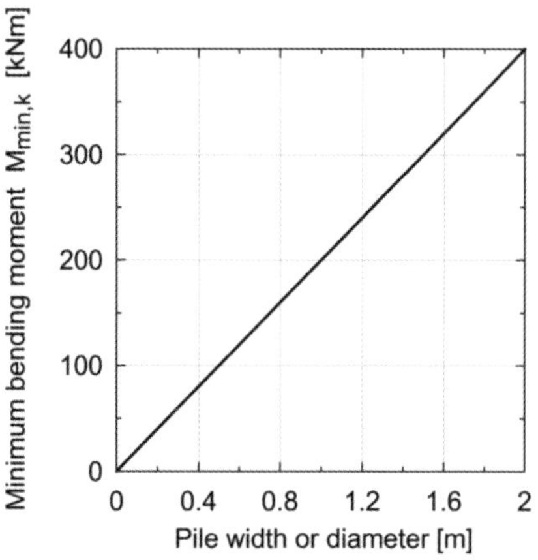

Figure 4.11 Characteristic minimum bending moment, after [50]

The data in Table 4.2 are with reference to the front piles in terms of the direction of the lateral pressure, whereby it is assumed that the spacing is $a \leq 4 \cdot a_s$ or $a \leq 4 \cdot D_s$ referred to the pile centres

(3) If lateral pressure after 4.5.2 needs to be taken into consideration, a characteristic minimum bending moment after Figure 4.11 resulting from lateral pressure should be taken into consideration in pile design for all directly affected and also for more distant piles. Balancing compression forces may not be adopted in this case.

4.5.6 Effects on piles

(1) The characteristic effects square to the pile axis resulting from lateral pressure can be calculated from the actions determined in 4.5.1 to 4.5.5 acting on bars with point-bearings or using the subgrade reaction modulus method.

(2) The design effects result from the characteristic effects using the governing partial factors.

4.6 Additional Effects on Raking Piles Resulting from Ground Deformations

4.6.1 Introduction

(1) Raking piles installed in areas comprising strata sensitive to settlements are often deformed in the course of construction by the settlement imposed by the placement of fills or other, additional surcharges. This imposed deformation causes additional effects in the raking piles in the form of settlement-induced bending effects ('settlement bending').

(2) Examples of actions causing settlement bending in raking piles are shown in Figure 4.12. The actions resulting from flow pressure or the earth pressure difference after Eq. (4.5) or (4.6) for the vertical surcharge components are in approximation determined for each pile's zone of influence of $3D_s \leq 3$ m $\leq$ pile distances. However, the component acting normal to the pile axis cannot become greater than the flow pressure determined using Eq. (4.5).

(3) Settlement bending in raking piles must be taken into consideration appropriately when analysing pile capacity.

(4) It is imperative that shear forces are adequately transferred in the pile connection to the superstructure.

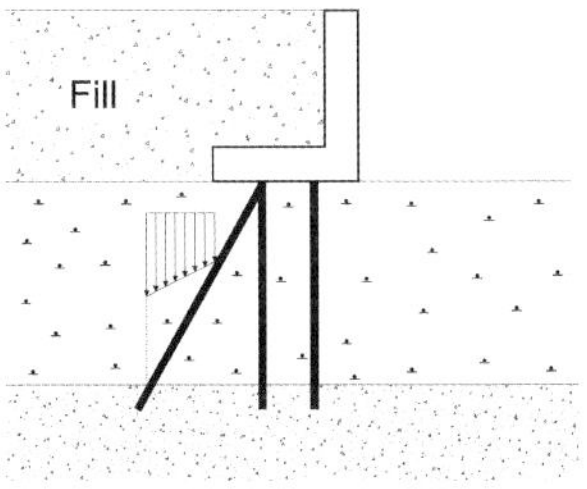

Figure 4.12 Actions on raking piles

4.6.2 Surcharges resulting from anchoring steel and micropiles

(1) Anchorages utilising steel piles and micropiles represent a special application of raking piles in accordance with 4.6.1. Depending on the kind of the structure, the ground and the installation method, ground deformations can occur which often cause significant additional strains in anchorages as result of the downdrag effects between the ground and the anchorage. Some examples of imposed deformations are shown in Figure 4.13. In addition to sufficient bearing capacity, the durability of the anchorages must also be ensured taking the surcharges into consideration.

(2) Grouted micropiles are predominantly used when anchoring of existing structures and penetrating obstructions is necessary. Typical anchoring elements for new structures are steel beams, installed as foldable anchors, grouted driven piles (RI piles) and steel beams of uniform section. These two systems are always differentiated when designing anchorages.

(3) With regard to steel beams, monitoring of backfilled quay walls has shown that pile deflections up to approximately 30 cm to 50 cm can occur as a result of the hydraulic filling process and/or the installation of deep foundation elements. With a limit bending of L/30, a value covering the usual installation impacts is recommended. L is the pile length from the pile connection to the base of the settlement-prone stratum. Only 70 % of the design capacity may be

adopted to cover the imposed surcharge effects. Design is carried out for the axial force only, taking into consideration a reduced ultimate bearing capacity after Eq. (4.13). Also the anchor connection and all individual components must be designed for the reduced ultimate bearing capacity.

$$\frac{\sigma_{E,d}}{\sigma_{R,d}} \leq 0{,}70 \tag{4.13}$$

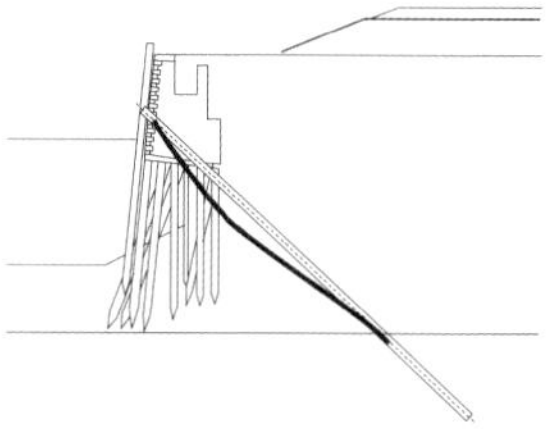

Consolidation settlements

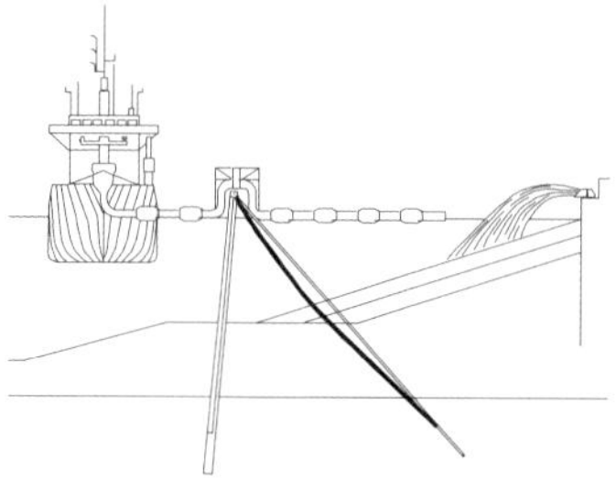

Ground deformations caused by the soil filling process

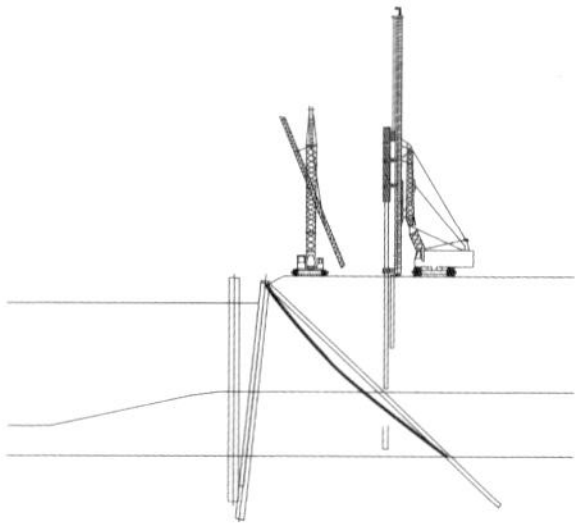

Compaction settlements

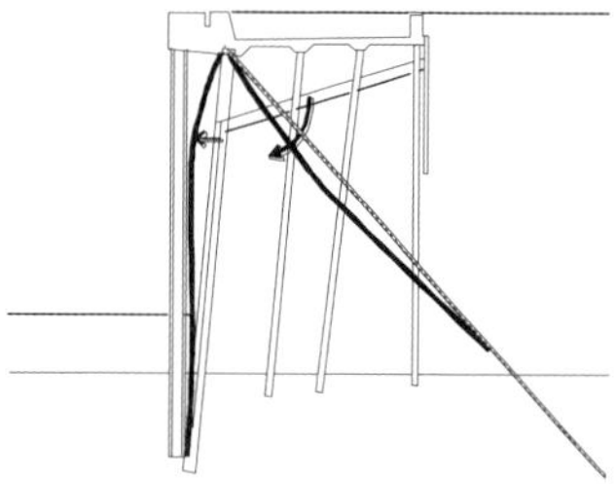

Ground deformations caused by sheet pile wall deflection

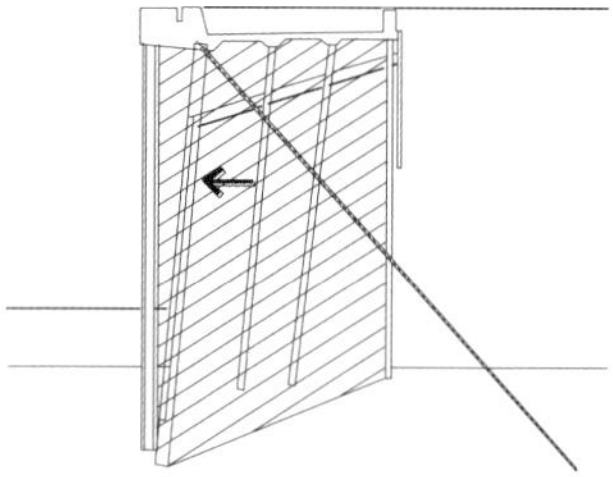

Block bearing behaviour

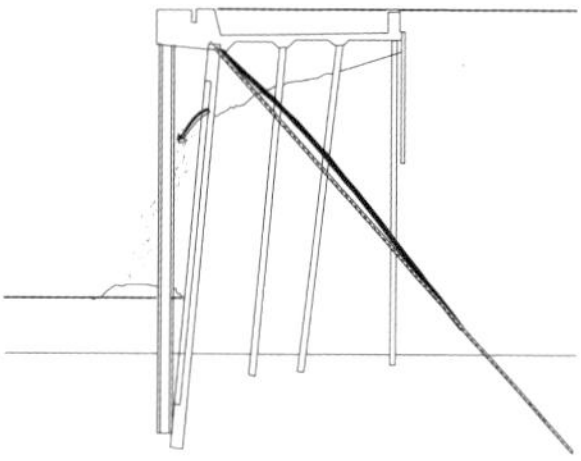

Heave caused by load relief

Figure 4.13 Situations where additional strain is caused on steel piles resulting from ground deformations, examples

Because elastic strain is already exceeded given a deflection of 50 cm lateral to the pile axis, permanent strains can occur in both the pile and the pile connection area. It is therefore recommended to reduce the ultimate bearing capacity of low-ductility components (e.g. welds) to 60% after Eq. (4.14).

$$\frac{\sigma_{E,d}}{\sigma_{W,R,d}} \leq 0{,}60 \tag{4.14}$$

(4) The durability of post-grouted micropiles is ensured by the steel tendon's corrosion protection, normally the grout body. Pile deflection as result of ground movements can therefore endanger durability. Experience on quay walls in northern Germany has shown that deflections should not exceed L/100, with L being the pile length from the pile connection to the base of the settlement-prone stratum. Elastic design is then carried out for the axial force only, taking into account a reduced ultimate bearing capacity after Eq. (4.15). High-strength steel, where there is generally little separation between elastic strain and elongation at failure, is commonly employed for the steel tendons of composite piles. For this reason, the "elastic-elastic" analysis method shall always be adopted to analyse this anchoring system. The anchor connection and all individual components must also be designed for the reduced ultimate bearing capacity. The formation of a hinged connection is recommended.

$$\frac{\sigma_{E,d}}{\sigma_{R,d}} \leq 0{,}85 \tag{4.15}$$

(5) If steel beams are used where excessive bending is involved, all action effects resulting from the respective surcharges must be determined for a more detailed design. The plastic cross-sectional and system reserves can be considered in this case and the yield strength is adopted for this purpose. The strain shall not exceed 5 % for a usual failure strain of approximately 20–25 %. It shall be checked that mentioned failure strains are applicable for the selected steel.

(6) Structural design specifications must be given. When forming a hinged pile connection care shall be taken that a hinge can form without other structural elements failing. Pile bending always results in an increase in the axial force as a consequence of the change in length. In some circumstances this is compensated for by the flexibility of the pile head. Particular attention on this effect is required when post-anchoring. The pile axial force decreases from the anchor connection to the anchor toe. Pile joints should therefore be located as deep as possible.

4.7 Foundation Piles in Slopes and at Retaining Structures

4.7.1 Foundation piles in slopes

(1) If piles are installed in an existing, sufficiently stable slope, a lateral pressure effect is not anticipated as long as the ground's stress state is not unfa-

vourably altered by pile installation or subsequent actions, e.g. additional surcharges near the top of the slope, or downslope excavations.

(2) If the slope is not sufficiently stable or the ground's stress state is unfavourably altered by subsequent actions, the resulting lateral effect on the piles must be taken into consideration. Its magnitude depends on the relative movement between the ground and the piles.

(3) If the utilisation factor $\mu \leq 1$ in the slope's global stability analysis is adhered to for the BS-P design situation after the EC 7-1 Handbook [44], experience shows that the ground deformations remain small and it is sufficient to adopt the active earth pressure as the lateral pressure. The following is adopted for calculating the zone of influence B of the earth pressure effect on each pile of the respective foundation:

a) for closely spaced piles

$$B = (B' + 3\,D_s)/n_G \tag{4.16}$$

where:

B' distance between the centres of the outer piles of the pile group;
n_G number of piles in the pile group;
D_s pile diameter or a_s for square piles.

b) for piles with larger spacing normal to the dip:

$$B = 3D_s \leq 3\text{ m} \tag{4.17}$$

c) or, if the influence of shear strength on the arching effect is taken into consideration after [17]:

$$B = D_s/K_a \tag{4.18}$$

where:

K_a earth pressure coefficients for $\beta = \delta = 0°$.

If in doubt the smallest of the values a) to c) has to be selected.

The earth pressure effect shall be adopted for the pile length above the governing slip surface as identified in a global stability analysis.

(4) If the slope's allowable utilisation factor is exceeded or if large relative movements between the ground and the piles are anticipated as a result of time-dependent ground deformations, e.g. "creeping slopes", a lateral pressure or downhill force (slope shear) on the piles must be adopted. The magnitude of this action shall be determined such way that the supporting effect of the piles guarantees global stability, i.e. reduces the utilisation factor to the value to be adhered to.

(5) In addition to the simplified approaches outlined above, 3.5 must also be observed.

4.7.2 Foundation piles at retaining structures

(1) If an abutment or a retaining wall founded on piles is backfilled, stress alterations in the ground below the wall towards the unloaded side are caused by the one-sided surcharge. This reduces the ground reaction in the upper region of the pile.

(2) The piles are generally subjected to a lateral pressure near the original ground surface, because the passive earth pressure is smaller than the actions.

(3) A pile support by the ground through subgrade reaction is only present at greater depths. The limit depth, i.e. the level above which the piles support the ground and experience horizontal loading, which they transfer to the ground in the zone below this level, can be determined using two methods as illustrated in Figure 4.14:

a) by global stability analyses, whereby the failure surface displaying the necessary stability or adhering to the allowable utilisation factor is governing, or
b) in simplification, by comparing the characteristic earth pressure in the planar case, whereby the active and passive earth pressure angles can be

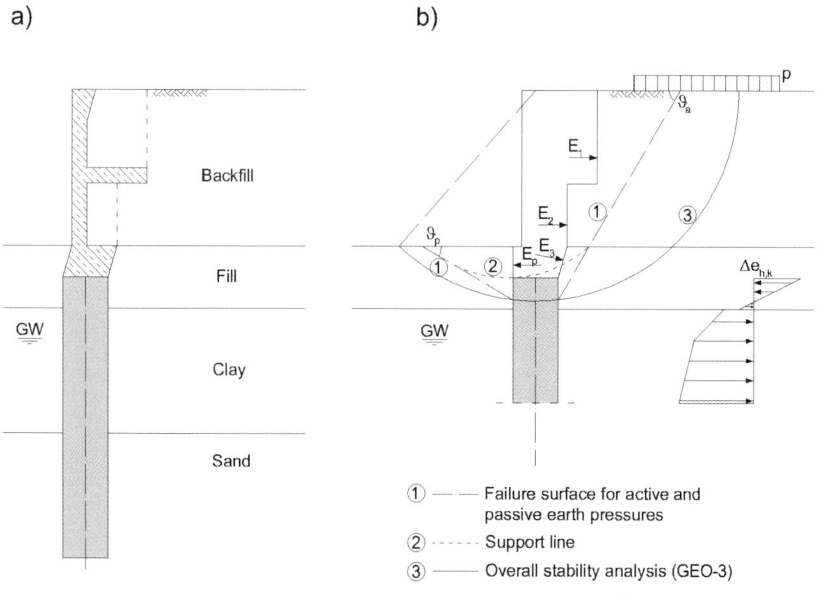

Figure 4.14 Piles at retaining structures, a) system, b) determining the limit depth with loading and supporting stresses; E_1 to E_3, E_p: forces in the decoupled system, see Figure 4.5

estimated using a pressure line to DIN 4084 (guide values: $\delta_a = +2/3\ \varphi'$, $\delta_p = -1/3\ \varphi'$), and the supporting characteristic passive earth pressure may only be taken into consideration at 50% in terms of deformations.

(4) Below the governing slip surface or where the mobilisable characteristic passive earth pressure exceeds the characteristic active earth pressure, the piles can be regarded as supported by the ground and a characteristic elastic support can be adopted, whereby the calculated contact stresses may not exceed any possible ground resistance, see the EC 7-1 Handbook [44] and 6.3.2 (1).

If the piles are also within the zone of influence of a downward slope, the influence of the inclination must be taken into consideration when determining the passive earth pressure and the modulus of subgrade reaction.

5 Bearing Capacity and Resistances of Single Piles

5.1 General

(1) The ultimate and serviceability limit state analyses for pile foundations are dealt with in Sections 1.2.3 and 6.

(2) Guidance and recommendations on determining pile resistances using different methods are given below. In addition, the peculiarities of each of the different analyses are described.

(3) It is pointed out that, as outlined also in the EC 7-1 Handbook [44], common German practice, in use for decades, for calculation of the pile resistances does not permit purely geotechnical methods, which utilise the shear strength or the stiffness modulus of the soil as input parameters, e.g. models of the bearing capacity at the pile base.

(4) Regardless of (3), it is within the permissions given in the general principles of the EC 7-1 Handbook [44] to allow analysis methods for determination of pile resistances for certain, distinct applications, if they are calibrated against pile test results under conditions comparable with the proposed application. This applies particularly to the increasing use of numerical methods, e.g. the finite-element method, see 5.11. However, realistic modelling of pile installation, e.g. of bored and displacement piles, and the associated alterations to the ground around the pile, has proven difficult.

(5) Because generally accepted methods for analysing pile resistances are not yet developed, the EC 7-1 Handbook [44] requires that pile resistances shall normally be determined on the basis of pile test results. It is also permissible to determine axial pile resistances on the basis of empirical data (5.4), which should, however, also be derived from pile load tests.

(6) In terms of load application, pile load tests can be differentiated into:

- static (Section 9) pile load tests and
- dynamic (Section 10) pile load tests.

(7) Pile load tests can be performed for a various purposes, e.g.:

- to determine the external capacity of single piles or pile groups, to define the basis of the design;
- to check the basis of the design;
- to check the suitability of a pile system in terms of the prevalent ground conditions;
- to verify adherence to given governing displacement limits in terms of the serviceability of structures;
- to check the internal capacity of piles;
- to test the quality of the execution.

Editors: German Geotechnical Society (*Deutsche Gesellschaft für Geotechnik e.V.*)

(8) Load tests shall be carried out in particular if:

- insufficient experience is available in terms of the pile execution methods or the prevalent ground conditions;
- highly variable ground conditions are encountered or;
- new execution methods should be tested.

(9) If piles are installed in ground or water judged to be aggressive after DIN 4030, appropriate investigations suiting the pile system used shall be carried out to determine whether the skin friction alters during the working life of the pile foundation. Depending on the result, skin friction might not be adopted, or only be adopted in part, when determining pile resistances.

5.2 Determining Pile Resistances from Static Pile Load Tests

5.2.1 General

(1) The objective of pile load tests is the determination of a characteristic resistance-settlement or resistance-heave curve in order to derive the pile resistances $R_{c,k}$ and $R_{t,k}$, for the ultimate limit state (ULS) and the serviceability limit state (SLS). The testing procedure is dealt with in Section 9.

(2) In tension pile load tests the reaction beam must be long enough to bridge any soil prism attached to the pile or the influence of the beam supports on pile capacity must be negligible.

(3) Pile load tests should be carried out on site, but the results of pile load tests carried out under comparable conditions may also be used to specify the characteristic resistance-settlement or resistance-heave curves.

(4) The following conditions must be adhered to satisfy transferability and comparability requirements:

- the same pile type and similar cross-sectional dimensions and embedment depths in the load-bearing ground;
- similar ground conditions, in particular of load-bearing strata, in terms of soil type and mean strength (penetration test results) and;
- comparability confirmed by a geotechnical expert.

(5) After the EC 7-1 Handbook [44] comparable load test results are treated as tested or measured data, $R_{c,m}$ and $R_{t,m}$, for deriving the characteristic resistances and the application of the correlation factors ξ_i.

(6) The EC 7-1 Handbook [44] assumes that resistance-settlement or resistance-heave curves (RSC/RHC) compiled from the tested or measured data, $R_{m,i}$, of one or more axial pile load tests are used. The characteristic pile resistance, $R_{c,k}$ or $R_{t,k}$ (R_{ult} = R(ULS)), is in turn derived from these test data, forming the basis for the ultimate limit state analysis and the characteristic RSC or RHC under service load conditions. From this, R(SLS) is determined in turn as the basis for the serviceability limit state analysis. The procedure described in

the EC 7-1 Handbook [44] for deriving a bearing capacity limit value for the pile resistance $R_{c,k}$ or $R_{t,k}$ (R_{ult} = R(ULS)) from $R_{m,i}$ is included in Annex A4.1.

Note: As noted in 3.1.1, the EC 7-1 Handbook [44] abbreviates the compression pile resistance in the ultimate limit state, R_c(ULS), to R_c, and the tension pile resistance, R_t(ULS), to R_t. In order to achieve clarity, and supplementary to the EC 7-1 Handbook [44], the ultimate limit state is referred to in these Recommendations on Piling as R_c, R_t or $R = R_{ult}$ = R(ULS), depending on the respective requirements. Pile resistances in the serviceability limit state are shortened to R = R(SLS) hereafter.

(7) Two areas shall be differentiated when deriving the characteristic pile resistances:

- characteristic pile resistances in the ultimate limit state, see 5.2.2;
- characteristic pile resistances in the serviceability limit state, see 5.2.3.

(8) Examples of how to determine the characteristic pile resistances and derive the characteristic resistance-settlement curves from pile load test results or measured data are given in Annex B1 and Annex B6.

5.2.2 Characteristic pile resistances in the ultimate limit state

(1) After the EC 7-1 Handbook [44], a correlation factor ξ_i must be introduced for the pile resistances in the ultimate limit state derived from the measured data $R_{c,m,i}$ or $R_{t,m,i}$. It takes account of pile execution influences and irregularities of the ground.

(2) The numerical values and the method to determine the correlation factors ξ_i are given in Annex A4.1

(3) If the ultimate bearing resistance is not obvious from the form of the measured RSC for compression piles, then

$$s_g = s_{ult} = 0{,}10 \cdot D_b \tag{5.1}$$

can, in approximation and for all pile systems, be adopted for the limit settlement s_g or s_{ult} (see EC 7-1 Handbook [44]), where

s_{ult} settlement in the ultimate limit state;
s_g limit settlement or failure settlement.

Note: Normally, s_{ult} and s_g are regarded as equal; s_{ult} formally designates the ultimate limit state analysis method in accordance with EC 7-1 Handbook [44]; s_g designates the settlement on pile failure.

(4) Pile base shapes other than circular shall be converted to equivalent pile base diameters D_{eq}.

5.2.3 Characteristic pile resistances in the serviceability limit state

(1) Qualified weighting of the individual load test results should be undertaken as the basis for analyses in the serviceability limit state, and the derivation of the characteristic resistance-settlement or resistance-heave curves. Depending on the boundary conditions it can also be necessary to derive upper and lower bounds of the characteristic RSC/RHC from the pile load test results or the measured data for the serviceability limit state analysis, also see 6.4.

(2) To acquire a smooth characteristic or ultimate pile resistance curve, the measured characteristic resistance-settlement or resistance-heave curves must be realistically converted to one or more characteristic resistance-settlement or resistance-heave curves under service loads. This shall normally be done by the geotechnical expert or geotechnical designer. In the standard case, the mean of the measured RSC/RHCs acquired from several load tests, or the test curve acquired from a pile load test for the characteristic RSC/RHC under service load should be selected and increases or reductions Δs_k should be applied based on the options shown in 6.4, Figure 6.1. The increases and reductions depend on the anticipated scatter in pile bearing capacities as a function of the pile type and the ground conditions, as well as the structural boundary conditions of the structure.

(3) Where pile systems displaying only minor settlements under service loads are employed, e.g. some displacement pile systems in load-bearing ground, stability in the serviceability limit state is often also provided if the piles are sufficiently stable in the ultimate limit state.

5.3 Determining Pile Resistances from Dynamic Pile Load Tests

(1) In accordance with the EC 7-1 Handbook [44], under certain circumstances the compressive pile resistances may also be derived from dynamic pile load tests as described in Section 10.

(2) The correlation factors ξ_i and the model factors η_D given in the EC 7-1 Handbook [44] must be adopted, whereby the correlation factors shall be increased by a value $\Delta\xi$ between 0 and 0.4, depending on the type of calibration used for the dynamic pile load tests, see 10.6. The basic values of the correlation factors $\xi_{0,5}$ or $\xi_{0,6}$ depend on the number n of the dynamic pile load tests. The procedure described in the EC 7-1 Handbook [44] for deriving a characteristic ultimate pile resistance, $R_{c,k}$, from the measured or test data $R_{c,m,i}$ is included in Annex A 4.2.

(3) Calibration against static pile tests is required to ensure that the damping factors for determining the dynamic component of the total resistance using direct methods are correctly selected. Use of the extended method with com-

plete modelling represents current best practice for determining pile resistances. This method shall preferentially be adopted.

(4) If predominantly non-cohesive soils are prevalent in the load-bearing strata in which the piles are embedded, and the testing institute can prove that extensive experience from, dynamic pile tests is available for the region, the characteristic pile resistances R_k, may be derived even if no static pile load tests have been carried out on the site. In this case R_k may be deducted from the tested or measured data from dynamic pile tests R_m with the correlation factors $\xi_{5,6}$ and $\Delta\xi$ for the case 'calibration of dynamic pile load tests with static pile load tests carried out on similar construction projects'. The reason for this is the multitude of comparative dynamic and static pile load tests available for non-cohesive ground conditions in the whole of northern Germany and, resultantly, the availability of reliable calibrations and experience.

Note: These provisions only apply to prefabricated driven piles to 2.2.2 when the procedure is expressively agreed to by the geotechnical expert and when pile driving reports are compiled during installation as the scatter in terms of pile installation for these piles is of only minor relevance compared e.g. to bored piles.

(5) For dynamic load tests on piles in cohesive soils, in accordance with the EC 7-1 Handbook [44], the following procedure shall be followed:

a) For piles in soils sensitive to creep and in unsaturated cohesive soils, the results of dynamic pile load tests shall always be calibrated against static pile load tests from the same site.
b) In saturated cohesive soils excessive porewater pressures can increase the capacities measured in dynamic pile load tests. Dynamic pile load tests may therefore not be adopted to determine the characteristic pile capacity if the pile base is situated in or the governing skin friction is mobilised in such soils.
c) In deviation to the provisions in a) and b), dynamic pile load tests may be carried out in cohesive soils if reliable, regional, empirical data are available and their applicability is expressly confirmed by a geotechnical expert for the respective case.

(6) When interpreting the results of dynamic pile load tests, in addition to static load test results, the information given in 5.4 on empirical data should also be considered, see (7).

(7) When deriving characteristic pile resistances from dynamic pile load tests the following steps should normally be followed:

a) The testing institute employed for the dynamic pile load testing provides the test results in a test report containing the test or measured data $R_{c,m,i} = R_{c,stat}$ differentiated into $(R_{c,m})_{mean}$ and $(R_{c,m})_{min}$ in accordance with 10.8.
b) A plausibility check of the measured values is necessary in order to minimise uncertainties connected with the dynamic load testing procedure. To

this end, the geotechnical expert or geotechnical designer shall compare the mean dynamic pile load test values $(R_{c,m})_{mean}$ to the upper bounds of the empirical data for the pile system in accordance with 5.4 or with local experience. After carrying out the plausibility check the geotechnical expert or the geotechnical designer shall confirm or modify the mean value $(R_{c,m})_{mean}$.

c) The geotechnical expert or geotechnical designer then converts the test data or the modified values in accordance with a) and b) into characteristic pile resistances $R_{c,k}$ on a project-specific basis for use in stability analyses. In this process, he shall take the stipulations in the EC 7-1 Handbook [44] and Annex 4.2 into consideration.

d) If the plausibility check after c) reveals significant differences and considerable assessment uncertainties, it is recommended that the test data (raw data) be independently examined and evaluated again, e.g. by means of a separate evaluation employing complete modelling. The independent examination and evaluation is of particular importance for conditions falling into Geotechnical Category GC 3.

The procedure for deriving characteristic pile resistances from dynamic pile load tests is in principle shown in Figure 5.1.

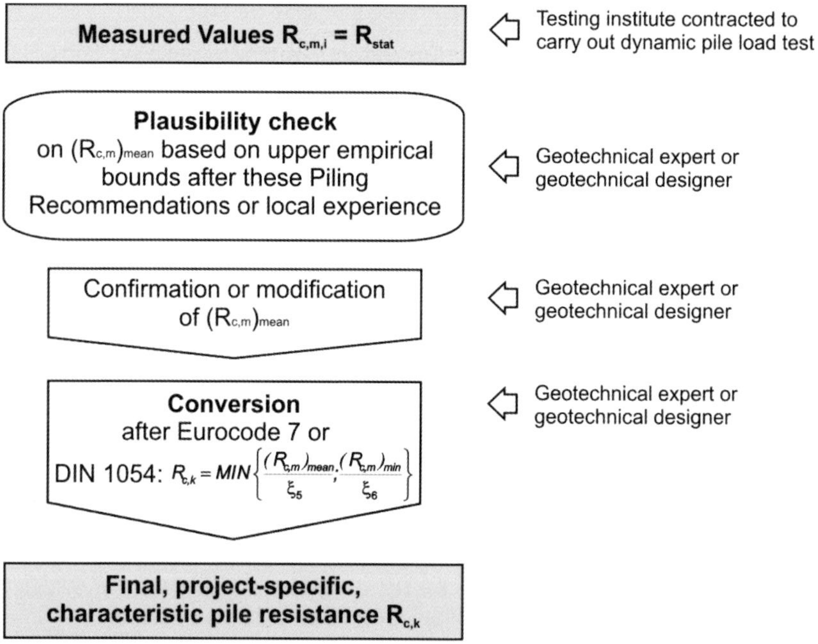

Figure 5.1 Steps to determine characteristic pile resistances from measured or test data of dynamic pile load tests

(8) If dynamic pile load tests are used where the results acquired from test data $R_{c,m,i}$ are available as complete resistance-settlement curves (RSC), the guidance in 5.2.3 and 6.4 shall be observed or applied accordingly to derive the characteristic RSC in the serviceability limit state.

(9) An example for determining the characteristic pile resistances from the measured dynamic load test data is contained in Annex B2. Annexes C1 and C2 include examples for deriving the static measured pile resistance values $R_{c,m,i} = R_{c,stat}$ from the dynamic measurements

(10) If both static and dynamic pile load tests are carried out on a site, two cases are differentiated after the EC 7-1 Handbook [44]:

- If, after applying the correlation factors, the characteristic pile resistances acquired by evaluating the static pile load tests alone are larger than those acquired by evaluating the dynamic pile load tests, only the results of the static pile load tests may be adopted.
- If, after applying the correlation factors, evaluation of the dynamic load tests result in greater bearing capacities, higher characteristic resistances than those from the static pile load tests may only be adopted when if this is comprehensible and justified. The increased resistances must be confirmed for the specific case by the geotechnical expert or the geotechnical designer.

5.4 Axial Pile Resistances Based on Empirical Data

5.4.1 General

(1) The EC 7-1 Handbook [44] allows axial pile resistances to be derived from empirical data, in addition to determining pile resistances from both static and dynamic pile load tests, see 5.2 and 5.3.

(2) Of the methods described in the EC 7-1 Handbook [44], 7.6.2.3, under the heading 'Ultimate compressive resistance determined from ground test results', only the method using Eq. (5.2) below should be adopted in Germany, see EC 7-1 Handbook [44], Nationally Determined Parameters (NDP) to 7.6.2.3 (5)P and NDP to 7.6.3.3 (4). The method is known in Germany as 'Determining axial pile resistances based on empirical data' and is dealt with below. The 'ground tests' (geotechnical investigations) shall be performed such that it is possible to reliably assign the characteristic empirical data, which have been derived from load tests, for the pile end bearing capacity $q_{b,k}$ and pile skin friction $q_{s,k}$ results. To this end the EC 7-1 Handbook [44] gives the following fundamental equations:

$$R_{b,k} = A_b \cdot q_{b,k} \tag{5.2a}$$

$$R_{s,k} = \sum_i A_{s,i} \cdot q_{s,i,k} \tag{5.2b}$$

$$R_{c,k} = R_{b,k} + R_{s,k} \tag{5.2c}$$

(3) However, in principle, the EC 7-1 Handbook [44] allows all data derived from experience to be used, in addition to the empirical data dealt with below, assuming their utility for the proposed case can be appropriately demonstrated.

(4) The applicability of the empirical data given in the subsequent paragraphs, as well as other, general data derived from experience must be confirmed by a geotechnical expert or a geotechnical designer for the proposed application.

(5) The standard procedure for determining the characteristic resistances of tension piles is the static pile load test. A deviation from this rule should be made in exceptional circumstances only and if expressly confirmed by a geotechnical expert for the particular case.

5.4.2 Guidance for the application

(1) The national empirical axial pile resistance data forming the basis for applying the EC 7-1 Handbook [44] are summarised below.

(2) Analogous to 5.4.1 (3) and (4), other empirical data may also be adopted if their validity can be appropriately demonstrated.

(3) After the EC 7-1 Handbook [44], the characteristic values of the pile resistances shall be specified, in principle, on the basis of the results of geotechnical investigations in accordance with the EC 7-2 Handbook [45] and other information. For the construction of pile foundations normally a geotechnical expert shall be involved.

(4) The empirical data are summarised below as ranges of axial pile resistances for the various pile systems (see Section 2). Characteristic values of the pile resistances $R_{c,k}$ for the proposed application may be derived from these empirical data if a geotechnical expert or a geotechnical designer is involved and the provisions of the EC 7-1 Handbook [44] are met. The given empirical data should normally only be used to derive characteristic compression pile resistances, see 5.4.1 (5) and 5.4.10.

(5) The given empirical data refer to both non-cohesive and cohesive soils. See 3.3 for classifications. The EC 7-1 Handbook [44] and 3.3 also include guidance on classifying coarse-grained, fine-grained and mixed-grained soils under the headings of non cohesive and cohesive soils to DIN 18196. In the context of determining the characteristic pile resistances, consulting a geotechnical expert is recommend, in particular where mixed-grained soils are present.

(6) The following recommendations should be understood by the client, the geotechnical expert or designer, and the structural engineer such that more favourable pile capacities and resistances may be adopted as a benefit for increased extend of project-related geotechnical investigations for classifying and describing the respective ground, also see 5.4.3.

(7) Classification of the prevalent soils for the respective case as necessary to apply the empirical pile resistance data summarised in the following

tables shall be based on the EC 7-2 Handbook [45] and, in particular, should address:

- Stratification;
- Grain size distribution;
- Density, strength and consistency conditions;
- Shear parameters and stiffness.

In addition to direct site investigations, e.g. through boreholes and laboratory testing, the results of indirect investigations should be available, e.g. static or dynamic probing (CPT or DPH), borehole dynamic probing (BDP), vane tests, etc., also see 3.2.

5.4.3 Application principles and limitations of tabled data

(1) Empirical pile end bearing capacity and pile skin friction data for the various pile systems are summarised in tables below as a function of the respective ground conditions. The numerical values listed in the tables refer to the CPT cone resistances q_c in non-cohesive soils and the undrained shear strength c_u of cohesive soils.

(2) The EC 7-1 [44] and EC 7-2 [45] Handbooks state that soil parameters scatter considerably as a result of the geological formation conditions and history. This applies especially to pile capacity and pile resistances in the ultimate and serviceability limit states, see Figure 5.2, because in addition to the ground-related scatter, considerable influences can result from the execution. When specifying characteristic soil properties, normally 'conservative mean values' are adopted. See EC 7-1 Handbook [44], and 5.2 and 5.3 for details of the procedure for specifying pile resistances from the results of pile load tests.

(3) The empirical pile base resistance q_b and pile skin friction q_s data ranges summarised in the tables below are based on numerous and predominantly static pile load test results, which were analysed for this purpose. The evaluation strategies and principles are described in [29], [57], [85] and [158]. As outlined in [57], empirical pile load test evaluations were made and related to statistical values, differentiated into 10%, 20% and 50% quantiles as input for

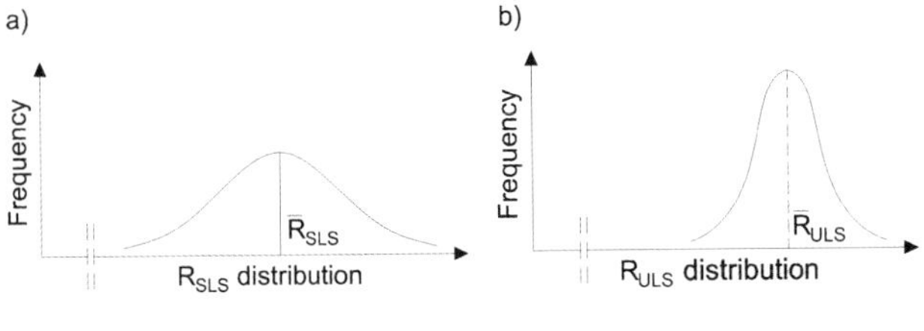

Figure 5.2 Qualitative distribution of pile resistances in the serviceability (a) and ultimate limit states (b)

the tables. This allows the user to also assess the probability and the risk of pile resistance deviations below of the tabled values for a specific project application.

(4) The following tables contain a range of empirical values for quantiles from 10 % to 50 % as shown in Figure 5.3. This means that, as a result of scatter, around 10 % of the in-situ pile resistances can be below the lower value in the tables, and around 50 % below the upper value (mean value).

(5) Concerning the magnitude of the range of table values it is expressly pointed out that the quantile range in Figure 5.3 represents an orientation only. The stated boundaries of around 10 % to 50 % can vary depending on the adopted load test results and the resulting distribution and scatter, because they are based on the adopted load test result population and the boundary conditions of the ground.

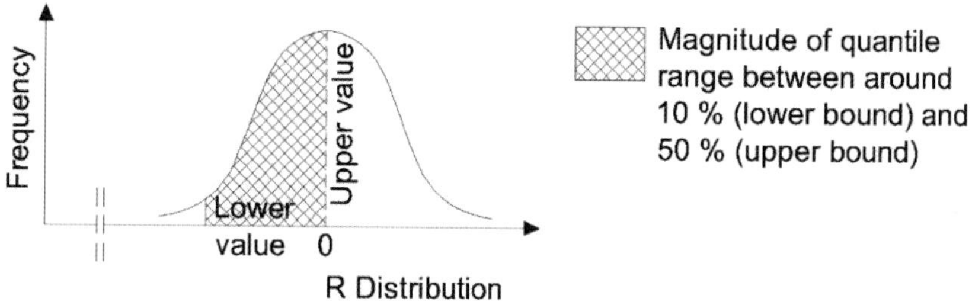

Figure 5.3 Fractiles for the ranges of pile resistance table data from empirical data, compared to load test results

(6) Normally and under condition that the site investigation has been carried out in line with the EC 7-2 Handbook [45] and 3.2, and that the provisions of 5.4.2 are observed, the lower table values (minimal values) should be adopted.

(7) The lower table values can also be adopted for the preliminary design if the final, project-related site investigation results as mentioned in (6) are not yet available.

(8) Pile resistances above the lower (minimal) values, interpolated between the lower and the upper table values, may only be selected for the specific application by the designer if they are expressly confirmed by a geotechnical expert. Local conditions and experience, and the actual situation on the ground must be taken into consideration. The local conditions and experience, and the proposed application, must also be taken into consideration.

(9) If, in terms of pile type and ground conditions, comparable load test results are available, they can be adopted to determine pile resistances as outlined in the EC 7-1 Handbook [44]. Comparability must be confirmed by a geotechnical expert or geotechnical designer, also see 5.2.1 (4) and (5).

(10) The soil strength range given in the table data includes mean CPT cone resistances q_c = 7,5 to 25 MN/m² for non-cohesive soils and undrained shear strengths $c_{u,k}$ = 100 to 250 kN/m² for cohesive soils related to the end bearing capacity, and $c_{u,k}$ = 60 to 250 kN/m² related to skin friction.

(11) Skin friction and support from soil layers where q_c < 7,5 MN/m² or $c_{u,k}$ < 60 kN/m² may only be taken into consideration if confirmed by the geotechnical expert or geotechnical designer and the different deformations of the layers for the mobilisation of the skin friction are considered.

(12) In layered ground with soil layers exceeding or falling short of the requirements stipulated in (11), the same rule applies. In addition, the provisions of 4.4 must be observed.

(13) For the serviceability limit state analysis acc. to 6.4, the measured and derived characteristic resistance-settlement curves should be modified for the serviceability state, if necessary, and applying 5.2.3 as appropriate.

(14) Guidance on comparing the empirical table data to measured data from static and dynamic pile load tests: the tabled characteristic values of pile base resistance and skin friction were derived from pile load test results by statistical application of the quantiles mentioned above to the measured data. Correlation factors ξ, see Annexes 4.1 and 4.2, were not applied. They are instead already incorporated in the partial factors for empirical values as stipulated in the EC 7-1 Handbook [44], Table A2.3 and Annex A3.2. This is done by a model factor η_E in accordance with the EC 7-1 Handbook [44], NDP to 7.6.2.3 (8) and NDP to 7.6.3.3 (6). A fact to be realised when comparing the characteristic table data to characteristic values derived from data measured during pile load tests. Also see 5.3 (6).

5.4.4 Prefabricated driven piles

5.4.4.1 General

(1) Information on the characteristic base resistance and skin friction of prefabricated driven piles derived from empirical data acquired in recent investigations based on [57], and guidance on their application, are provided below. After DIN EN 12699:2001-05 prefabricated driven piles comprise prefabricated concrete, steel, timber and cast-iron piles, see 2.2.2.

(2) The following table data do not apply to timber and cast-iron piles. The capacity of timber piles may be assessed on the basis of DIN 4026:1975-08.

(3) The elements of the characteristic resistance-settlement curve for prefabricated displacement piles (prefabricated driven piles) corresponding to the proposals in [57] are shown in Figure 5.4 for a settlement up to $s_{ult} = s_g$. See 5.2.2 (3) for details of s_{ult} and s_g.

(4) Differentiation is made between the settlement-dependent pile base resistance R_b (s) and the pile shaft resistance R_s (s).

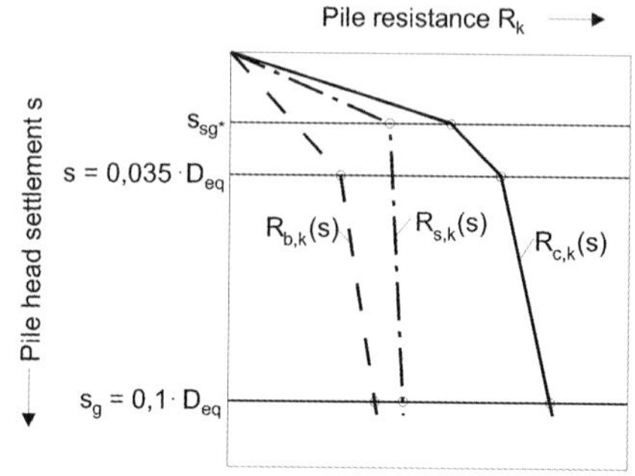

Figure 5.4 Elements of the characteristic resistance-settlement curve for prefabricated driven piles

(5) The limit settlement for $R_{b,k}$ ($s_{ult} = s_g$) applies in analogy to Eq. (5.1):

$$s_{ult} = s_g = 0{,}10 \cdot D_{eq} \tag{5.3}$$

where:

D_{eq} equivalent diameter of the pile base [m].

For square and rectangular, prefabricated piles and steel sections, the equivalent pile base diameter is calculated using Eq. (5.4):

$$D_{eq} = 1{,}13 \cdot a_s \tag{5.4a}$$

for square piles and

$$D_{eq} = 1{,}13 \cdot a_s \cdot \sqrt{a_L / a_s} \tag{5.4b}$$

for rectangular piles and steel sections.

where:

a_s side length of a pile with square cross-section or length of the smaller side of a rectangular, prefabricated pile or steel section;

a_L length of the longer side of the cross-section of a rectangular, prefabricated pile or steel section. For steel sections the respective dimensions of the profile shall be taken.

(6) When mobilising the ultimate limit state, the characteristic settlement for $R_{s,k}$ (s_{sg^*}) is calculated using Eq. (5:5); use MN as dimension of $R_{s,k}$ (s_{sg^*}):

$$s_{sg^*}\,[\text{cm}] = 0{,}5 \cdot R_{s,k}\,(s_{sg^*})\,[\text{MN}] \le 1\,[\text{cm}] \tag{5.5}$$

(7) The characteristic axial pile resistance is determined from:

$$R_{c,k}(s) = R_{b,k}(s) + R_{s,k}(s) = \eta_b \cdot q_{b,k} \cdot A_b + \sum_i \eta_s \cdot q_{s,k,i} \cdot A_{s,i} \qquad (5.6)$$

where:

A_b	nominal pile base area, for steel section piles as shown in Figure 5.5;
$A_{s,i}$	nominal pile shaft area in stratum i; for steel section piles as shown in Figure 5.5 with the effective circumference U;
$q_{b,k}$	characteristic value of the base resistance, derived from Tables 5.1 and 5.3;
$q_{s,k,i}$	characteristic value of the skin friction in stratum i, derived from Tables 5.2 and 5.4;
η_b	base resistance model factor from Table 5.5;
η_s	skin friction model factor from Table 5.5;
$R_{c,k}(s)$	settlement-dependent, characteristic compression pile resistance;
$R_{b,k}(s)$	settlement-dependent, characteristic pile base resistance;
$R_{s,k}(s)$	settlement-dependent, characteristic shaft resistance;
s_{sg*}	characteristic settlement at which mobilisation of the ultimate skin friction for the settlement-dependent, characteristic pile shaft resistance begins;
s_{sg}	limit settlement for the settlement-dependent, characteristic pile shaft resistance ($s_{sg} = s_g$).

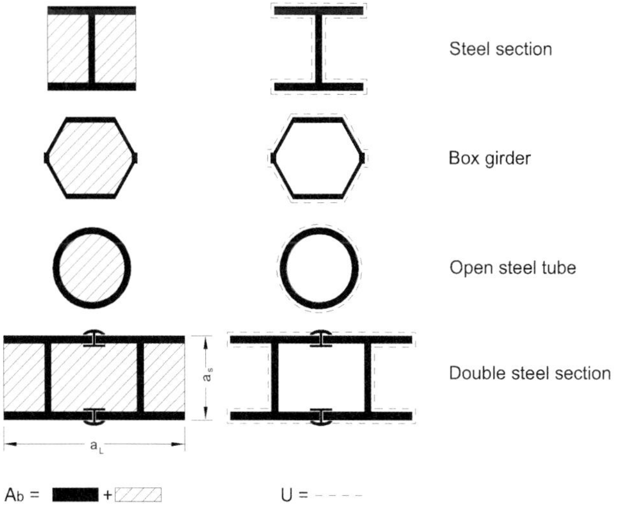

Figure 5.5 Nominal values of pile base areas and pile shaft areas of steel sections

5.4.4.2 Empirical values of base resistance and skin friction of prefabricated driven piles

(1) The empirical base resistance and skin friction data given in Table 5.1 to Table 5.4 apply to the following piles, provided they are embedded at least 2,50 m into a load-bearing layer and the model factors listed in Table 5.5 are taken into account:

- prefabricated, reinforced concrete and prestressed concrete driven piles where D_{eq} = 0,25 to 0,50 m;
- closed-ended steel tube piles with diameters up to 800 mm;
- open-ended steel tube and hollow box piles with diameters between 300 mm and 1 600 mm;
- steel sections with flange widths between 300 mm and 500 mm, and section heights between 290 mm and 500 mm, and
- steel box piles.

The table data apply to prefabricated driven piles. The table values must be reduced when piles are installed through vibration, because substantial capacity reductions have been reported for isolated cases of vibrated piles, e.g. in [92]. The amount of the reduction must be confirmed by the geotechnical expert. For the base resistances the reduction is dispensed with if the pile is driven into the load-bearing stratum for the final $8 \cdot D_{eq}$ (in m).

(2) The validity of the table values depends on:

- the mean cone resistance q_c of the CPT in non-cohesive soil and
- the shear strength of the undrained soil $c_{u,k}$ for cohesive soils.

(3) When specifying the governing mean cone resistance q_c of the CPT or the characteristic undrained shear strength $c_{u,k}$ differentiation shall be made between:

- the zone governing the pile base resistance (from $1 \cdot D_{eq}$ above to $4 \cdot D_{eq}$ below the pile base) and
- the zone governing the skin friction (mean value for the affected stratum).

If the ground stratification has a greater impact on the CPT cone resistance or the undrained shear strength, two or more mean skin friction zones must be specified separately.

(4) When applying the values given in Table 5.1 and Table 5.3 it is assumed that:

- the thickness of the load-bearing stratum below the pile base is not less than five times the equivalent pile base diameter, but at least 1,50 m and;
- $q_c \geq 7{,}5$ MN/m² or $c_{u,k} \geq 100$ kN/m² are confirmed for this zone.

Regardless of this, it is recommended that in non-cohesive soils the pile bases are founded in zones where $q_c \geq 10$ MN/m².

(5) If the above geometrical values given are not met, analysis of safety against a punching failure is required. In addition, it must then be verified that the underlying ground does not substantially impair the settlement behaviour.

(6) If Tables 5.1 to 5.5 are applied in approximation to mixed-grained soils, the definitions for cohesive and non-cohesive soils in accordance with the EC 7-1 Handbook [44] or 3.3 may be adopted. If the table values are applied to mixed-grained soils, particular attention should be paid to 5.4.2 (5).

(7) Also see 5.4.3 for the conditions for applying the table values.

Table 5.1 Empirical data ranges for the characteristic base resistance $q_{b,k}$ for prefabricated, reinforced concrete and prestressed concrete driven piles in non-cohesive soils

Relative pile head settlement s/D_{eq}	Pile base resistance $q_{b,k}$ [kN/m²]		
	mean CPT cone resistance q_c [MN/m²]		
	7,5	15	25
0,035	2 200–5 000	4 000–6 500	4 500–7 500
0,100	4 200–6 000	7 600–10 200	8 750–11 500
Intermediate values may be linearly interpolated.			

Table 5.2 Empirical data ranges for the characteristic skin friction $q_{s,k}$ for prefabricated, reinforced concrete and prestressed concrete driven piles in non-cohesive soils

Settlement	Pile skin friction $q_{s,k}$ [kN/m²]		
	mean CPT cone resistance q_c [MN/m²]		
	7,5	15	25
s_{sg^*}	30–40	65–90	85–120
$s_{sg} = s_g = 0,1\ D_{eq}$	40–60	95–125	125–160
Intermediate values may be linearly interpolated.			

Table 5.3 Empirical data ranges for the characteristic base resistance $q_{b,k}$ for prefabricated, reinforced concrete and prestressed concrete driven piles in cohesive soils

Relative pile head settlement s/D_{eq}	Pile base resistance $q_{b,k}$ [kN/m²]		
	shear strength $c_{u,k}$ of the undrained soil [kN/m²]		
	100	150	250
0,035	350–450	550–700	800–950
0,100	600–750	850–1 100	1 150–1 500
Intermediate values may be linearly interpolated.			

Table 5.4 Empirical data ranges for the characteristic skin friction, $q_{s,k}$, for prefabricated, reinforced concrete and prestressed concrete driven piles in cohesive soils

Settlement	Pile skin friction $q_{s,k}$ [kN/m²]		
	shear strength $c_{u,k}$ of the undrained soil [kN/m²]		
	60	150	250
s_{sg^*}	20–30	35–50	45–65
$s_{sg} = s_g = 0{,}1D_{eq}$	20–35	40–60	55–80
Intermediate values may be linearly interpolated.			

Table 5.5 Model factors for pile base resistance and skin friction η_b or η_s for prefabricated driven piles when adopting the values from Tables 5.1 to 5.4

Pile type		η_b	η_s
Reinforced concrete and prestressed concrete		1,00	1,00
Steel section[1)] ($h \le 0{,}50$ m and $h/b_F \le 1{,}5$)	$s = 0{,}035 \cdot D_{eq}$	$0{,}61 - 0{,}30 \cdot h/b_F$	0,60
	$s = 0{,}10 \cdot D_{eq}$	$0{,}78 - 0{,}30\ h/b_F$	
Double steel section		0,25	0,60
Open-ended steel tube and hollow box ($0{,}3\ \text{m} \le D_b \le 1{,}60$ m)		$0{,}95 \cdot e^{-1{,}2 \cdot D_b}$	$1{,}1 \cdot e^{-0{,}63 \cdot D_b}$
Closed steel tube ($D_b \le 0{,}80$ m)		0,80	0,60
[1)] h = height of steel girder section, b_F = flange width of steel girder section			

Note: Only a small pile load test database is available for the values in Tables 5.3 and 5.4 for prefabricated driven piles in cohesive soils. The empirical data were therefore carefully selected and normally return smaller pile resistances than those given across-the-board e.g. in the old DIN 4026:1975-08. Moreover, the values given in DIN 4026 were based on a lower global safety factor of $\eta \approx 1{,}5$ compared to today's provisions.

(8) An example for determining the characteristic resistances of prefabricated driven piles is included in Annex B4.

5.4.4.3 Empirical data on the bearing capacity of open-ended steel tubes and hollow boxes

(1) When driving open-ended steel tubes or hollow boxes, high stresses can develop in the penetrating soil through compaction and clogging inside the bottom of the steel section. This mobilises an increased internal pile skin friction. The soil penetrating the section is called a plug, regardless of any developed stresses.

(2) Regardless of the stress effects within the open pile section and the resulting skin friction mentioned in (1), a simplified procedure for determining the characteristic pile resistances of open-ended steel tubes and hollow box sections from empirical data may be adopted, such that in approximation a closed pile base (plug) is numerically adopted for the applications given in Table 5.5. By adopting these assumptions and the model factors given in Table 5.5 for open-ended steel tubes and hollow boxes, the characteristic pile resistance can be determined using the methods described in 5.4.4.2 and 5.4.4.3. The correlation factors apply to pile diameters 0,3 m $\leq D \leq$ 1,6 m. Additional guidance on methods for determining pile resistances for pile diameters $D >$ 1,6 m are included in [85].

(3) For open-ended, non-circular sections, the empirical data ranges used in Tables 5.1 to 5.5 may also be adopted in approximation if no other information is available. An equivalent diameter D_{eq} for an imaginary tube section can be determined using Equation (5.4) via the respective, equivalent pile base area.

(4) Additional guidance on load transfer within a plug can be found in [85] and [86], together with a summary of related publications.

(5) The internal pile skin friction decreases with increasing pile diameter and decreasing density of the soil. High dilatancy of non-cohesive soils, on the other hand, impacts positively on the value of the internal pile skin friction. Excessively high pile driving energy leads to the plug being pushed through the pile and, consequently, reduces the pile capacity. A reduction in internal pile skin friction is also anticipated if other pile installation methods are used, like pressing or vibrating. If the pile base is enlarged the driving energy can possibly be reduced; however, the internal skin friction is also reduced.

5.4.4.4 Experience with prefabricated piles in rock and very dense or cemented soils

(1) When driving prefabricated reinforced concrete, prestressed concrete, steel or timber piles into cemented soil or rock, as well as into very dense and cemented soils, special conditions must be observed depending on the soil or rock properties:

- Prefabricated piles can not or only with limitations be driven into rock or cemented soil.
- Driving into unaltered or partially weathered rock is not normally possible (Table 3.6). When reaching the rockhead, the driving resistance achieves a transitionless maximum. Under these conditions, loads can only be transferred via base resistance. The applicability of the pile system under these conditions should be checked by a geotechnical expert on a case-by-case basis.
- In cemented soil and in rock with a weathering zone the achievable embedment length is primarily a function of the degree of weathering. This is

especially the case for variable strength rock types. Depending on the degree of weathering, the achievable embedment depth can be between a few centimetres and several metres. In such cases, the pile resistance should be verified by load tests. Generally, an abrupt increase in driving resistance can be identified when slightly weathered rock strata are reached.

(2) For prefabricated full displacement piles driven onto or into rock, often limit resistances are achieved which are in the order of the 'internal' capacity (structural resistance) of the piles. Consequently, these determine the pile resistance. A minimum embedment depth as for piles in soils is not required.

5.4.5 Cast-in-place concrete piles

5.4.5.1 General

(1) The elements of the characteristic resistance-settlement curve are shown in Figure 5.4 for settlements up to $s_{ult} = s_g$. See 5.2.2 (3) for details of s_{ult} and s_g.

The equivalent diameter D_{eq} is replaced here by the diameter of the base plate, D_b, for Simplex piles and the diameter of the drive tube, D_s, for Franki piles.

(2) For Simplex piles, the settlement-dependent pile base resistance $R_b(s)$ and the shaft resistance $R_s(s)$ are differentiated..

(3) For $R_{b,k}$ ($s_{ult} = s_g$) the limit settlement after Eq. (5.1) applies..

(4) When mobilising the ultimate limit state the characteristic settlement for $R_{s,k}$ (s_{sg*}) applies; use MN as dimension of $R_{s,k}$ (s_{sg*}):

$$s_{sg*}\,[\mathrm{cm}] = 0{,}5 \cdot R_{s,k}\,(s_{sg*})\,[\mathrm{MN}] \leq 1\,[\mathrm{cm}] \tag{5.7}$$

(5) The characteristic axial pile resistance of Simplex piles shall be determined from:

$$R_{c,k}(s) = R_{b,k}(s) + R_{s,k}(s) = q_{b,k} \cdot A_b + \sum_i q_{s,k,i} \cdot A_{s,i} \tag{5.8}$$

where:

A_b	nominal pile base area; governing is the diameter of the base plate D_b;
$A_{s,i}$	nominal pile shaft area in stratum i; governing is the diameter of the drive tube D_s;
$q_{b,k}$	characteristic value of the tip resistance, derived from Table 5.6;
$q_{s,k,i}$	characteristic value of the pile skin friction in stratum i, derived from Tables 5.7 and 5.8;
$R_{c,k}(s)$	settlement-dependent, characteristic compressive pile resistance;
$R_{b,k}(s)$	settlement-dependent, characteristic pile base resistance;
$R_{s,k}(s)$	settlement-dependent, characteristic shaft resistance;
s_{sg*}	characteristic settlement for mobilising the ultimate skin friction for the settlement-dependent characteristic pile shaft resistance, where $R_{s,k}(s_{sg*}) = R_{s,k}(s_{sg})$;

s_{sg} limit settlement for the settlement-dependent, characteristic pile shaft resistance ($s_{sg} = s_g$).

(6) The characteristic axial pile resistance of Franki piles in the ultimate limit state shall be determined from:

$$R_{c,k} = R_{b,k} + R_{s,k} = R_{b,k} + \sum_i q_{s,k,i} \cdot A_{s,i} \quad (5.9)$$

where:

$R_{b,k}$ characteristic pile base resistance after Figures 5.6 to 5.11;

$A_{s,i}$ nominal pile shaft area in stratum i higher than 0,80 m above the driving depth with governing diameter of the drive tube D_s;

$q_{s,k,i}$ characteristic value of the pile skin friction in stratum i, derived from Tables 5.10 and 5.11.

5.4.5.2 Empirical values of base resistance and skin friction of Simplex piles

(1) The table values apply to cast-in-place concrete piles with driven drive tube. The table values shall be reduced for vibrated piles, see 5.4.4.2 (1).

(2) The validity of the table values depends on:

- the mean cone resistance, q_c, of the CPT in non-cohesive soil and
- the shear strength of the undrained soil, $c_{u,k}$, in cohesive soils.

(3) When specifying the governing mean cone resistance q_c, of the CPT or the characteristic undrained shear strength $c_{u,k}$ the following zones shall be differentiated:

- the zone governing the end bearing capacity (from $1 \cdot D_{eq}$ above to $4 \cdot D_{eq}$ below the pile base) and
- the zone governing the skin friction (mean value for the affected stratum).

If the ground stratification has a great influence on the CPT cone resistance or the undrained shear strength, two or more mean skin friction zones must be specified separately.

(4) Condition for application of the values in Table 5.6 is that:

- the thickness of the load-bearing stratum below the pile base is not less than five times the pile base diameter, but at least 1,50 m and
- $q_c \geq 7{,}5$ MN/m² is confirmed for this zone..

Regardless of this, it is recommended that the pile bases are founded in zones where $q_c \geq 10$ MN/m² in non-cohesive soils.

(5) If the above geometrical values are not met, analysis of safety against a punching failure is required. In addition, it must then be verified that the underlying ground does not substantially impair settlement behaviour.

(6) If Tables 5.6 to 5.8 are applied in approximation to mixed-grained soils, the definitions for cohesive and non-cohesive soils in accordance with the

EC 7-1 Handbook [44] or 3.3 may be adopted. If the table values are applied to mixed-grained soils, particular attention should be paid to 5.4.2 (5).

Table 5.6 Empirical data ranges for the characteristic base resistance $q_{b,k}$ for Simplex piles in non-cohesive soils

Relative pile head settlement s/D_b	Pile base resistance $q_{b,k}$ [kN/m²]		
	mean CPT cone resistance q_c [MN/m²]		
	7,5	15	25
0,035	2 200–5 000	4 000–6 500	4 500–7 500
0,100	4 200–6 000	7 600–10 200	8 750–11 500
Intermediate values may be linearly interpolated.			

Table 5.7 Empirical data ranges for the characteristic skin friction $q_{s,k}$ for Simplex piles in non-cohesive soils

Settlement	Pile skin friction $q_{s,k}$ [kN/m²]		
	mean CPT cone resistance q_c [MN/m²]		
	7,5	15	25
s_{sg^*}	55–70	105–135	130–165
$s_{sg} = s_g = 0{,}1D_b$	55–70	105–135	130–165
Intermediate values may be linearly interpolated.			

Table 5.8 Empirical data ranges for the characteristic skin friction $q_{s,k}$ for Simplex piles in cohesive soils

Settlement	Pile skin friction $q_{s,k}$ [kN/m²]		
	Shear strength $c_{u,k}$ of the undrained soil [kN/m²]		
	60	150	250
s_{sg^*}	25–40	45–65	60–85
$s_{sg} = s_g = 0{,}1D_b$	25–40	45–65	60–85
Intermediate values may be linearly interpolated.			

(7) Also see 5.4.3 for conditions for applying the table values.

Note: The data base for pile base resistances $q_{b,k}$ in cohesive soils is still too small to allow empirical data to be derived.

5.4.5.3 Empirical values of base resistance and skin friction of Franki piles

(1) The characteristic feature in terms of the axial pile resistance in this pile system is the enlarged pile base relative to the pile shaft. This is achieved by internal driving of almost dry concrete from the bottom of the drive tube.

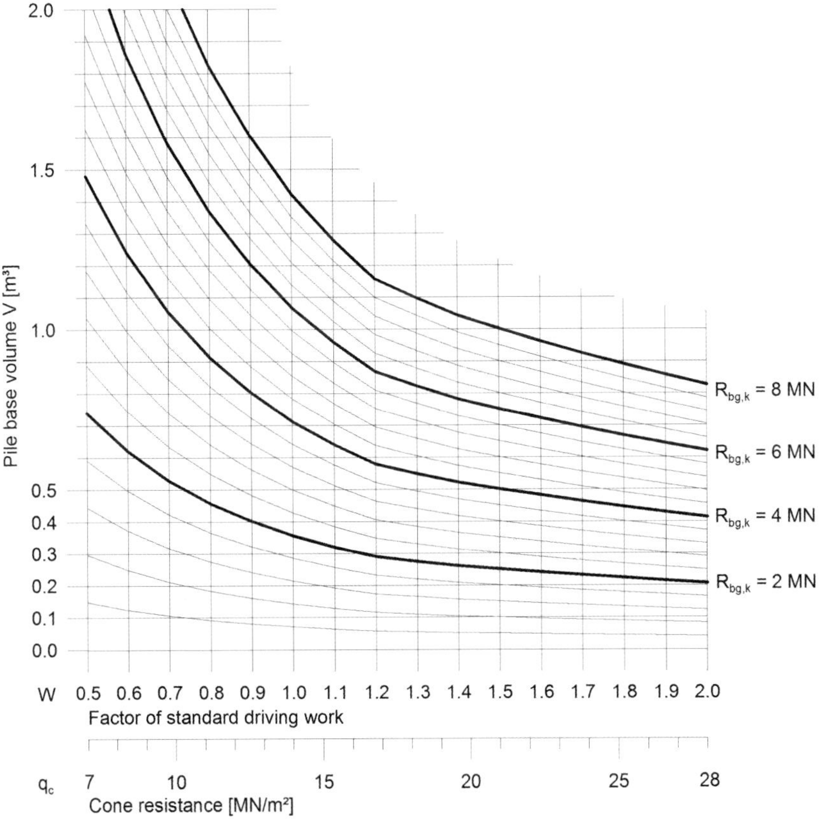

Figure 5.6 Lower bound empirical data for base resistances (ULS) and necessary base volume of Franki piles in non-cohesive soils

(2) Dimensioning the necessary base volume is carried out as a function of the required pile base resistance and the driving work expended when driving the drive tube for the final two metres. Base design nomograms have been established for this purpose, see Figures 5.6 to 5.11, which also relate the pile base design to the locally determined bearing capacity of the ground.

(3) The expended driving work W_{actual} is defined relative to a standard driving work W_{norm} which is a function of the diameter of the driven tube and the mass of the hammer. The values of the standard driving work are given in Table 5.9. The quotient W_{actual}/W_{norm} is defined as the standard driving work component W and is the input variable for the base design nomograms in Figures 5.6 to 5.11.

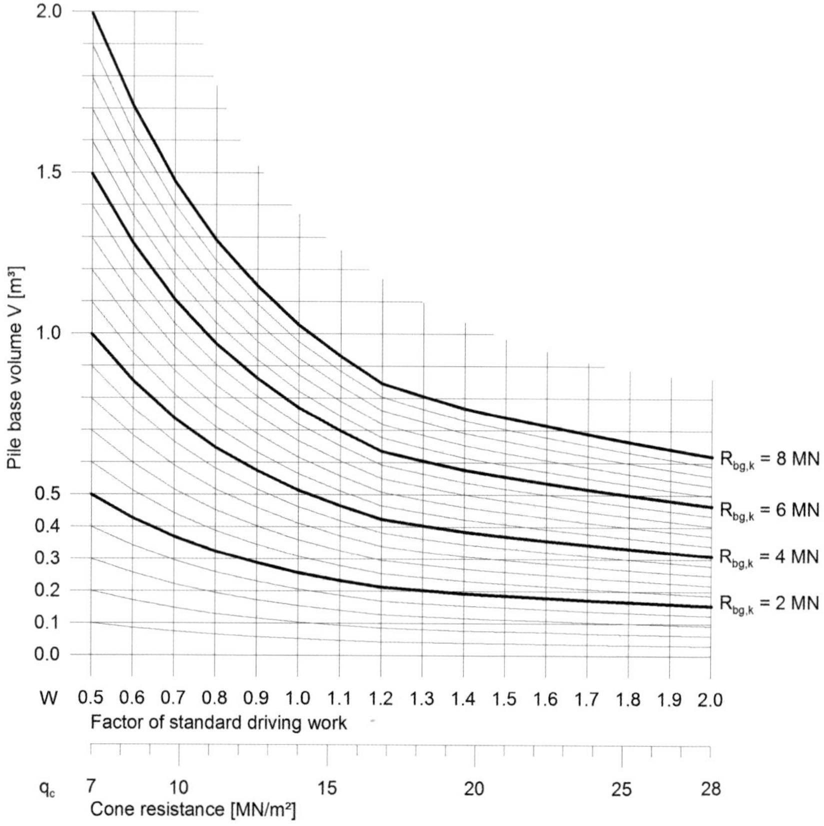

Figure 5.7 Upper bound empirical data for pile base resistances (ULS) and necessary base volume of Franki piles in non- cohesive soils

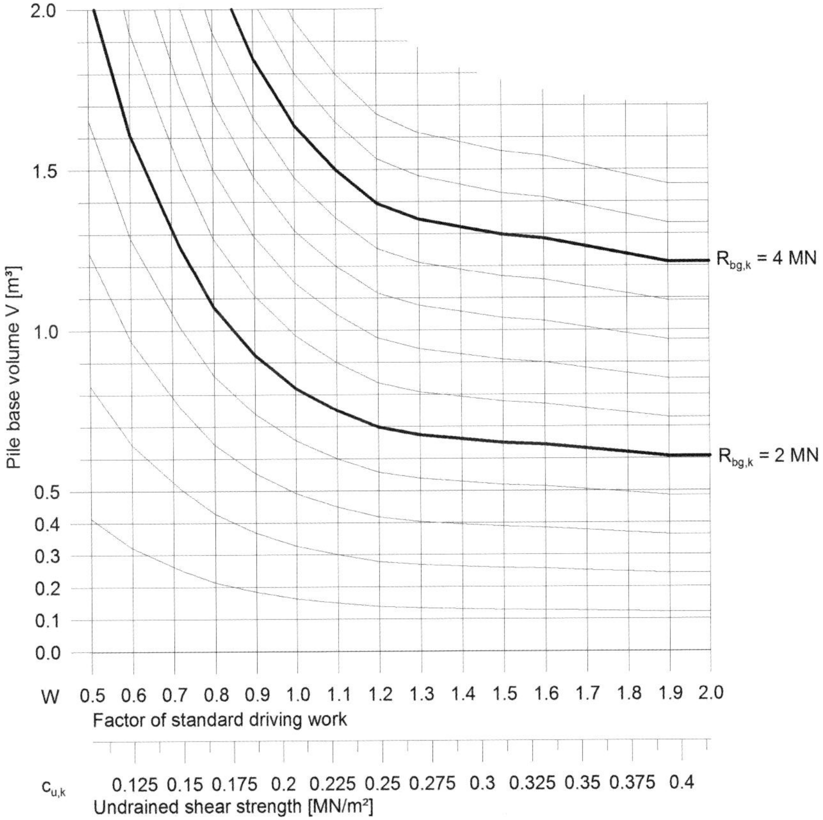

Figure 5.8 Lower bound empirical data for pile base resistances (ULS) and necessary base volume of Franki piles in cohesive soils

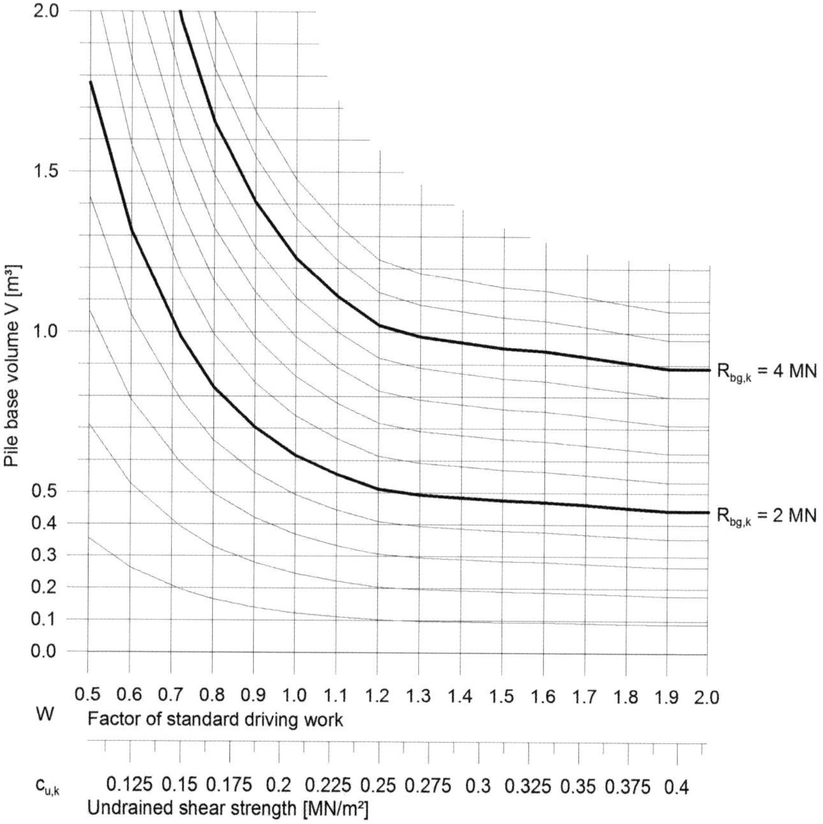

Figure 5.9 Upper bound empirical data for pile base resistances (ULS) and necessary base volume of Franki piles in cohesive soils

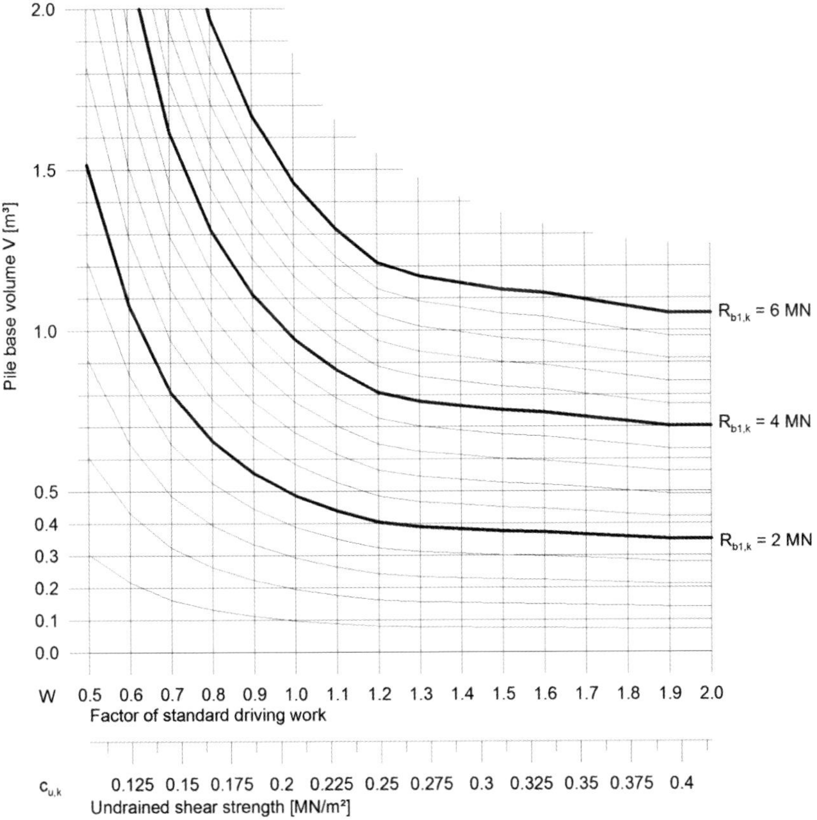

Figure 5.10 Lower bound empirical data for pile base resistances (ULS) and necessary base volume of Franki piles in boulder clay

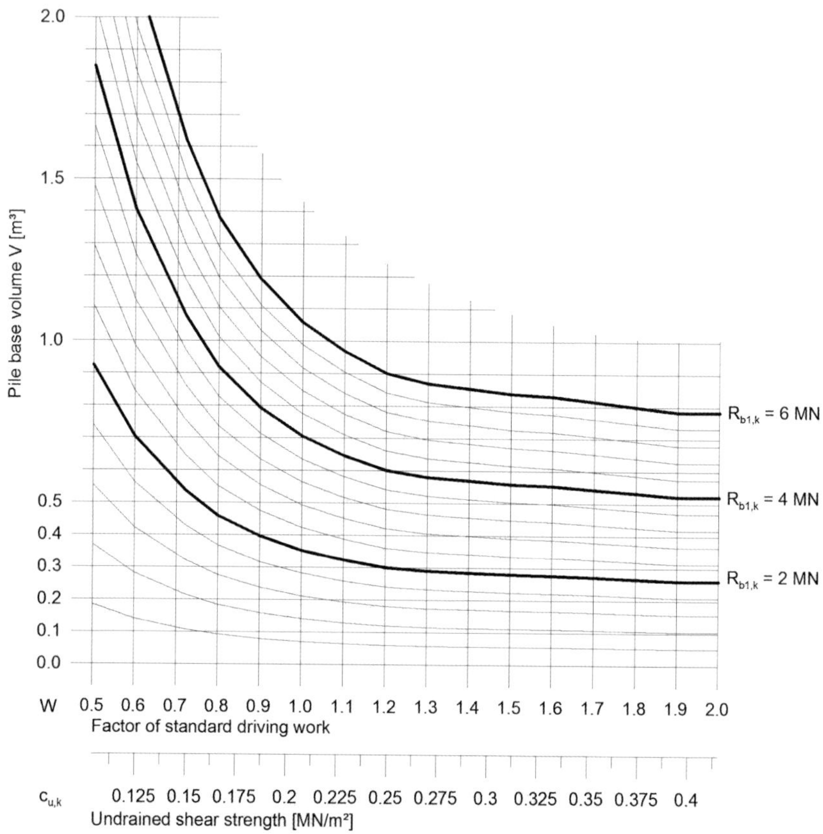

Figure 5.11 Upper bound empirical data for pile base resistances (ULS) and necessary base volume of Franki piles in boulder clay

Table 5.9 Standard driving work W_{norm} for vertical Franki piles

Tube diameter D_s [cm]	Hammer weight [kN]	Drop height [m]	Number of blows/2 m	Standard driving work W_{norm} [kNm]
42	22,0	6,5	125	17 875
51	30,0	6,5	125	24 375
56	37,5	6,5	125	30 469
61	45,0	6,5	125	36 563

(4) When applying Table 5.10 it is assumed that the mean cone resistance governing the pile skin friction is defined separately for two or more zones, if ground stratification has a major impact on cone resistance.

Table 5.10 Empirical data ranges for the characteristic skin friction $q_{s,k}$ for Franki piles in non-cohesive soils

Mean cone resistance q_c of the CPT [MN/m²]	Ultimate limit state value $q_{s,k}$ of pile skin friction [kN/m²]
7,5	70–95
15	115–150
≥ 25	135–180
Intermediate values may be linearly interpolated.	

Table 5.11 Empirical data ranges for the characteristic skin friction $q_{s,k}$ for Franki piles in cohesive soils

Shear strength $c_{u,k}$ of the undrained soil [kN/m²]	Ultimate limit state value $q_{s,k}$ of pile skin friction [kN/m²]
60	35–45
150	55–70
≥ 250	70–90
Intermediate values may be linearly interpolated.	

(5) Conditions for the use of the nomograms are:

- a governing zone ($2 \cdot D_S$ or at least 1,0 m above and up to $3 \cdot D_S$ or at least 1,50 m below the driving depth) is taken into consideration for the pile end bearing capacity;
- the proportion of the standard driving work is $W = \frac{W_{actual}}{W_{norm}} \geq 0,5$ for the final two driving metres;
- the drive tube is withdrawn 0,8 m for the formation of the pile base;
- skin friction is only adopted from 0,8 m above the driving depth in accordance with Tables 5.10 and 5.11.

Note: Because a large number of static pile load test data are available, pile capacity data for North German boulder clay (mixed grained soils) have also been included for these pile systems.
When assigning mixed-grained soils to either non-cohesive or cohesive soils, the regional geological conditions, e.g. pre-consolidation, must be

taken into consideration, in addition to grain size distribution and plastic properties. Also influences arising from pile installation on the bearing capacity of the soils, e.g. pile type, dynamic actions, water, etc. must be taken into account. A boulder clay, of which glacial preconsolidation has been proven, may, e.g. and in individual cases, be classified as a non-cohesive soil despite its cohesive character as stated in both the EC 7-1 Handbook [44] and in DIN 18196. The prerequisite for this is a reliable, regional geological correlation by the geotechnical expert.

(6) Using this pile system, prior to pile installation, ground improvement can be implemented by means of gravel compaction in the pile base and shaft zones. In such cases the driving work for re-driving the drive tube subsequent to successful gravel compaction is the governing factor for dimensioning the base volume.

(7) When installing raked piles the expended driving work is reduced by the factor f:

- pile inclination: vertical to 10:1 $\rightarrow$ f = 1,00;
- pile inclination to 18:1 $\rightarrow$ f = 0,95;
- pile inclination to 16:1 $\rightarrow$ f = 0,90;
- pile inclination to 14:1 $\rightarrow$ f = 0,85.

(8) Additional analysis of the serviceability limit state (SLS) is normally not necessary, because the settlements in the serviceability limit state for this pile system are very low and in general lie empirically in the region of 0,5 to 1,0 cm.

(9) An example for determining the pile resistances and for pile design is included in Annex B7.

5.4.6 Bored piles

5.4.6.1 General

(1) The elements of the characteristic resistance-settlement curve for bored piles are shown in Figure 5.12 for settlement up to $s_{ult} = s_g$. See 5.2.2 (3) for details of s_{ult} and s_g.

(2) The settlement-dependent pile base resistance $R_b(s)$ and the pile shaft resistance $R_s(s)$ are differentiated.

(3) The limit settlement applies in analogy to Eq. (5.1) for $R_{b,k}$ ($s_{ult} = s_g$):

$$s_g = 0{,}10 \cdot D_b \tag{5.10}$$

where:

D_b diameter of the pile base in m.

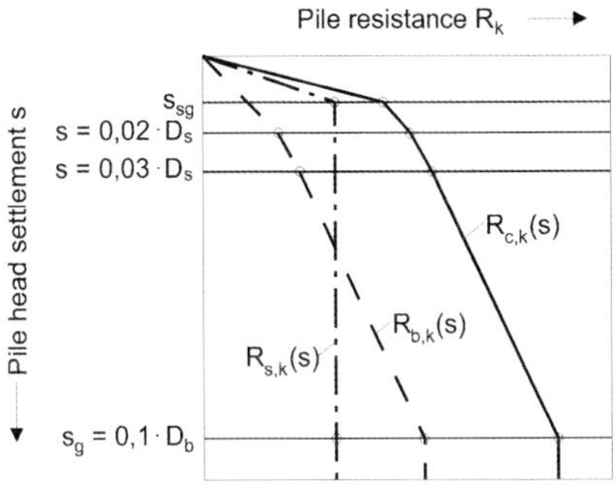

Figure 5.12 Elements of the characteristic resistance-settlement curve for bored piles

(4) The governing reference variable for settlement in the serviceability limit state is the pile shaft diameter D_s. On piles without an enlarged base the pile diameter D (here: $D = D_s = D_b$) is adopted as the reference variable for settlement at the ultimate limit state. The pile diameter D_b is the governing reference variable for piles with an enlarged base. The diaphragm wall thickness D shall be adopted for diaphragm wall elements barettes).

(5) The limit settlement applies for the characteristic pile shaft resistance $R_{s,k}(s_{sg})$ in MN in at ultimate limit state:

$$s_{sg}\,[\text{cm}] = 0{,}5 \cdot R_{s,k}\left(s_{sg}\right)[\text{MN}] + 0{,}5\,[\text{cm}] \le 3\,[\text{cm}] \tag{5.11}$$

(6) The characteristic axial pile resistance is determined from

$$R_{c,k}(s) = R_{b,k}(s) + R_{s,k}(s) = q_{b,k} \cdot A_b + \sum_i q_{s,k,i} \cdot A_{s,i} \tag{5.12}$$

where:

A_b	nominal value of the pile base area;
$A_{s,i}$	nominal value of the pile shaft area in stratum i;
$q_{b,k}$	characteristic value of the base resistance, derived from Tables 5.12 and 5.14;
$q_{s,k,i}$	characteristic value of the skin friction in stratum i, derived from Tables 5.13 and 5.15;
$R_{c,k}(s)$	settlement-dependent, characteristic compressive pile resistance;
$R_{b,k}(s)$	settlement-dependent, characteristic base resistance;
$R_{s,k}(s)$	settlement-dependent, characteristic shaft resistance;
s_{sg}	limit settlement for the settlement-dependent characteristic shaft resistance.

5.4.6.2 Empirical values of base resistance and skin friction of bored piles

(1) The empirical data for pile base resistance and skin friction given in Tables 5.12 to 5.15 apply to bored piles from D_s or D_b = 0,30 to 3,00 m, which embed at least 2,50 m into a load-bearing stratum and depend on:

- the mean cone resistance q_c of the CPT with depth in non-cohesive soil and
- the shear strength of the undrained soil $c_{u,k}$ for cohesive soils.

Note: The magnitude of the lower table values (minimum values) were first adopted in DIN 4014:1990-03 on the basis of the investigations by [29].

(2) When specifying the governing mean cone resistance q_c of the CPT or the characteristic undrained shear strength $c_{u,k}$ differentiation shall be made between:

- the zone governing the base resistance from $1 \cdot D_b$ above and $4 \cdot D_b$ below the pile base for pile diameters up to D_b = 0,6 m, and $1 \cdot D_b$ above and $3 \cdot D_b$ below the pile base for diameters greater than D_b = 0,6 m and
- the zone governing the skin friction (mean value for the affected stratum);

If ground stratification has a great influence on the CPT cone resistance or the undrained shear strength, two or more mean pile skin friction zones must be specified separately.

(3) Condition for the application of the values of Tables 5.12 and 5.14 are:

- the thickness of the load-bearing layer below the pile base is not less than 3 times the pile base diameter, but at least 1,50 m and
- $q_c \geq 7{,}5$ MN/m² or $c_{u,k} \geq 100$ kN/m² is confirmed in this zone.

Regardless of this, founding the pile bases in zones where $q_c \geq 10$ MN/m² is recommended.

(4) If the above geometrical values are not met, analysis of safety against a punching failure is required. In addition, it must then be verified that the underlying ground does not substantially impair settlement behaviour.

(5) If Tables 5.12 to 5.15 are applied in approximation to mixed-grained soils, the definitions for cohesive and non-cohesive soils in accordance with the EC 7-1 Handbook [44] or 3.3 may be adopted. If the table values are applied to mixed-grained soils, particular attention should be paid to 5.4.2 (5).

(6) Also see 5.4.3 for the conditions for applying the table values.

(7) An example for determining the characteristic resistances of bored piles is included in Annex B3.

Table 5.12 Empirical data ranges for the characteristic base resistance $q_{b,k}$ for bored piles in non-cohesive soils

Relative settlement of the pile head s/D_s or s/D_b	Pile base resistance $q_{b,k}$ [kN/m²] mean CPT cone resistance q_c [MN/m²]		
	7,5	15	25
0,02	550–800	1 050–1 400	1 750–2 300
0,03	700–1 050	1 350–1 800	2 250–2 950
0,10 ($\hat{=}$ s_g)	1 600–2 300	3 000–4 000	4 000–5 300
Intermediate values may be linearly interpolated. For bored piles with enlarged base the values shall be reduced to 75 %.			

Table 5.13 Empirical data ranges for the characteristic skin friction $q_{s,k}$ for bored piles in non-cohesive soils

Mean CPT cone resistance q_c [MN/m²]	Ultimate limit state value $q_{s,k}$ of pile skin friction [kN/m²]
7,5	55–80
15	105–140
≥ 25	130–170
Intermediate values may be linearly interpolated.	

Table 5.14 Empirical data ranges for the characteristic base resistance $q_{b,k}$ for bored piles in cohesive soils

Relative settlement of the pile head s/D_s or s/D_b	Pile base resistance $q_{b,k}$ [kN/m²] Shear strength $c_{u,k}$ of the undrained soil [kN/m²]		
	100	150	250
0,02	350–450	600–750	950–1 200
0,03	450–550	700–900	1 200–1 450
0,10 ($\hat{=}$ s_g)	800–1 000	1 200–1 500	1 600–2 000
Intermediate values may be linearly interpolated. For bored piles with a flared base the values are reduced to 75 %.			

Table 5.15 Empirical data ranges for the characteristic skin friction $q_{s,k}$ for bored piles in cohesive soils

Shear strength $c_{u,k}$ of the undrained soil [kN/m²]	Ultimate limit state value $q_{s,k}$ of pile skin friction [kN/m²]
60	30–40
150	50–65
≥ 250	65–85
Intermediate values may be linearly interpolated.	

5.4.6.3 Empirical data for base resistance and skin friction of piles in rock and cemented soils

(1) The data base for deriving characteristic pile resistances from pile load tests in rock and cemented soil is comparatively small and does not allow statistical deduction of empirical values. This is due to the specific difficulties when recording and describing the rock mechanics properties of soft and weathered rock. Therefore, the execution of pile load tests is strongly recommended, see [23] and [100].

Tables 5.16 to 5.19 summarise the available empirical data and allow the rough estimation of the range of anticipated pile capacities, on condition that the common methods for investigation and classification of rock and soft rock are adopted and a geotechnical expert is consulted. This includes such features as compressive rock strength, discontinuities, degree of weathering, decomposition and disintegration status, as well as mineralogy and texture.

(2) On condition of:

- minimum 0,5 m embedment of the bored piles into rock of an unconfined compressive strength $q_{u,k} \geq 5$ MN/m² or;
- minimum 2,50 m embedment into rock of an unconfined compressive strength $q_{u,k} \leq 0{,}50$ MN/m²

the values for the ultimate limit state (ULS) are given in Table 5.16 for the pile base resistance and the skin friction as a function of the unconfined compressive strength $q_{u,k}$ of the rock samples.

(3) Conditions for applying the empirical values given in Table 5.16 are that:

- the rock is homogeneous and present with sufficient thickness;
- the spatial orientation of the rock surface and of the discontinuities do not promote failure mechanisms;
- no open discontinuities are present, nor discontinuities filled with easily deformable materials and;
- a reduction in strength as a result of drilling cannot occur, e.g. due to ingress of water in claystone and marlstone.

Table 5.16 Empirical data ranges for the characteristic base resistance $q_{b,k}$ and characteristic pile skin friction $q_{s,k}$ for bored piles in rock

Unconfined compressive strength $q_{u,k}$ [MN/m²]	Ultimate limit state value $q_{b,k}$ of base resistance [kN/m²]	Ultimate limit state value $q_{s,k}$ of skin friction [kN/m²]
0,5	1 500–2 500	70–250
5,0	5 000–10 000	500–1 000
20,0	10 000–20 000	500–2 000
Intermediate values may be linearly interpolated.		

(4) It shall be checked for the specific case if the anticipated pile settlements in the rock justify the adoption of skin friction in overburden layers. If in doubt, skin friction in such layers shall not be adopted.

(5) Alternatively, the ultimate limit state value of the base resistance $q_{b,k}$ in rock or cemented soil may be estimated on the basis of the weathering state and the degree of mineral bonding as shown in Table 5.17. The pile skin friction ultimate limit state value $q_{s,k}$ may be adopted at one tenth of the values of Table 5.17.

Table 5.17 Reference values for the characteristic base resistance $q_{b,k}$ for bored piles in rock or soft rock

Weathering state and degree of mineral bonding*)	Ultimate limit state value $q_{b,k}$ of pile base resistance [kN/m²]		
	Rock type		
	Massive igneous rocks and metamorphites, e.g. granite, gabbro, basalt, gneiss	Conglomerates, breccias, sandstone, limestone, dolomites	Marlstone, siltstone, claystone
Unaltered, very good mineral bonding	16 000	11 000	8 000
Partially weathered, good mineral bonding	9 000	6 000	4 000
Moderately weathered, moderate mineral bonding	4 000	3 000	values for soils apply
Decomposed or completey altered, poor or no mineral bonding	values for soils apply.		

*) Mineral bonding to DIN 4022

(6) The values in Table 5.17 apply to wide spacing of discontinuities of 1,0 m and more. For smaller spacing, the values shall be reduced by at least 25 %.

(7) Conditions for the application of the values of Table 5.17 are that:

- the piles are at least 3 m long;
- the embedment depth into the rock corresponds to at least half of the pile diameter, and;
- no open discontinuities are present, nor discontinuities filled with easily deformable materials.

(8) For piles in siltstone, mudstone and sandstone the ultimate limit state values (ULS) given in Table 5.18 and 5.19 for the pile base resistance $q_{b,k}$, and for the skin friction $q_{s,k}$ can be adopted on the basis of rock type-specific empirical data after [30] and [102]. The requirements stipulated in (6) apply.

(9) Piles in compact rock normally only display minor settlement, meaning that analysis of the serviceability limit state can be dispensed with if the ultimate limit state is adhered to. In contrast to this, the empirical data available

Table 5.18 Empirical data for the characteristic base resistance $q_{b,k}$ and characteristic skin friction $q_{s,k}$ for bored piles in siltstone and claystone

<table>
<tr><th rowspan="2">Rock type</th><th rowspan="2">Strength acc. to EC 7-1 Handbook [44]</th><th colspan="3">Degree of weathering</th><th colspan="2">Guide parameter</th><th colspan="2">Pile resistances</th></tr>
<tr><th colspan="2">to [153]</th><th>to [33]</th><th>$q_{u,k}$ [MN/m^2]</th><th>w_n [%]</th><th>$q_{s,k}$ [kN/m^2]</th><th>$q_{b,k}$ [kN/m^2]</th></tr>
<tr><td rowspan="2">Rock</td><td>Hard to very hard</td><td>Unaltered</td><td>V0</td><td>VU</td><td>> 100</td><td>4–8</td><td>800</td><td>8 000</td></tr>
<tr><td>Hard</td><td>Partially weathered</td><td>V1</td><td rowspan="2">VA</td><td>> 50</td><td>5–10</td><td>400</td><td>4 000</td></tr>
<tr><td rowspan="5">Soft rock</td><td>Moderately hard</td><td rowspan="2">Moderately weathered</td><td rowspan="2">V2</td><td>12,5–50</td><td rowspan="4">8–16</td><td rowspan="2">300</td><td rowspan="2">3 500</td></tr>
<tr><td rowspan="2">Moderately weak</td><td rowspan="3">VE</td><td rowspan="2">5–12,5</td></tr>
<tr><td rowspan="2">Weathered</td><td rowspan="2">V3</td><td rowspan="2">200</td><td rowspan="2">2 500</td></tr>
<tr><td>Weak</td><td>1,25–5</td></tr>
<tr><td>Very weak</td><td>Highly weathered</td><td>V4</td><td>VZ</td><td>< 1,25</td><td>14–20</td><td>90</td><td>1 600</td></tr>
<tr><td>Completely weathered rock</td><td>Crumbly/soil</td><td>Totally weathered</td><td>V5</td><td>Soil</td><td>< 0,6</td><td>18–30</td><td>60</td><td>1 000</td></tr>
</table>

for the dependence on deformation of the mobilisation of skin friction and base resistance of bored piles installed in rock and soft rock display broad scatter, see [131]. However, the pile settlements required to mobilise the ultimate skin friction are normally considerably lower than in cohesive soils. If the deformation-dependent mobilisation of the pile resistance (equivalent spring stiffness) influences the design of the foundation and the structure, static pile load tests are recommended to determine the resistance-settlement behaviour of the bored piles.

(10) The empirical data given in Tables 5.16 to 5.19 are shown as a function of differing input parameters. A geotechnical expert with experience of piling in rock should always be involved when selecting suitable characteristic pile resistances from the tables. Also 5.4.1 (4) shall be observed.

Table 5.19 Empirical data for the characteristic base resistance $q_{b,k}$ and characteristic skin friction $q_{s,k}$ for bored piles in sandstone

<table>
<tr><th rowspan="2">Rock type</th><th rowspan="2">Strength acc. to EC 7-1 Handbook [44]</th><th>Degree of weathering</th><th>Guide parameter</th><th colspan="2">Pile resistances</th></tr>
<tr><th>To [33]</th><th>$q_{u,k}$ [MN/m^2]</th><th>$q_{s,k}$ [kN/m^2]</th><th>$q_{b,k}$ [kN/m^2]</th></tr>
<tr><td rowspan="2">Rock</td><td>Hard to very hard</td><td>VU</td><td>> 100</td><td>700</td><td>8 000</td></tr>
<tr><td>Hard</td><td rowspan="3">VA</td><td>> 50</td><td>500</td><td>6 000</td></tr>
<tr><td rowspan="5">Soft rock</td><td>Moderately hard</td><td>12,5–50</td><td rowspan="2">200–400</td><td rowspan="2">4 000</td></tr>
<tr><td rowspan="2">Moderately weak</td><td rowspan="2">5–12,5</td></tr>
<tr><td rowspan="2">VE</td><td rowspan="2">100–200</td><td rowspan="2">2 500</td></tr>
<tr><td>Weak</td><td>1,25–5</td></tr>
<tr><td>Very weak</td><td>VZ</td><td>< 1,25</td><td>80</td><td>1 600</td></tr>
<tr><td rowspan="2">Completely weathered rock</td><td>Crumbly/soil</td><td rowspan="2">Soil</td><td>< 0,6</td><td>60</td><td>1 200</td></tr>
<tr><td>Cohesive/soil</td><td>< 0,6</td><td>40</td><td>800–1 000</td></tr>
</table>

5.4.6.4 Diaphragm wall elements (barettes)

(1) The empirical data compiled in Tables 5.12 to 5.19 may also be adopted for diaphragm wall elements (barettes) if the cross-sections and dimensions defined in 2.2.1.7 are adhered to.

(2) If skin friction should be utilised, any prefabricated elements placed within the diaphragm wall may not come into contact with the ground.

5.4.6.5 Bored pile walls and diaphragm walls

(1) The empirical data reproduced in Tables 5.12 to 5.19 may also be adopted for contiguous or secant pile walls and diaphragm walls if only the effective, net areas contacting the ground are adopted for determining base resistance and skin friction. A rectangular equivalent wall of, in plan view, equal area is adopted as the basis for the design.

(2) If the walls are exposed, making differential earth pressures effective, e.g. as a result of active earth pressure behind the wall and passive earth pressure in front of the wall, skin friction may only be adopted for the wall areas with an earth pressure angle $\delta \leq 0$, also see [26].

5.4.7 Partial displacement piles

(1) The group of bored displacement piles is divided into partial displacement piles (2.2.1.4, 2.2.1.5) to DIN EN 1536 and screw piles (full displacement bored piles, 2.2.4 and 5.4.8) to DIN EN 12699. In the literature and in common language this describes piles systems installed using rotary methods, whereby the ground is either partially or completely displaced.

(2) Pile systems where the configuration of the drive tube is such that soil is transported from the load-bearing embedment zone and/or redistributed into overlying, weaker soils, belong to the partial displacement pile group. They include systems using drive tubes and possessing a continuous flight auger for only part of their length (2.2.1.5).

(3) Partial displacement piles can also be constructed using a continuous flight auger (2.2.1.4). The degree of soil displacement depends on numerous factors; among others the inside diameter D_i of the hollow stem in relation to the outside diameter of the auger D_a, the design of the auger, the soil type and ground stratification and, not least, the quality of the installation. DIN EN 1536 contains information on installing partial displacement piles.

(4) Accordingly, the various types of partial displacement piles differ in many features, making universally applicable conclusions in terms of empirical pile resistance data only possible to a limited degree.

(5) The values contained in Tables 5.12 to 5.15 can be adopted to determine characteristic resistances based on empirical data. For specific cases these values may be increased by the geotechnical expert if:

- the ratio of the diameter of the hollow stem to the outside diameter of the auger $\geq 0{,}6$ and
- the volume of the transported soil is less than 70 % of the pile volume.

When specifying the increase factor relative to the tabled values:

- the uniformity and the density of non-cohesive soils;
- the consistency of cohesive soils, and
- regional experience with the adopted pile system

must be taken into consideration. No increase is allowed in very poorly graded or loose, non-cohesive soils, and in soft, cohesive soils.

Long-term, regional, empirical data for partial displacement piles in medium-dense sands below the groundwater table and with uniformity coefficients between $U = D_{60}/D_{10} \geq 2{,}0$ and $U \leq 2{,}5$ are available for Northern Germany. Assuming piles are correctly installed a general increase factor of 1,15 has proven useful. Larger uniformity coefficients can indicate larger bearing capacities.

(6) When specifying the governing mean cone resistance q_c of the CPT or the characteristic undrained shear strength $c_{u,k}$ differentiation shall be made between:

- the zone governing the pile base resistance (from $1 \cdot D_s$ above to $4 \cdot D_s$ below the pile base) and
- the region governing pile skin friction (mean value for the affected stratum).

If the ground stratification has a greater influence on the CPT cone resistance or the undrained shear strength, two or more mean pile skin friction zones must be specified separately.

(7) All other conditions for adopting the data given in Tables 5.12 to 5.15 are similar to those for bored piles as outlined in 5.4.6.2.

(8) The elements of the characteristic resistance-settlement curve for bored piles up to a settlement of $s_{ult} = s_g$ as shown in Figure 5.12 also govern partial displacement piles. The characteristic shaft resistance $R_{s,k}(s_{sg})$, characteristic axial pile resistance $R_{c,k}(s)$ and characteristic pile base resistance $R_{b,k}(s)$ are also determined using Eq. (5.11) or Eq. (5.12), analogous to bored piles, also see 5.4.6.1.

5.4.8 Screw piles

5.4.8.1 General

(1) Screw piles are also referred to as full displacement bored piles (2.2.4).

(2) The term 'boring' represents ground loosening and extraction. The designation "full displacement bored pile" is therefore somewhat contradictory, because terminologically boring and full displacement are mutually exclusive and could be misjudged. Although the designation has now established itself in the professional literature, for pile systems in which a drive tube is installed statically, i.e. rotated and/or pressed, vibration-free and the soil, in particular in the embedment zone in the load-bearing ground, is completely displaced, are primarily known as screw piles. DIN EN 12699 contains information on the execution of screw piles. Because of the static, rotating installation method used, screw piles differ from other, dynamically installed, full displacement pile systems, e.g. prefabricated displacement piles.

5.4.8.2 Empirical values of base resistance and skin friction of screw piles

(1) Screw piles are systems in which the soil is completely displaced, with the exception of minor uplift at the ground surface. Moreover, no soil redistribution from the embedment zone in the load-bearing ground to overlying, weaker strata may occur. Empirical base resistance and skin friction data are given for the Atlas pile and Fundex pile systems.

(2) When specifying the governing mean cone resistance q_c of the CPT or the characteristic undrained shear strength $c_{u,k}$ differentiation shall be made between:

- the zone governing the base resistance (from $1 \cdot D_s$ above to $4 \cdot D_s$ below the pile base) and;
- the zone governing the skin friction (mean value for the affected stratum).

If the ground stratification has a greater influence on the CPT cone resistance or the undrained shear strength, two or more mean pile skin friction zones must be specified separately.

(3) Conditions for the application of the data of Tables 5.20, 5.22 and 5.24 are that:

- the thickness of the load-bearing layer below the pile base is not less than three pile diameters, but at least 1,50 m and;
- $q_c \geq 7{,}5$ MN/m² or $c_{u,k} \geq 100$ kN/m² is confirmed for this zone.

Regardless of this, founding the pile bases in zones where $q_c \geq 10$ MN/m² is recommended.

(4) The elements of the characteristic resistance-settlement curve for bored piles up to a settlement of $s_{ult} = s_g$ as shown in Figure 5.12 are also relevant for screw piles. The determination of the characteristic shaft resistance $R_{s,k}(s_{sg})$, of

Table 5.20 Empirical data ranges for the characteristic base resistance $q_{b,k}$ for Atlas piles in non-cohesive soils

Relative pile head settlement s/D_s	Pile base resistance $q_{b,k}$ [kN/m²]		
	mean CPT cone resistance q_c [MN/m²]		
	7,5	15	25
0,02	950–1 400	1 650–2 300	2 650–3 450
0,03	1 200–1 850	2 150–2 950	3 350–4 450
0,10 ($\hat{=}$ s_g)	2 750–4 000	4 750–6 500	6 000–8 000

Intermediate values may be linearly interpolated.
D_s: relative to external helix diameter

the characteristic pile base resistance $R_{b,k}(s)$ and of the characteristic axial pile resistance $R_{c,k}(s)$ also follows in analogy to bored piles, using Eq. (5.11) or Eq. (5.12), see 5.4.6.1.

Table 5.21 Empirical data ranges for the characteristic skin friction $q_{s,k}$ for Atlas piles in non-cohesive soils

Mean CPT cone resistance q_c [MN/m²]	Ultimate limit state value $q_{s,k}$ of pile skin friction [kN/m²]
7,5	85–105
15	160–200
≥ 25	200–245
Intermediate values may be linearly interpolated. Pile shaft area A_{si}: relative to the external helix diameter D_s.	

Table 5.22 Empirical data ranges for the characteristic base resistance $q_{b,k}$ for Atlas piles in cohesive soils

Relative pile head settlement s/Ds	Pile base resistance $q_{b,k}$ [kN/m²]		
	Shear strength $c_{u,k}$ of the undrained soil [kN/m²]		
	100	150	250
0,02	600–800	900–1 250	1 300–1 950
0,03	750–950	1 050–1 500	1 650–2 350
0,10 ($\hat{=}$ s_g)	1 350–1 750	1 800–2 500	2 200–3 250
Intermediate values may be linearly interpolated. D_s: relative to external helix diameter			

Table 5.23 Empirical data ranges for the characteristic skin friction $q_{s,k}$ for Atlas piles in cohesive soils

Shear strength $c_{u,k}$ of the undrained soil [kN/m²]	Ultimate limit state value $q_{s,k}$ of pile skin friction [kN/m²]
60	40–60
150	75–95
≥ 250	95–120
Intermediate values may be linearly interpolated. Pile shaft area A_{si}: relative to the external helix diameter D_s.	

Table 5.24 Empirical data ranges for the characteristic base resistance $q_{b,k}$ for Fundex piles in non-cohesive soils

Relative pile head settlement s/D_b	Pile base resistance $q_{b,k}$ [kN/m²] for a mean CPT cone resistance q_c [MN/m²]		
	7,5	15	25
0,02	1 300–1 900	2 500–3 100	3 650–4 350
0,03	1 650–2 500	3 250–3 950	4 650–5 550
0,10 ($\hat{=}$ s_g)	3 800–5 500	7 200–8 800	8 300–10 000
Intermediate values may be linearly interpolated. D_b: Diameter of base plate			

Table 5.25 Empirical data ranges for the characteristic skin friction $q_{s,k}$ for Fundex piles in non-cohesive soils

Mean CPT cone resistance q_c [MN/m²]	Ultimate limit state value $q_{s,k}$ of pile skin friction [kN/m²]
7,5	35–50
15	85–115
≥ 25	115–145
Intermediate values may be linearly interpolated.	

Note: The data base for pile resistances from pile tests in cohesive soils is still too small to allow empirical data to be derived.

(5) An example for determining the characteristic resistances of Fundex piles is included in Annex B5.

5.4.9 Grouted displacement piles and micropiles

5.4.9.1 General

(1) If, in exceptional cases, no pile load tests are performed on pressure-grouted piles, vibro-injection piles, grouted micropiles ($D_s \leq 0{,}30$ m) or tubular grouted piles, the characteristic pile resistance $R_{c,k}$ or $R_{t,k}$ in the ultimate limit state may be determined for compression and tension piles using:

$$R_{c,k} = R_{t,k} = R_{s,k} = \sum_i q_{s,i,k} \cdot A_{s,i} \qquad (5.13)$$

where:

$A_{s,i}$ nominal value of the pile shaft area in stratum i;

$q_{s,i,k}$ characteristic value of the pile skin friction in stratum i, derived from Tables 5.26 and 5.32;

$R_{c,k}$ or $R_{t,k}$ characteristic compressive or tensile pile resistance for the ultimate limit state;

$R_{s,k}$ characteristic pile shaft resistance for the ultimate limit state.

See 5.4.10 for additional conditions.

5.4.9.2 Empirical values of skin friction of pressure-grouted piles

(1) Empirical values of pile skin friction in the ultimate limit state are provided in Tables 5.26 and 5.27.

Table 5.26 Empirical data ranges for the characteristic skin friction $q_{s,k}$ for pressure-grouted piles in non-cohesive soils

Mean CPT cone resistance q_c [MN/m²]	Ultimate limit state value $q_{s,k}$ of pile skin friction [kN/m²]
7,5	105–135
15	180–230
≥ 25	225–275
Intermediate values may be linearly interpolated.	

Table 5.27 Empirical data ranges for the characteristic skin friction $q_{s,k}$ for pressure-grouted piles in cohesive soils

Shear strength $c_{u,k}$ of the undrained soil [kN/m²]	Ultimate limit state value $q_{s,k}$ of pile skin friction [kN/m²]
60	40–50
150	80–90
≥ 250	95–105
Intermediate values may be linearly interpolated.	

(2) The circumference of the pile shoe shall be adopted as the circumference of the grout body.

(3) For compression piles, under conditions below, in addition to skin friction, base resistance may be adopted like that for prefabricated driven piles:

- the thickness of the load-bearing layer below the pile base is not less than 3- times the pile base diameter, but at least 1,50 m and
- $q_c \geq 7{,}5$ MN/m² is confirmed in this zone.

Regardless of this, founding the pile bases in zones where $q_c \geq 10\ MN/m^2$ is recommended.

(4) The driving progress must be adapted to the capacity of the grout mixing and pumping unit so that the void formed by soil displacement is completely filled with grout.

5.4.9.3 Empirical values of skin friction of vibro-injection piles

(1) Empirical pile skin friction data in the ultimate limit state for non-cohesive soils are given in Table 5.28.

(2) The circumference of the pile shoe shall be adopted as the circumference of the grout body.

Table 5.28 Empirical data ranges for the characteristic skin friction $q_{s,k}$ for vibro-injection piles in non-cohesive soils

Mean CPT cone resistance q_c [MN/m²]	Ultimate limit state value $q_{s,k}$ of pile skin friction [kN/m²]
7,5	90–115
15	150–195
≥ 25	180–220
Intermediate values may be linearly interpolated.	

(3) An additional pile end bearing capacity may only be derived from pile load tests.

(4) The vibration progress must be adapted to the capacity of the grout mixing and pumping unit so that the void formed by soil displacement is completely filled with grout.

5.4.9.4 Empirical values of skin friction of grouted micropiles

(1) Empirical values of pile skin friction in the ultimate limit state are given in Tables 5.29 and 5.30.

Table 5.29 Empirical data ranges for the characteristic skin friction $q_{s,k}$ for grouted micropiles ($D_s \leq 0,30$ m) in non-cohesive soils

Mean CPT cone resistance q_c [MN/m²]	Ultimate limit state value $q_{s,k}$ of pile skin friction [kN/m²]
7,5	135–175
15	215–280
≥ 25	255–315
Intermediate values may be linearly interpolated.	

Table 5.30 Empirical data ranges for the characteristic skin friction $q_{s,k}$ for grouted micropiles ($D_s \leq 0{,}30$ m) in cohesive soils

Shear strength $c_{u,k}$ of the undrained soil [kN/m²]	Ultimate limit state value $q_{s,k}$ of pile skin friction [kN/m²]
60	55–65
150	95–105
≥ 250	115–125
Intermediate values may be linearly interpolated.	

(2) The largest outside diameter of the drilling tool, the drilling pipe or the casing shall be taken to determine the circumference of the grout body. If piles are installed with external flushing, the pile shaft diameter may be adopted as the outside casing diameter plus 20 mm.

Note: When adopting the empirical data given in Table 5.30 it must be noted that the data base from load tests of grouted micropiles in cohesive soils is small. Depending on the specific situation, higher capacities might be anticipated.

(3) Pile end bearing capacity may not be adopted in addition.

5.4.9.5 Empirical values of skin friction in tubular grouted piles

(1) Empirical values of pile skin friction in the ultimate limit state are given in Tables 5.31 and 5.32.

Table 5.31 Empirical data ranges for the characteristic skin friction $q_{s,k}$ for tubular grouted piles in non-cohesive soils

Mean CPT cone resistance q [MN/m²]	Ultimate limit state value $q_{s,k}$ of pile skin friction [kN/m²]
7,5	170–210
15	255–320
≥ 25	305–365
Intermediate values may be linearly interpolated.	

Note: When adopting the empirical data given in Tables 5.31 and 5.32 it must be noted that the data base from load tests of tubular grouted piles is small. Depending on the specific situation, higher capacities might be anticipated.

Table 5.32 Empirical data ranges for the characteristic skin friction $q_{s,k}$ for tubular grouted piles in cohesive soils

Shear strength $c_{u,k}$ of the undrained soil [kN/m²]	Ultimate limit state value $q_{s,k}$ of pile skin friction [kN/m²]
60	70–80
150	115–125
≥ 250	140–150
Intermediate values may be linearly interpolated.	

(2) The largest outside diameter of the drilling tool, the drilling pipe or the casing shall be taken to determine the circumference of the grout body. If piles are installed with external flushing, the pile shaft diameter may be adopted as the outside casing diameter plus 20 mm.

(3) Pile end bearing capacity may not be adopted in addition.

5.4.9.6 Bond stress in grouted displacement piles

(1) DIN SPEC 18538 stipulates that the design values of the bond stresses $\tau_{R,d,1}$ and $\tau_{R,d,2}$ are adopted for analysis of the bond in the pile shaft/grout interface of grouted displacement piles to DIN EN 12699:

a) $\tau_{R,d,1}$ = 0,2 N/mm² uniformly for all surfaces for grouted, driven grouted and vibro-injection piles with H-sections;
b) $\tau_{R,d,2} = e_0 \cdot \mu/1{,}1$ for all other sections.

where:

e_0 at-rest earth pressure at the respective depth;
μ friction coefficient, set as $\mu = 0{,}5$ for rough rolled steel sections.

A cube compressive strength of $f_{c,cube} \geq 25$ N/mm² for the grout and a minimum grout thickness of 25 mm are assumed.

(2) Annex A6 contains informative more up-to-date values for the bond stress, determined during a recent research project, see [61], [62].

5.4.10 Applying the empirical data to tension piles

(1) As outlined already in 5.4.1 (5), pile load tests should always be performed when dealing with tension piles. Also the EC 7-1 Handbook [44] allows the estimation of tension pile resistances from empirical data in exceptional cases only.

(2) If, in well-founded, exceptional cases, tension piles resistances based on empirical data are adopted, the derived characteristic skin friction values must

be confirmed by a geotechnical expert or geotechnical designer for the specific situation. In this case it shall also be checked if the empirical data given in the above tables should be further and considerably reduced for deriving tension pile resistances, e.g. by applying appropriate calibration factors.

(3) When determining a characteristic resistance-heave curve based on empirical data, the limit heave $s_{sg,t}$ may be approximately determined using:

$$s_{gt} = 1{,}30 \cdot s_{sg} \tag{5.14}$$

where s_{sg} or s_{sg*} is adopted after Eq. (5.5) or (5.7), and (5.11), or accordingly for other pile types.

(4a) Where the empirical values to 5.4.9 are adopted in the design, the design length of the grout body of pressure-grouted piles or vibro-injection piles should be limited to a maximum of 15 m and that for tubular grouted piles or grouted micropiles to a maximum of 12 m, as this corresponds to the current range of experience. If greater grout body lengths are designed, the respective skin friction values shall be derived from load tests or shall be confirmed by the geotechnical expert

5.5 Bored Piles with Enlarged Bases

(1) It can be necessary to increase the pile base resistance of bored piles by enlarging the base, for both compressive and tensile loads.

(2) The voids required in the base area for this purpose may only be formed in sufficiently stable ground.

(3) Any base enlargements formed either manually or by machine should be as concentric to the pile shaft as possible.

(4) If the pile base is below the groundwater table the void must be supported by positive water pressure, or by other fluids, in accordance with usual bored pile execution.

(5) The ratio of enlarged base height to base protrusion may not fall below the following values:

- in non-cohesive soils 3 : 1;
- in cohesive soils 2 : 1.

(6) Compression and tension piles shall be differentiated when assessing the load-bearing capacity. After investigations published in [111], the following applies, in particular, to bored piles:

a) A zone of reduced skin friction develops above the enlarged base of compression piles which probably results from loosening of the soil. This is compensated further above, probably as a result of arching stresses. Consequently, for an approximate analysis the mean skin friction values as given in 5.4.6 may also be adopted above base enlargements.

b) Also (longer) bored piles with enlarged bases transfer loads predominantly via the pile shaft, similar to piles without enlarged bases, unless the pile is founded on a particularly hard soil stratum or on rock. Base enlargements are therefore generally only technically and economically expedient in firm soils and on weak rocks.

5.6 Additional Methods Using the EC 7-1 and EC 7-2 Handbooks

(1) In addition to static or dynamic pile load tests the EC 7-1 Handbook [44], and in places the EC 7-2 Handbook [45], also list:

- empirical or analytical methods and;
- observation of the behaviour of a similar pile foundation;

to determine the resistance of axially loaded piles. See EC 7-1 Handbook [44], NDP to 7.6.2.3 (5)P and NDP to 7.6.3.3 (4)P for national usage.

(2) 5.4.1 (3) and (4) must be observed for these cases.

5.7 Pile Resistances for Grouted Shafts and Bases

(1) Because the resistances of the pile shaft and pile base dependent on the relative displacements, the proportion of skin friction in the serviceability limit state is considerable. Grouting of the pile shaft therefore increases the pile capacity and can simultaneously reduce settlements, see [103].

(2) Shaft and base grouting can reduce negative influences on the bearing capacity, in particular when bored piles are installed in soils of low density or soft-consistency, or, disturbance of the surrounding ground result from pile installation.

(3) Pile base grouting is carried out to prestress the subsoil or to increase the capacity of the pile base subsequent to pile construction.

(4) Shaft grouting is carried out using grouting pipes connected to the reinforcement, such that the position of the pipes cannot change when installing the reinforcement and placing the concrete. Simple sleeved pipes with only one valve are normally used for single stage grouting. Provided appropriate configuration and flushing, regrouting can be carried out a number of times. Alternatively, the shaft can also be grouted using the tube-a-manchette method, using double packers at varying depths.

(5) The following information complementing that on the ground conditions, is essential for shaft grouting [70]:

- type of grouting pipes;
- single or multiple stage grouting;
- specific shaft surface area per sleeve (valve);

- grout quantity per sleeve (valve);
- maximum grouting pressure;
- pumping rate in l/min;
- procedure for recording of grouting data.

(6) Base grouting can be carried out using flexible box structures at the pile base, or at 90° crossed grouting pipes or valves at the base of the pile.

(7) In addition to detailed information on the ground conditions, the following governing parameters, are required for base grouting:

- type of grouting equipment (flexible box structures, simple sleeved pipes, or other);
- grout quantity;
- maximum grouting pressure;
- maximum heave of pile head during grouting;
- type of heave monitoring;
- procedure for recording of grouting data.

(8) Where purposeful grouting of the shafts of bored piles is carried out, the empirical pile skin friction data for grouted micropiles given in 5.4.9.4 can be adopted for preliminary design or the values given for bored piles in 5.4.6.2 be increased by approximately 50% [103]. Further information on the enhancement of pile capacity by shaft and base grouting is provided in [112].

5.8 Resistances of Piles Under Lateral Loads

(1) Soil resistances lateral to the pile axis are generally given as subgrade reaction moduli, in particular for long, flexible piles.

(2) The resistances shall be differentiated for two different analysis methods:

a) determination of the effects within the pile, e.g. bending moments, see (3) and (4);
b) determination of the horizontal deformations of a pile foundation, see (5).

(3) In addition to back-calculations from horizontal load tests, the subgrade reaction moduli outlined in the EC 7-1 Handbook [44] may also be determined from the relationship: $k_{s,k} = E_{s,k}/D_s$, see 6.3.2, Eq. (6.8), because the magnitude of the estimated subgrade reaction moduli plays only a minor role on pile effects. Not regulated in the EC 7-1 Handbook [44] is, however, the effect of the shape of progression with depth of the modulus of subgrade reaction which can be decisive, even when using the simplified approach. See e.g. [59], for guidance and additional information and literature references.

Note: The procedure specified in (3) above represents a simplified approach, and can be applied regardless of displacements. For more accurate analysis, the modulus of subgrade reaction should be determined from horizontal pile load tests as a function of deflections and be adopted corresponding to the deflections resulting from the effects on the complete structure.

(4) Estimates of deformations based on the relationship given in (3), also see 6.3.2, Eq. (6.8), are generally associated with major uncertainties. If horizontal deformation analyses are necessary, the EC 7-1 Handbook [44] demands they be based on pile load tests or on the results of comparable pile load tests.

(5) See [26] for subgrade reaction approaches used for piled walls for excavation pits.

(6) An example of laterally loaded piles is included in Annex B9.

5.9 Pile Resistances Under Dynamic Actions

(1) Where piles are, in contrast to static actions, subjected to considerable cyclic, dynamic or impact actions, markedly changed pile behaviour must be anticipated as a function of the respective boundary conditions. The EC 7-1 Handbook [44] requires that this behaviour shall be taken into consideration in the analysis and design of pile foundations.

(2) Additional information on definitions, bearing capacity and pile resistances can be found in 4.1 (1), 4.2 (4) and Section 13.

5.10 Internal Pile Capacity

5.10.1 General

(1) The internal capacity of piles and their connections to the structure depends on the pile materials and shall be analysed in this context.

(2) The material properties, analysis formats and partial factors given in the respective design standards govern the determination of the resistances of the respective structural members. The following standards apply, together with the information on the respective parameters specified in the National Annexes:

- DIN EN 1992-1-1 for reinforced and unreinforced bored piles to DIN EN 1536 and driven, steel or prestressed concrete piles to DIN EN 12699;
- DIN EN 1993-1-1 and, in particular, DIN EN 1993-5 for driven steel piles to DIN EN 12699 and micropiles to DIN EN 14199, in which the steel member is predominantly designed to accept the effects on the structural element;
- DIN EN 1994-1-1 for steel piles with a closed, concrete-filled cross-section (e.g. micropiles to DIN EN 14199 with tubular cross-sections).

(3) The standards controlling internal capacity analyses allow test-supported design for both the resistance capacities of cross-sections and for stability analyses.

a) DIN EN 1990 and DIN EN 1992-1-1 or DIN EN 1993-1-1 provide general provisions for test-supported design.
b) Instructions on performing test-supported design can be taken from DIN EN 1997-1 and the respective pile execution standards.

(4) When analysing the internal pile capacity, the pile action effects are generally determined based on the EC 7-1 Handbook [44]. Both the equilibrium conditions and the compatibility, in particular in terms of the interactions between the structure, the pile and the ground, must be taken into consideration.

(5) With regard to designing piles in terms of internal capacity, the durability of the construction for the proposed service life, in particular, must be demonstrated.

a) For steel piles, in particular, the reduction in the effective cross-section as a result of corrosion must be taken into consideration. DIN EN 1993-5 specifies corrosion rates as a function of the aggressive medium (various soil types and water) and the pile exposure.
b) The durability of steel or prestressed concrete piles shall be ensured through sufficient concrete cover of the reinforcement and by the concrete quality. The general reinforcement rules for in-situ concrete and prefabricated concrete elements given in Sections 8 and 10 of DIN EN 1992-1-1 must be observed.

(6) In addition to the information provided in DIN EN 1536, DIN EN 12699 and DIN EN 14199, supplementary guidance and provisions on material parameters and structural element resistances are provided in the DIN SPEC standards complementing the European pile execution standards, see 1.1 (2).

(7) Laterally loaded piles with circular cross-sections can also be designed after [5]. An example analysis is included in Annex B9.

5.10.2 Allowable cross-section stresses

(1) The design value of an effect in any cross-section may not exceed the corresponding design value of the allowable stress. If several types of effect occur simultaneously, this requirement also applies to the combination of these effects, see EC 7-1 Handbook [44].

(2) For driven steel piles to DIN EN 12699 and micropiles to DIN EN 14199, in which the steel member is predominantly designed to accept the effects on the structural element, the provisions of DIN EN 1993-1-1 and DIN EN 1993-5 apply.

a) DIN EN 1993-1-1 provides suitable information on the cross-section values to be adopted for tensile and compressive loads, and the formats for analysing normal forces, shear forces and bending loads. The interaction of the action effects must be taken into consideration in the analysis using suitable equations.
b) The provisions of the National Annex DIN EN 1993-1-1/NA also apply.

(3) For steel or prestressed concrete piles (bored piles to DIN EN 1536 or displacement piles to DIN EN 12699), the analysis of the allowable stresses in the cross-sections follows the provisions of DIN EN 1992-1-1.

a) The analysis formats with which the combination of normal force and bending moment are analysed for piles, are described in DIN EN 1992-1-1 and must be adhered to.
b) Specific construction provisions for bored piles are included in DIN EN 1992-1-1. In addition, the specifications in the execution standard DIN EN 1536 apply.
c) Additional provisions and requirements for prefabricated driven piles are formulated in DIN EN 1992-1-1. They must be observed in addition to the provisions of DIN EN 12699.
d) The provisions of the National Annex DIN EN 1992-1-1/NA also apply.

5.10.3 Resistance of piles against buckling failure in soil strata with low lateral support, and buckling analysis

(1) Buckling of compression piles is normally prevented by the supporting action of the ground lateral to the pile. Sufficiently stiff and strong soils therefore present no buckling hazard.

(2) However, this does not necessarily apply in soft soils.

(3) The EC 7-1 Handbook [44] states the following about buckling failure and buckling analysis:

a) The EC 7-1 Handbook [44], 7.8, requires that slender piles, when passing through water or thick deposits of very weak soils shall be checked against buckling. Without referring to slenderness it goes on to state that no buckling safety analysis is required if the piles are contained in soils with a characteristic, undrained shear strength $c_{u,k} > 10\ kN/m^2$.
b) In the EC 7-1 Handbook [44], A, Note on (5), it is pointed out that buckling can also occur in micropiles installed in soils with shear strengths $c_{u,k} > 10\ kN/m^2$.
c) DIN EN 1536 and DIN EN 12699 do not contain guidance on any limits from which stability analyses need to be performed.
d) DIN EN 14199 states that a buckling analysis should be performed for micropiles installed in a soil with characteristic undrained shear strength less than 10 kN/m², taking any imperfections into consideration.

(4) In the investigations carried out e.g. by [10], [93], [94] and [156], it was shown that buckling failure can also occur in slender piles in soil strata offering little lateral support, given unfavourable boundary conditions, even if the limits mentioned in (3) are adhered to. Also see [149] and [106] for micropiles.

(5) Piles in at least firm, cohesive soils and those in non-cohesive soils generally do not need to be analysed for buckling.

(6) Until further notice, the approaches given in Annex A5 for determining the resistance of piles against buckling failure may be adopted for piles requiring a buckling safety analysis. Alternatively, the capacity of a compression pile suspected of being unstable may be analysed in the course of test-supported design using static pile load tests, also see 5.10.1 (3).

(7) A buckling hazard must always be considered in compression piles passing through water. Guidance on a simplified check of the buckling length of partially free-standing steel piles can be found in DIN EN 1993-5. The method can also be adopted for partially free-standing, reinforced concrete or prestressed concrete piles.

(8) Specification of the governing parameters required to analyse buckling safety depends greatly on boundary conditions dictated by the ground situation, pile installation and the performance of the pile system (bond of steel, cement,, ground). The option of test-supported design should therefore be considered if the results of the numerical analyses contradict building practice experience, also see 5.10.1 (6).

5.11 Numerical Analyses of the Capacity of Single Piles

(1) As outlined in 5.1 (4), numerical methods may also be employed to determine pile resistances if the analysis models have been calibrated on comparable pile load test results.

(2) It should be noted that realistic pile installation modelling using numerical methods is generally difficult, and is inexact for certain pile systems in the context of practical construction. This is because the ground properties in the vicinity of the pile can be strongly influenced by pile installation. Numerical analyses works best for bored piles.

(3) Notes on mesh generation and modelling can be found e.g. in [146].

(4) The plausibility of characteristic pile resistances determined using numerical methods should be examined on the basis of pile resistances determined from the empirical data given in 5.4, and be confirmed by the geotechnical expert for the specific application.

6 Stability Analyses

6.1 Introduction

(1) The fundamentals of the analyses using the partial safety factor approach in accordance with the provisions of the EC 7-1 Handbook [44] are outlined in 1.2 and 3.1.1 (4).

(2) The conditions required to analyse the stability and serviceability of pile foundations are summarised in their entirety again below, even if this means repeating provisions of the EC 7-1 Handbook [44] and texts of the relevant standards. However, this aims to facilitate the use of the new provisions for pile foundations based on the EC 7-1 Handbook [44].

6.2 Limit State Equations

(1) According to the EC 7-1 Handbook [44], the characteristic actions F_k, H_k and $M_k$imposed on the piles by the superstructure or the ground, or the effects E_k,, shall be converted to design values for the ultimate limit state and serviceability limit state analyses using Analysis Method 2 (STR and GEO-2) after Eq. (6.1 a–c) or Eq. (6.2), adopting the appropriate partial factors (see Annexes A2 and A3).

$$F_d = F_{G,k} \cdot \gamma_G + F_{Q,rep} \cdot \gamma_Q \tag{6.1 a}$$

$$H_d = H_{G,k} \cdot \gamma_G + H_{Q,rep} \cdot \gamma_Q \tag{6.1 b}$$

$$M_d = M_{G,k} \cdot \gamma_G + M_{Q,rep} \cdot \gamma_Q \tag{6.1 c}$$

or

$$E_d = E_{G,k} \cdot \gamma_G + E_{Q,rep} \cdot \gamma_Q \tag{6.2}$$

(2) The analysis of the pile is satisfied if the limit state equations (6.3 a–c) and (6.4) are adhered to for the ultimate and serviceability limit states.

$$F_d \le R_{d,F} \tag{6.3 a}$$

$$H_d \le R_{d,H} \tag{6.3 b}$$

$$M_d \le R_{d,M} \tag{6.3 c}$$

and

$$E_d \le R_d \tag{6.4}$$

Recommendations on Piling – EA Pfähle.
Editors: German Geotechnical Society (*Deutsche Gesellschaft für Geotechnik e.V.*)

6.3 Bearing Capacity Analysis

6.3.1 Axially loaded piles

(1) The following procedure applies for analysing the 'external' capacity (load transfer to the ground) of an axially loaded, single pile of a pile foundation for the governing design situation in the ultimate limit state after the EC 7-1 Handbook [44]:

a) The characteristic, axial actions F_k, see 4.1 and 4.2, at the pile head are determined as foundation loads of the chosen system. The system consists of the selected pile type and dimensions, the pile head and superstructure. The foundation loads comprise the loads imposed by the structure and, as applicable, negative skin friction, and are separately determined as persistent and transient situations.
b) The design values F_d after 6.2 (1) are determined from the characteristic axial actions F_k on the pile.
c) Adopting the characteristic axial pile resistances determined after Section 5, the design values of the pile resistances in the ultimate limit state result from:

$$R_{c,d} = R_{c,k}/\gamma_t \quad \text{for compression pile resistance} \tag{6.5 a}$$

$$R_{t,d} = R_{t,k}/\gamma_{s,t} \quad \text{for tension pile resistance} \tag{6.5 b}$$

where γ_t or $\gamma_{s,t}$, are adopted from Annex A3.2 or the EC 7-1 Handbook [44], Table A2.3.

The partial factors apply equally to both the base and the shaft resistance.

(2) Using the determined axial design actions and resistances, it must be demonstrated that the piles fulfil the limit state conditions for all governing load cases and load combinations, see EC 7-1 Handbook [44] or 6.2 (2):

$$F_{c,d} \leq R_{c,d} \text{ or } \sum F_{c,d} \leq \sum R_{c,d} \quad \text{for compression pile resistances} \tag{6.6 a}$$

$$F_{t,d} \leq R_{t,d} \text{ or } \sum F_{t,d} \leq \sum R_{t,d} \quad \text{for tension pile resistances} \tag{6.6 b}$$

Note: In addition to the designations used in Equations (6.5) and (6.6) for the ultimate limit state of the piles as specified in the EC 7-1 Handbook [44], the Recommendations on Piling also include the abbreviations $R_g = R_{ult} = R(ULS)$, depending on context, also see Annex A1.

(3) If the limit state conditions are not fulfilled, the pile dimensions must be increased accordingly. If an uneconomical safety excess exists and should be lowered, the dimensions, e.g. the pile length, may be reduced accordingly. The analysis shall be repeated in both cases or be completed by iteration.

(4) When determining the design values of tensile loads, after 8.1.2 (2) and the EC 7-1 Handbook [44], a simultaneously acting characteristic compressive load resulting from favourable permanent actions shall be taken into consideration only by the partial factor $\gamma_{G,inf}$. Contrary to this, design of compression piles with simultaneously acting tensile load shall use the same GEO 2 method as for all foundation structures. All characteristic loads shall be summarised (compressive and tensile), taking their algebraic signs into account, and the resulting characteristic compressive loads, reduced by the tensile component, shall be converted to design values for the actions γ_G and γ_Q.

(5) The self-weight of the piles may be ignored in cases of compressive loads. For tensile loads the pile's self-weight may be adopted as a simultaneously acting compressive load in accordance with 8.1.2 (2).

(6) Safety against structural failure ('internal pile capacity' shall be analysed in accordance with 6.3.3, taking 5.10 into consideration.

(7) Annexes B1 and B7 contain examples of the analysis of the external capacity of axially loaded single piles.

(8) See 5.9 and 13 for details of the different performance of piles under cyclic actions.

(9) For grouted tension pile systems (grouted micropiles to DIN EN 14199 and grouted displacement piles to DIN EN 12699) in accordance with the EC 7-1 Handbook [44], a model factor η_M shall be taken into consideration for the calculation of the design values and Eq. (6.5 b) becomes:

$$R_{t,d} = R_{t,k} / (\gamma_{s,t} \cdot \eta_M) \qquad (6.5\ c).$$

The model factor is $\eta_M = 1{,}25$, regardless of the pile inclination. Eq. (6.5 c) also applies if, in well-founded, exceptional cases, no pile load test data are available and the pile resistances of grouted pile systems are derived from empirical data.

6.3.2 Laterally loaded piles

(1) The bearing capacity of long, flexible piles in the governing design situation in the ultimate limit state (ULS) situations STR and GEO needs not be analysed if the piles are embedded in the ground and the horizontal characteristic effect in the DS-P design situation is no more than 3 % of the vertical effect, or 5 % in the DS-T design situation. Proceed as follows for all other cases:

- Define the input values for determining ground reactions, e.g. a modulus of subgrade reaction after 5.8;
- Determine the characteristic action effects or the characteristic stresses using the characteristic values of actions and the characteristic values of the subgrade reaction moduli;

- Convert the characteristic action effects or the characteristic stresses to design values of effects by multiplying by the partial factors of actions after Annex A2 or EC 7-1 Handbook [44], Table A2.1 and 6.2 (1);
- Verify that the characteristic normal stress $\sigma_{h,k}$ along the pile shaft between the pile and the ground does not exceed the characteristic passive earth pressure $e_{ph,k}$ calculated for the two dimensional situation:

$$\sigma_{h,k} \leq e_{ph,k} \tag{6.7}$$

- The analysis using Eq. (6.7) assumes in approximation that only a stress comparison is required, which involves the normal stresses $\sigma_{h,k}$ between the pile and the ground, calculated using the modulus of subgrade reaction method, and the two dimensional characteristic passive earth pressure stresses $e_{ph,k}$ in front of the pile, calculated after the EC 7-1 Handbook [44]. The EC 7-1 Handbook does not stipulate that a displacement-dependent mobilisation of $e_{ph,k}$ be considered, also see [64]. The passive earth pressure stresses $e_{ph,k}$ are introduced here only as maximum normal stress limits $\sigma_{h,k}$. The numerical reduction to the two dimensional passive earth pressure (and not to the three-dimensional case) provides safety reserves with depth, which depend on the utilisation. The analysis relates primarily to calculations using the approximation approach after 5.8 for determining internal forces:

  $$k_{s,k} = E_{s,k}/D \tag{6.8}$$

  If pile load test results are available, realistic, if necessary non-linear, modulus of subgrade reaction variables and distributions should be derived from them for the ground resistance, as a function of the load ranges.
- Verify that the lateral ground resistance $B_{h,d}$ was not adopted at greater than allowed by the design value of the three-dimensional passive earth pressure $E^{r}_{ph,d}$ for the corresponding component of the embedment depth as far as the pile pivot (displacement zero point):

  $$B_{h,d} \leq E^{r}_{ph,d} \tag{6.9}$$

- Analyse safety against structural failure.

(2) Annex B9 contains an example for determining the characteristic effect and designing a laterally loaded, flexible, single pile.

(3) Proceed as follows to analyse short, rigid, single piles for the governing design situation in the ultimate limit state:

- Determine the resulting characteristic support forces and characteristic action effects for an estimated pile length and characteristic parameters, applying the equilibrium conditions.
- The characteristic support forces in the ground shall be converted to design forces by multiplying by the partial factors of actions from Annex A2, the EC 7-1 Handbook [44], Table A2.1 and 6.2(1), and be balanced against the

design values of the three-dimensional passive earth pressure. Methods that model the bearing capacity of the short, almost rigid piles may be adopted, e.g. the methods after [8], [148]. The final pile length shall be determined iteratively.

- When analysing safety against structural failure the characteristic action effects shall be converted to design values by multiplying by the partial factors for actions from Annex A2 or EC 7-1 Handbook [44] and Table A2.1 and 6.2(1).

(4) See 5.9 and 13 for details of dynamic actions.

6.3.3 Structural failure in piles

(1) Safety against structural failure (STR) must be demonstrated for all piles compliant with the type-specific regulations:

$$E_d \leq R_{M,d} \qquad (6.10).$$

(2) The governing design values of the effects E_d shall be determined as outlined in Sections 4, 6.2, 6.3.1 and 6.3.2.

(3) The material properties given in the respective construction standards shall be applied for the determination of the design values of the resistances $R_{M,d}$ of the structural elements.

6.4 Serviceability Analysies

6.4.1 Axially loaded piles

(1) If an appropriate examination reveals that the deformations of the pile foundation are relevant to the structure as a whole, an analysis of the serviceability limit state (SLS) must be performed for the governing design situations. Serviceability is demonstrated if the following condition is met:

$$F_d\,(SLS) = F_k \leq R_d\,(SLS) = R_k\,(SLS) \qquad (6.11)$$

Partial factors of $\gamma = 1.0$ are normally adopted for the actions and the resistances.

Analysis may also be performed using the allowable settlements. all. s_k as provided by the structural designer, assuming characteristic effects on the pile foundation in the serviceability limit state, as follows:

$$s_k \leq \text{all. } s_k \qquad (6.12)$$

Note: Where pile systems only display minor settlements under service loads, the serviceability limit state analysis can be covered in the analysis of the ultimate limit state, also see 5.2.3 (3).

(2) If the procedure described in (1) is adopted, it is first assumed that single pile performance is prevalent for the pile foundation structure as a whole, and

that an additional group effect influencing pile capacity can be ignored. Regardless of this, differential settlements Δs_k between the piles of a piled foundation structure can occur, even when single pile behaviour is anticipated. Such effects can result e.g. of heterogeneous ground conditions and/or influences from pile execution. After Figure 6.1 differentiation is to be made between anticipated

- minor differential settlements and
- substantial differential settlements.

(3) If only minor differential settlements are anticipated between individual piles in a structure, then the characteristic pile resistance shall be derived from an assessment of the pile load test results after 5.2 and 5.3, or based on empirical data after 5.4 using a specified, allowable characteristic settlement s_k as shown in Figure 6.1 a).

For the consideration of the ground-foundation-structure interaction, spring constants may be adopted for the piles, which are derived from the secant on the characteristic resistance-settlement or -heave curve, or from empirical data.

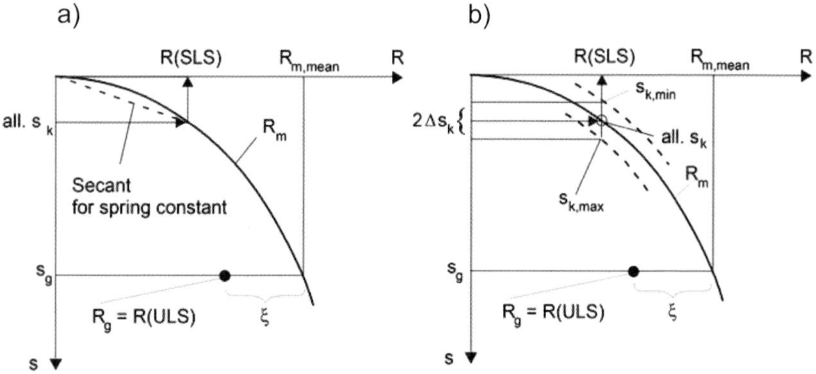

Figure 6.1 Possible method for derivation of characteristic resistances of isolated piles R(SLS) in the serviceability limit state from test data and data from resistance-settlement curves a) for anticipated minor differential settlements between isolated piles; b) for anticipated substantial differential settlements between isolated piles

(4) If substantial differential settlements are anticipated between the individual piles of a structure, proceed as in (3) for the single pile. Possible upper $s_{k,max}$ and lower bounds $s_{k,min}$ of the settlements s_k after Figure 6.1 b) shall be determined in the range of the resulting pile resistance R(SLS), adopting Eq. (6.13):

$$\Delta s_k = \kappa \cdot s_k \tag{6.13}$$

The factor κ depends on pile installation, ground stratification and the locations of the piles within the foundation structure, and should be specified in agreement with the geotechnical expert. Additional information can be found in [59].

(5) With regard to pile groups, the bearing capacity of the piles in the group can be additionally influenced by the group effect, also see Section 8.

(6) As a result of possible differential settlements between the piles or between individual groups as mentioned in (4), or within a pile group as mentioned in (5), for the characteristic pile resistance R(SLS), an ultimate limit state (ULS) or serviceability limit state (SLS) can result as a consequence of imposed effects in the pile head slab or the superstructure.

(7) Examination is also required, if the anticipated deformations of individual piles or of pile groups can result in a serviceability limit state of neighbouring structures, e.g. adjacent buildings or pipe inlets.

(8) Annexes B1 and B7 contain examples of the analysis of the serviceability limit state of axially loaded, single piles.

6.4.2 Laterally loaded piles

(1) The provisions of 6.4.1 shall be applied accordingly to laterally loaded piles.

(2) See 5.8 for additional guidance.

6.5 Pile Groups and Grillages

(1) See Section 8 for ultimate and serviceability limit state analyses of compression pile groups, tension pile groups and laterally loaded pile groups.

(2) Analysis of grillages is dealt with in Section 7.

(3) Examples of pile groups are provided in Annexes B11 to B13.

6.6 Piled Raft Foundations

(1) See 3.1.4 for details of bearing capacities and analyses of piled raft foundations.

(2) Both the ultimate and the serviceability limit states can be analysed on the basis of [46], or 8.3 and 8.4, whereby the EC 7-1 Handbook [44], A 7.6.2.8 A (1) must be observed.

7 Grillage Analysis

7.1 Analysis Models and Procedures

(1) Traditional grillage analysis is based on the simplified, static analysis model shown in Figure 7.1, as described in [128] and [137], for example, as follows:

- In structural terms the piles are hinged columns with bearings at the head and the toe.
- The toe bearing is immovable.
- The piles are adopted as linear elastic springs.
- The flexibility of the load-bearing ground is incorporated in the pile spring constant.
- The flexural stiffness of the capping slab or the superstructure is large in comparison to the pile stiffness ('rigid' pile head structure).
- The ground does not act on the piles themselves (through negative skin friction, for example).

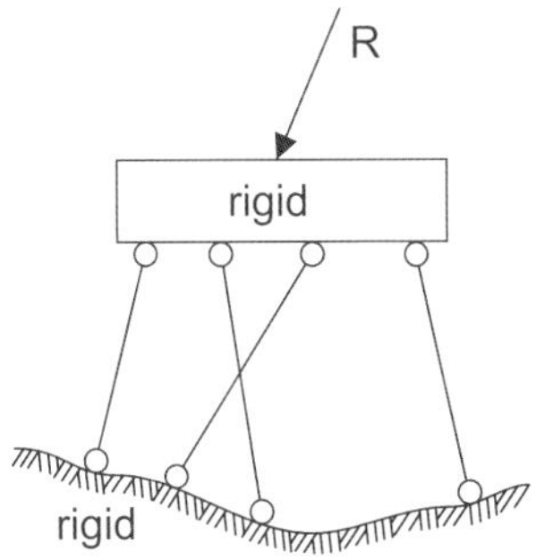

Figure 7.1 Grillage: System and boundary conditions

(2) Additional notes and the analysis equations for the methods discussed in (1) can be found in [137].

(3) Applying the described analysis route is generally very complex and extensive. However, for special cases such as:

- Foundations on vertical piles only;
- Double symmetrical grillages;
- Axis-symmetrical, three-dimensional grillages, for example for tower-like structures;

simple equations are given in [137], and are particularly useful for hand calculations at the draft stage.

Recommendations on Piling – EA Pfähle.
Editors: German Geotechnical Society (*Deutsche Gesellschaft für Geotechnik e.V.*)

(4) Because of the software applications now widespread in practice, grillages are increasingly analysed today using frame methods with embedded or non-embedded beams.

7.2 Non-linear Pile Bearing Behaviour in Grillage Analysis

(1) In grillage systems with differently utilised piles (asymmetrical pile systems in terms of geometry and/or loading), a larger pile bearing capacity can, in many cases, be demonstrated by taking the non-linear pile bearing behaviour into consideration.

(2) Notes on the impact on grillage analysis of the non-linearity of the piles' resistance-settlement curve can be found in [7], for example.

(3) In summary, comparative analyses, for example as in [124], have shown that in terms of the impact of the non-linearity of the piles' resistance-settlement curve when analysing grillages, it may be assumed that the maximum values of:

- the pile forces and
- depending on the actual load positions, the action effects in the pile capping slab

are generally reduced for piles loaded in compression.

8 Analysis and Verification of Pile Groups

8.1 Actions and Effects

8.1.1 Compression pile groups

(1) Actions on pile groups correspond to the actions on single piles as described in Section 4. However, it is generally that the distribution of total of the actions on individual piles is not distributed uniformly in a pile group.

(2) The effect on an individual group pile corresponds to the activated pile resistance, e.g. after 8.2.1.

(3) The effects on the pile head slab caused by the non uniform distribution of pile resistances must be taken into consideration. The forces acting on the pile capping slab resulting from the pile resistances and the associated settlements should be known for the slab design. Spring stiffnesses c_i can be adopted for each group pile for this purpose.

8.1.2 Tension pile groups

(1) As a principle, two cases shall be differentiated for analysis and determation of the actions on a tension pile group. One case refers to the pull-out resistance of the single pile in the GEO-2 limit state, and the other to the uplift of the entire pile group as a soil monolith in the UPL limit state, also see 8.3.2.

Note: The EC 7-1 Handbook [44] does not clearly stipulate whether the latter case, i.e. the analysis of the uplift of a soil monolith, needs also be performed for isolated piles, also see (5).

(2) If the piles are subjected exclusively to tensile loads, the design value of the effects required for analysis of the safety against pull-out is carried out as that for single piles, see Section 4. If characteristic tensile forces dominate in more unfavourable combinations of simultaneously acting compressive and tensile forces, the design value $F_{t,d}$ of the tensile effect shall be determined using:

$$F_{t,d} = F_{t,G,k} \cdot \gamma_G + F_{t,Q,rep} \cdot \gamma_Q - F_{c,G,k} \cdot \gamma_{G,inf} \quad (8.1)$$

where:

$F_{t,G,k}$ the characteristic value of the tensile effect on a pile or a pile group as a result of permanent actions;

γ_G the partial factor for permanent effects in the GEO-2 limit state;

$F_{t,Q,rep}$ the characteristic or representative value of the tensile effect on a pile or a pile group as a result of possible unfavourable, variable actions;

γ_Q the partial factor for unfavourable, varible effects in the GEO-2 limit state;

Editors: German Geotechnical Society (*Deutsche Gesellschaft für Geotechnik e.V.*)

$F_{c,G,k}$ the characteristic value of a simultaneously acting compressive effect as a result of permanent actions;

$\gamma_{G,inf}$ the partial factor for favourable, permanent, compressive effects in the GEO-2 limit state.

(3) When analysing the safety against pull-out of piles in the GEO-2 limit state shear forces T_k may be taken into consideration using:

$$F_{t,d} = F_{t,G,k} \cdot \gamma_G + F_{t,Q,rep} \cdot \gamma_Q - \left(F_{c,G,k} + T_k\right) \cdot \gamma_{G,inf} \tag{8.2}$$

The shear forces are treated as favourable, permanent compressive effects for determining the design value of the tensile effect. See (7), (8) and 8.3.2.2 (2) for additional guidance on consideration of shear forces T_k.

(4) For UPL limit state analysis after 8.3.2.2 of the attached block of soil the geometry, and therefore the weight $G_{E,k}$, of a block of soil attached e.g. through tension pile elements, may be determined using:

$$G_{E,k} = n_Z \cdot \left[l_a \cdot l_b \left(L - \frac{1}{3} \cdot \sqrt{l_a^2 + l_b^2} \cdot \cot\varphi \right) \right] \cdot \eta_Z \cdot \gamma \tag{8.3}$$

where, in addition to the previously defined variables:

$G_{E,k}$ the characteristic weight of the attached soil;
n_Z number of tension elements in a tension pile group;
L the length of the tension elements;
l_a the larger grid
l_b the smaller grid dimension;
γ the governing unit weight of the attached block of soil;
η_Z the calibration factor; $\eta_Z = 0{,}80$.

The associated geometrical model is shown in Figure 8.1 which also applies to edge piles. Where necessary, the unit weight γ is entirely or partially replaced by the effective weight density (under buoyancy) γ' of the soil.

(5) The geometry of the attached soil prism determined with Eq. (8.3) was derived using relationships described in [116]. However, the equation cannot be adopted for very large pile grid spacing. In this case the governing analysis is that described in 8.3.2.3 for the isolated pile. If only one isolated pile without an enlarged base is present, analysis of the attached block of soil in the UPL limit state may also be dispensed with.

(6) According to EC 7-1 Handbook [44], and [26] and (3), shear forces T_k may be taken into consideration together with other favourable actions on the superstructure or on the retaining walls of an excavation pit where the bottom is subject to uplift and anchored by piles. The analysis shall ensure that:

- force transfer from the wall into the base is provided by a suitable construction and
- the bottom slab is capable of uniformly distributing the transferred forces.

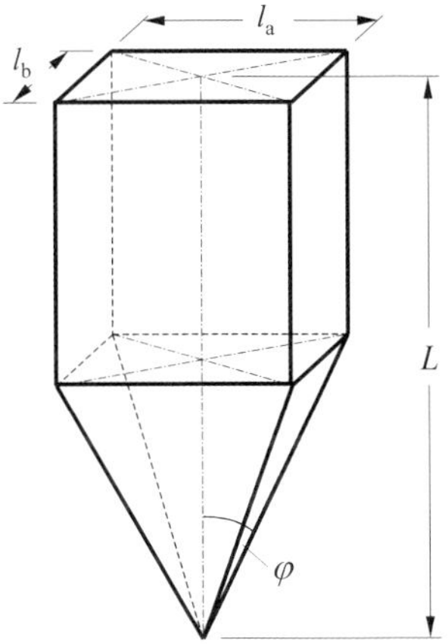

Figure 8.1 Geometry of the soil attached to an individual pile of a pile group

(7) For an excavation pit subject to uplift forces and anchored by piles the shear forces on the retaining wall should only be taken into consideration in a small strip bordering the walls. Otherwise, distribution of the load across the concrete bottom slab must be demonstrated, taking the stiffness of the slab and of the piles into consideration.

(8) An example analysis of a tension pile group is provided in Annex B12.

8.1.3 Laterally loaded pile groups

(1) The recommendations in 8.2.3 for laterally loaded pile groups relate to groups subjected to actions like foundation loads imposed by a superstructure. The actions are acting on the pile capping slab or above it. Geotechnical actions such as lateral pressures, see 4.5, are not considered here.

8.2 Bearing Capacity and Resistances of Pile Groups

8.2.1 Compression pile groups

8.2.1.1 Introduction

(1) In pile groups the individual piles normally participate to varying degrees in accepting the actions imposed on the group. Possible reduction of the load-bearing capacity (performance) of the single piles compared to the uninfluenced single pile must be taken into consideration, in cases where a group effect is governing.

(2) Concerning the load-bearing capacity (performance) of compression pile groups detailed knowledge is not yet available, in particular not for the various ground conditions and pile types.

(3) For determination of the generally higher group settlements in the serviceability limit state, in particular for long bored piles, as compared to single piles, the settlement-related group effect can be calculated using e.g. the approximation solution after [120], also see [88] and [115].

(4) Below, an approximation method using nomograms after [124] and [125] is proposed. It is derived from extensive 3D-FEM parameter studies and takes settlement- and resistance-related pile group behaviour into consideration. The method should be preferentially adopted for determining the settlement behaviour in and the analysis of the serviceability limit state of pile groups. The method is suited also to derive characteristic pile spring stiffnesses, specific to the position within the pile group, as basis for determining characteristic effects in the pile capping slab or in the superstructure and for structural analyses, see 8.3.1.2. Whether or not the fundamental conditions of the method are adhere to in a project-related application should be confirmed on a case-by-case basis be a geotechnical expert or the geotechnical designer.

(5) Examples of the application of the method after (4) can be found in Annex B11.

8.2.1.2 Group effect in terms of the settlements of bored pile groups

(1) The mean settlement s_G of a bored pile group corresponds to the settlement of a single pile to which the group factor G_s is applied, as a result of the mean action F_G on the group piles.

$$s_G = s_E \cdot G_s \tag{8.4}$$

where:

s_G mean settlement of a pile group;
s_E settlement of a comparable single pile;
G_s settlement-related group factor for the mean settlement of a pile group.

(2) The settlement-related group factor G_s for determining the mean settlement of a pile group subject to a centrally acting, vertical action is given by:

$$G_s = S_1 \cdot S_2 \cdot S_3 \tag{8.5}$$

where:

S_1 factor concerning the influence of the soil type and the group geometry (pile length L, pile embedment depth in load-bearing ground d, pile centre distances a as shown in Figure 8.2 and Figure 8.3);

S_2 group size influence factor as shown in Figures 8.4, 8.5 and 8.6;
S_3 pile type influence factor (for the group effect of displacement piles and grouted micropiles, see 8.2.1.4 and 8.2.1.5).

(3) The nomograms in Figure 8.2 to Figure 8.6 to determine the settlement-related group effect are differentiated for cohesive and non-cohesive soils. The

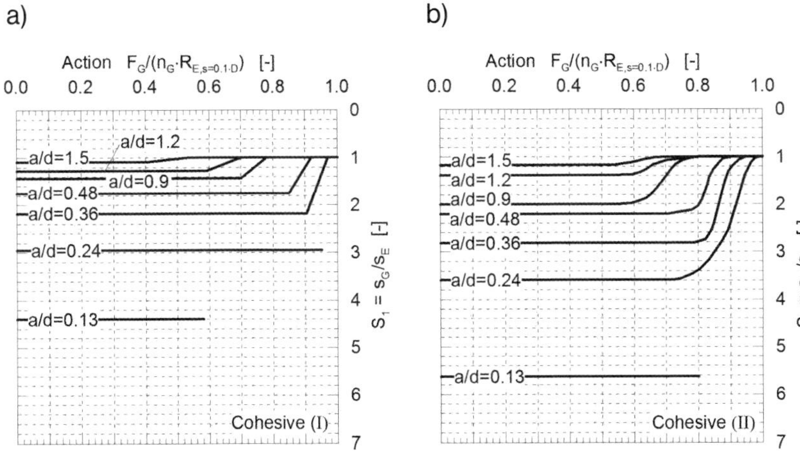

Figure 8.2 Nomograms for determining the mean settlement of a bored pile group for cohesive soils: a) cohesive (I), E_s = 5–15 MN/m² b) cohesive (II), E_s = 15–30 MN/m²

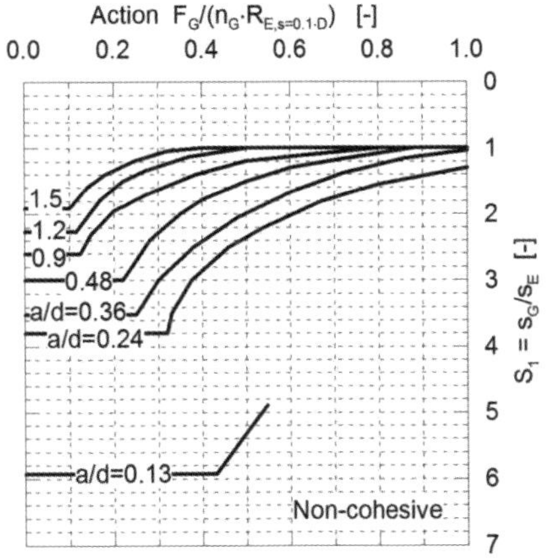

Figure 8.3 Nomograms for determining the mean settlement of a bored pile group for non-cohesive soils, $E_s \geq$ 25 MN/m²

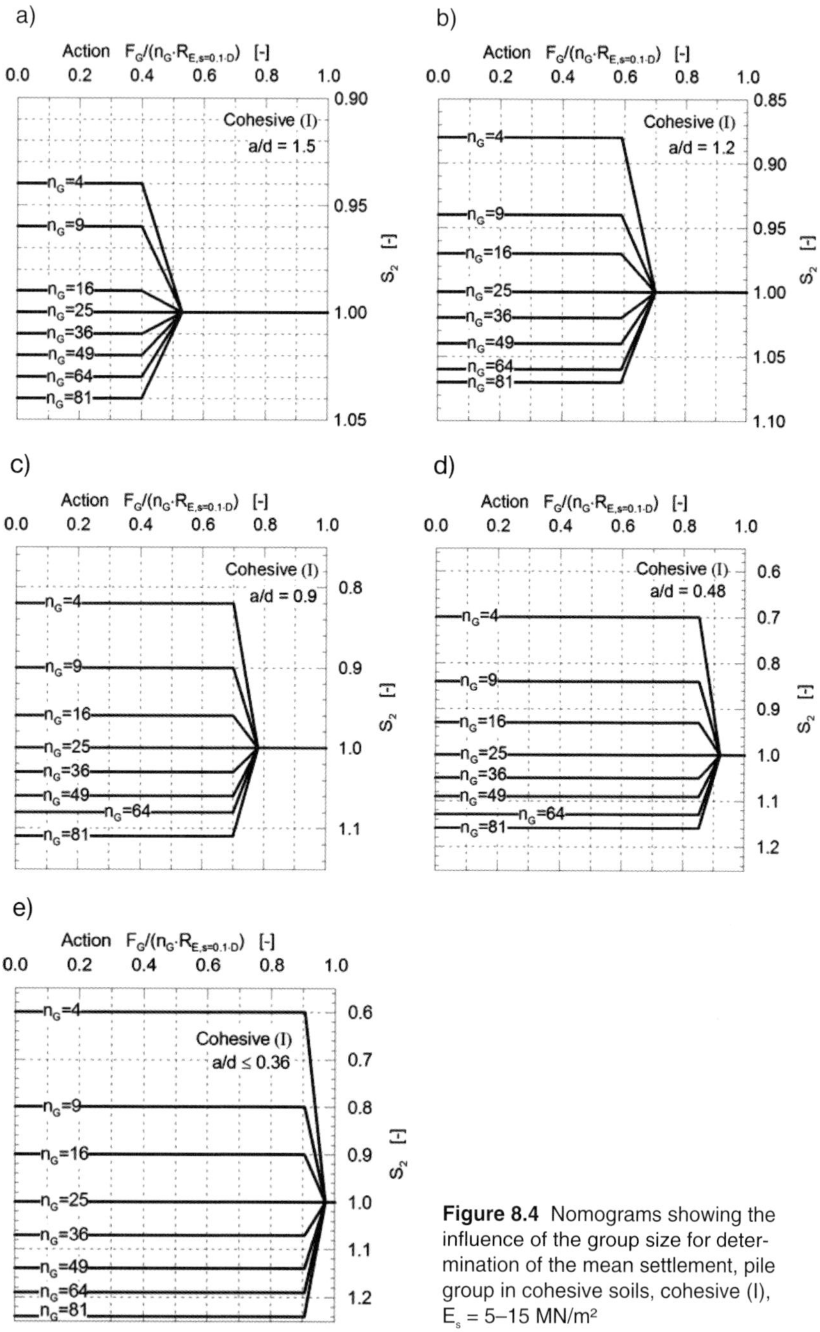

Figure 8.4 Nomograms showing the influence of the group size for determination of the mean settlement, pile group in cohesive soils, cohesive (I), E_s = 5–15 MN/m²

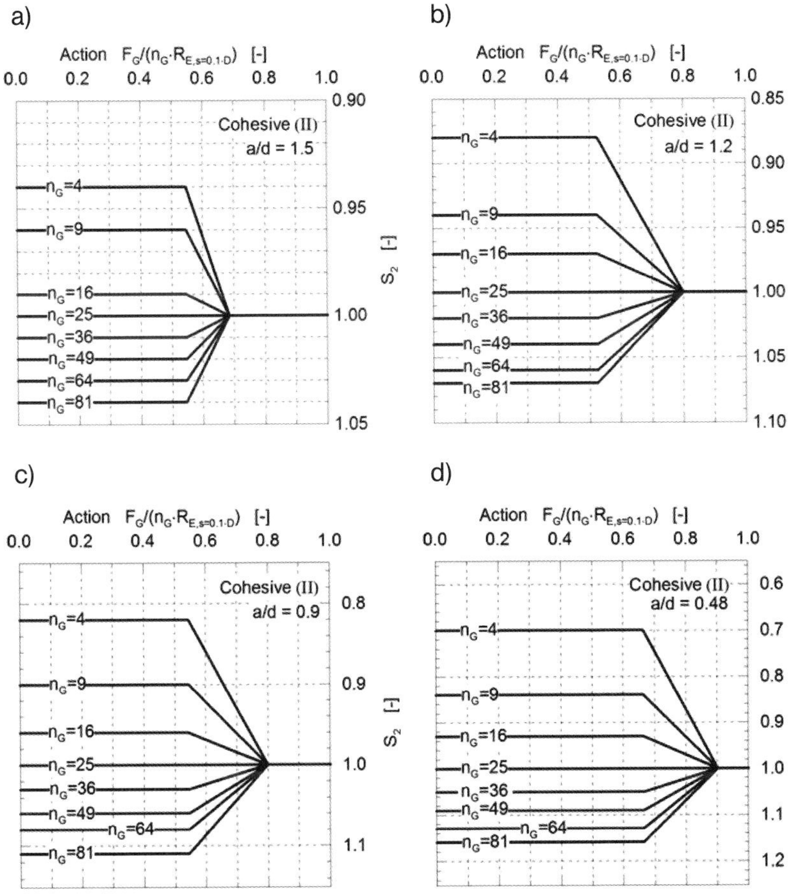

Figure 8.5 Nomograms showing the influence of the group size for determination of the mean settlement of a pile group in cohesive soils, cohesive (II), E_s = 15–30 MN/m²

stiffness modulus, representing the governing soil parameter for this method, was, in approximation, set as follows for these soils types (application limits):

- cohesive I: $E_s = 5-15$ MN/m²;
- cohesive II: $E_s = 15–30$ MN/m²;
- non-cohesive: $E_s \geq 25$ MN/m².

Soils with stiffness moduli $E_s < 5$ MN/m² are regarded as not load-bearing strata, in which no significant pile resistance is activated.

Note: If, for cohesive or non-cohesive soils, other stiffness moduli are prevalent, the corresponding stiffness modulus group may be selected in approximation, regardless of the soil type.

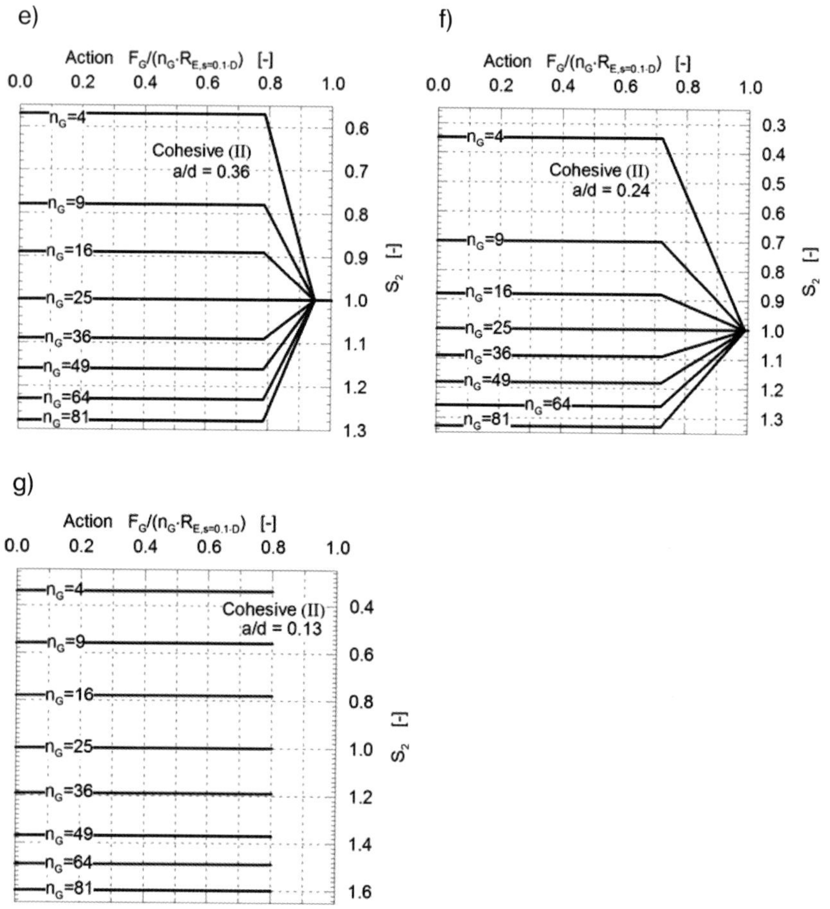

Continued Figure 8.5 Nomograms showing the influence of the group size for determination of the mean settlement of a pile group in cohesive soils, cohesive (II), E_s = 15–30 MN/m²

(4) The pile capping slab or superstructure is assumed as almost rigid, i.e. the differential settlements within the pile group are negligible for the performance.

(5) The nomogram method is initially related to square bored pile groups. However, it can in approximation also be adopted for regular, non-square, pile groups. Like for square groups, the number of piles in the group n_G is adopted as the input variable for the nomograms in Figures 8.2 to 8.6. The following abbreviations are used:

a	pile centre distances in the pile group;
d	pile embedment depth in the load-bearing strata;
D	pile diameter;
F_G	vertical action on the whole pile group;
L	pile length;
$R_{E,s=0,1 \cdot D}$	pile resistance of a single pile for a settlement $s = 0,1 \cdot D$;
s_E	settlement of a single pile;
s_G	mean settlement of a pile group;
n_G	number of piles in a pile group.

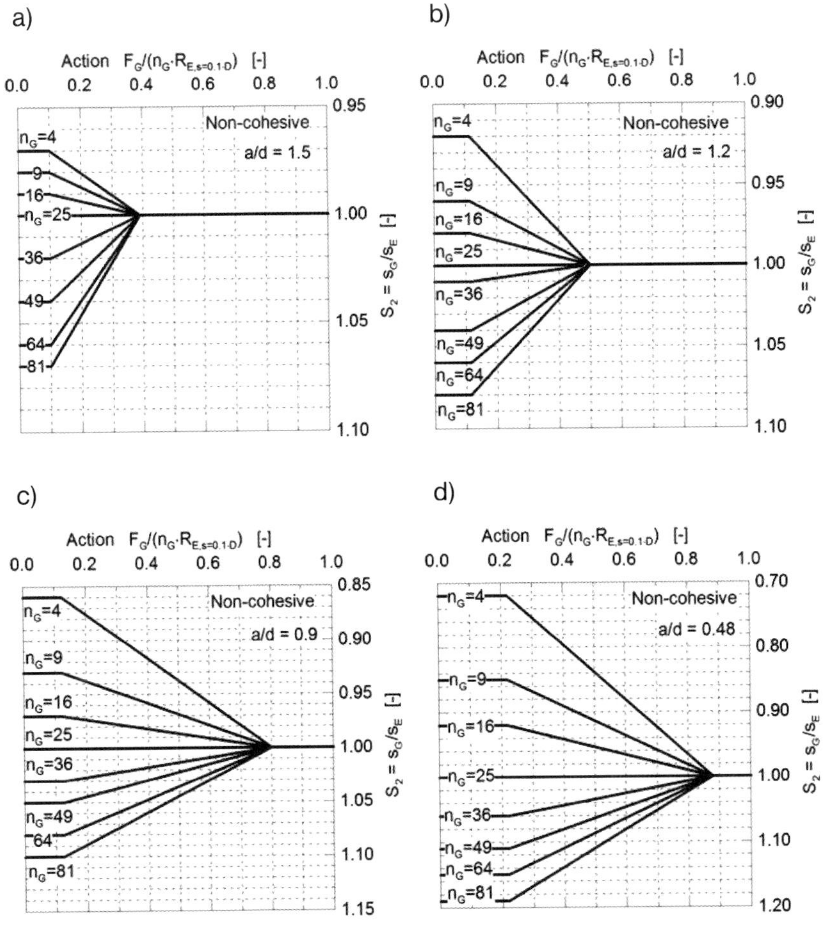

Figure 8.6 Nomograms s the influence of the group size for determination the mean settlement of a pile group in non-cohesive soils, $E_s \geq 25$ MN/m²

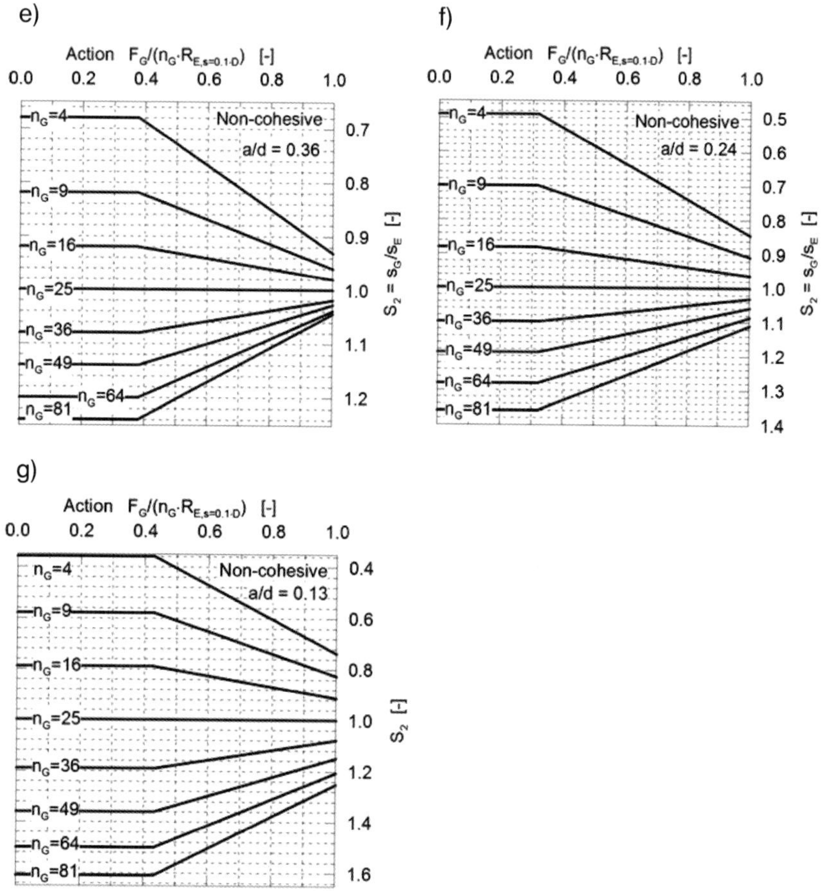

Contrinued Figure 8.6 Nomograms showing the influence of the group size for determination the mean settlement of a pile group in non-cohesive soils, $E_s \geq 25$ MN/m²

(6) The nomograms were developed on the basis of investigations and parameter studies using the finite element method, see [124], on bored piles with diameters D = 0,3 to 1,50 m and pile lengths L = 9 to 24 m. They should therefore only be applied to similar geometries. There are no limitation on pile centre distances, although applications should be limited to a/D ≥ 2.

(7) In accordance with 8.2.1.3 (6), the determined settlements should be adapted once again, if necessary, if the sum of the group pile forces is not in conformance with the total action on the pile group.

(8) Moment actions predominantly produce rotations of the pile foundation, which can be taken into consideration using Eq. (8.6). The factor η_M is a meas-

ure for the magnitude of the moment action relative to the vertical action and simultaneously accounts for the group geometry over the number of piles, and the pile centre distances, thus:

$$\eta_M = \frac{M \cdot n_G}{F_G \cdot n \cdot a} \tag{8.6}$$

where:

η_M factor for the relative magnitude of the moment action;
M moment action;
n_G number of piles in the group, where $n_G = n \cdot n$;
F_G vertical total action on the pile group;
n number of piles in a pile row;
a pile centre distances.

The additional settlements resulting from the moment action can be determined for all group piles via the rotation tan δ of the pile capping slab, see Figure 8.7.

(9) Groups of displacement piles normally display a deviating settlement behaviour when comparing with bored pile groups. Because of preloading of the group piles, resulting e.g. from the driving energy and stressing effects between the piles, smaller settlements than for bored pile groups should be anticipated for displacement pile groups, at least close to the serviceability limit state, see 8.2.1.4.

8.2.1.3 Resistances in (bored) group piles

(1) The resistances of individual bored piles in a group, taking the group effect into consideration, are given by the pile resistance of a single pile to which the group factor is applied for a settlement equal to the mean settlement of the pile group:

$$R_{G,i} = R_E \cdot G_{R,i} \tag{8.7}$$

where:

$R_{G,i}$ group pile resistance (pile i in the group);
R_E resistance of a comparable single pile;
$G_{R,i}$ resistance-related group factor for pile i in the group.

(2) The group factor $G_{R,n}$ for determining the resistances of group piles below a centrally acting, vertical total action is given by:

$$G_{R,i} = \lambda_1 \cdot \lambda_2 \cdot \lambda_3 \tag{8.8}$$

where:

λ_1 influence factor for soil type, group geometry (pile length L, embedment depth in load-bearing stratum d, pile centre distance a, respective settlement s as shown in Figures 8.8 to 8.10);

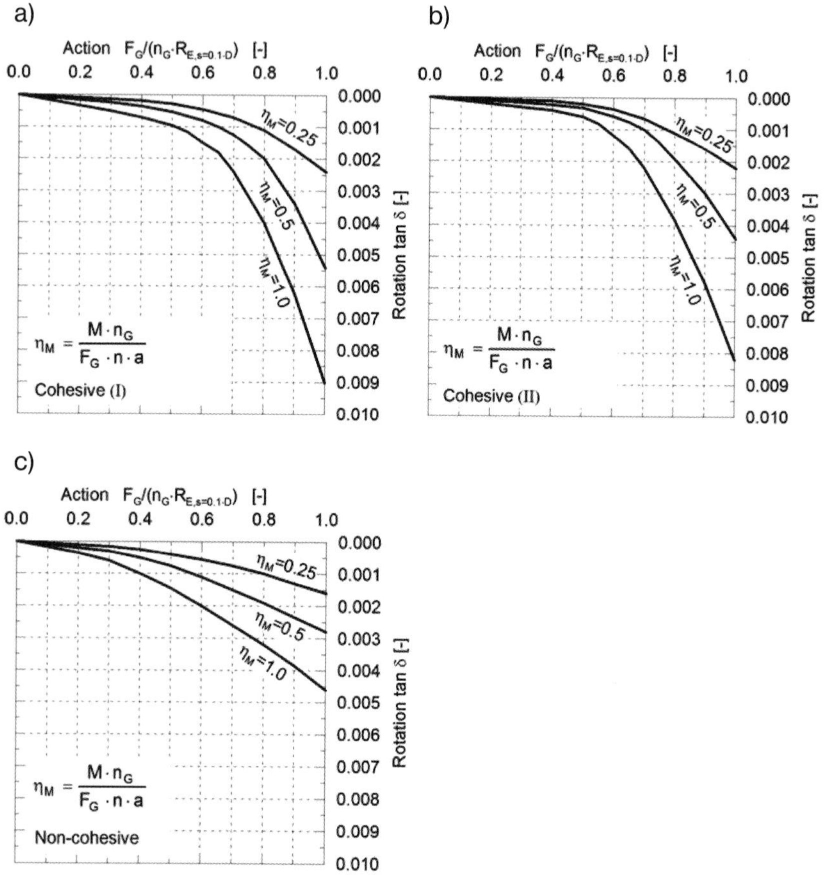

Figure 8.7 Nomograms for determining the rotation of the pile capping slab as a result of moment actions for a) cohesive soils (I), E_s = 5–15 MN/m², b) cohesive soils (II), E_s = 15–30 MN/m², c) non-cohesive soils, $E_s \geq$ 25 MN/m²

λ_2 influence factor for group size as shown in Figure 8.11;
λ_3 influence factor for pile type (for the group effect of displacement piles and grouted micropiles, see 8.2.1.4 and 8.2.1.5).

(3) 8.2.1.2 (3) to (6) also apply for the nomograms shown in Figure 8.8 to Figure 8.11.

(4) Only the four piles at the corners of a pile foundation are regarded as *corner piles*. *Edge piles* are the piles arranged in the respective outer rows, between the corner piles. *Inner piles* are all piles surrounded on all sides by neighbouring piles. The corner and edge pile definitions are retained even in

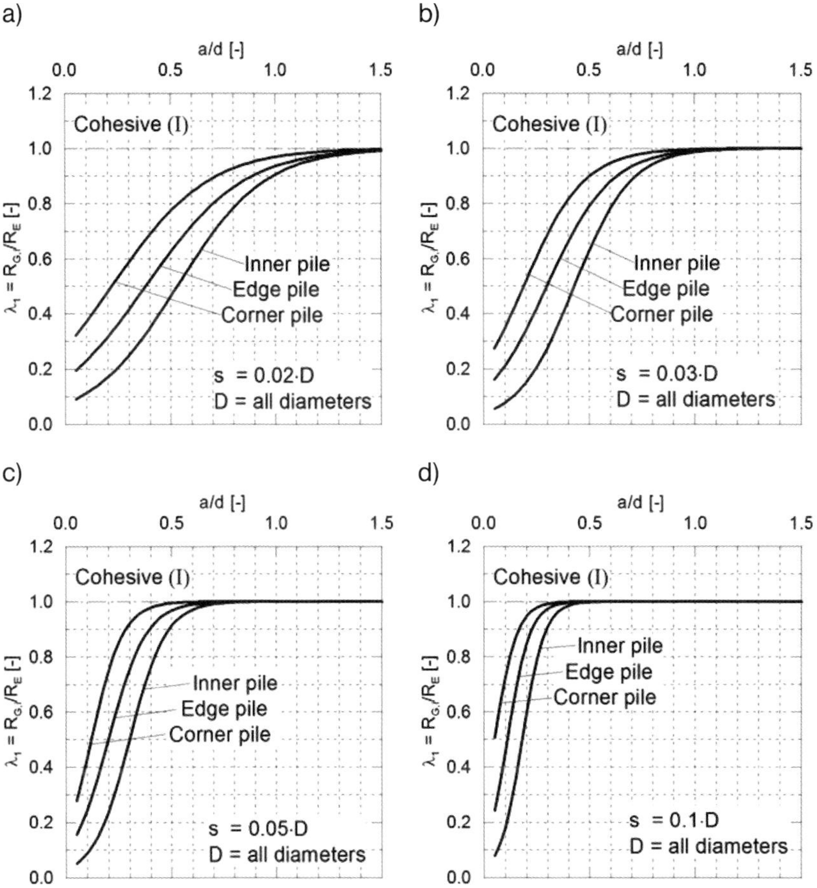

Figure 8.8 Nomograms for consideration of the resistance-related group effect of bored pile groups in cohesive soils (I) for E_s = 5–15 MN/m² and a relative settlement of a) s = 0,02 · D, b) s = 0,03 · D, c) s = 0,05 · D, d) s = 0,1 · D

large pile groups and are not extended to cover several rows, see 3.1.3, Figure 3.2.

(5) Use of the nomograms shown in Figures 8.8 to 8.10 should be limited to pile groups containing a maximum number of 9 piles per side.

(6) If the sum of all pile resistances in the group deviates substantially from the value of the total action on the group, iterative correction of the settlement determined after 8.2.1.2 is recommended. If the pile resistances are underestimated, the settlement must be increased; if the pile resistances are overestimated, the settlement must be decreased.

(7) With regard to square pile groups with 3 × 3 and 4 × 4 piles, Figure 8.11 gives the influence factor λ_2 for taking the group size into consideration for inner piles. By definition, smaller pile groups have no inner piles as discussed in (4). In pile groups with 5 and more piles per side, the influence factor should be adopted as $\lambda_2 = 1.0$.

(8) In non-square pile groups the number of piles on the longer side shall be adopted as the input variable for Figure 8.11.

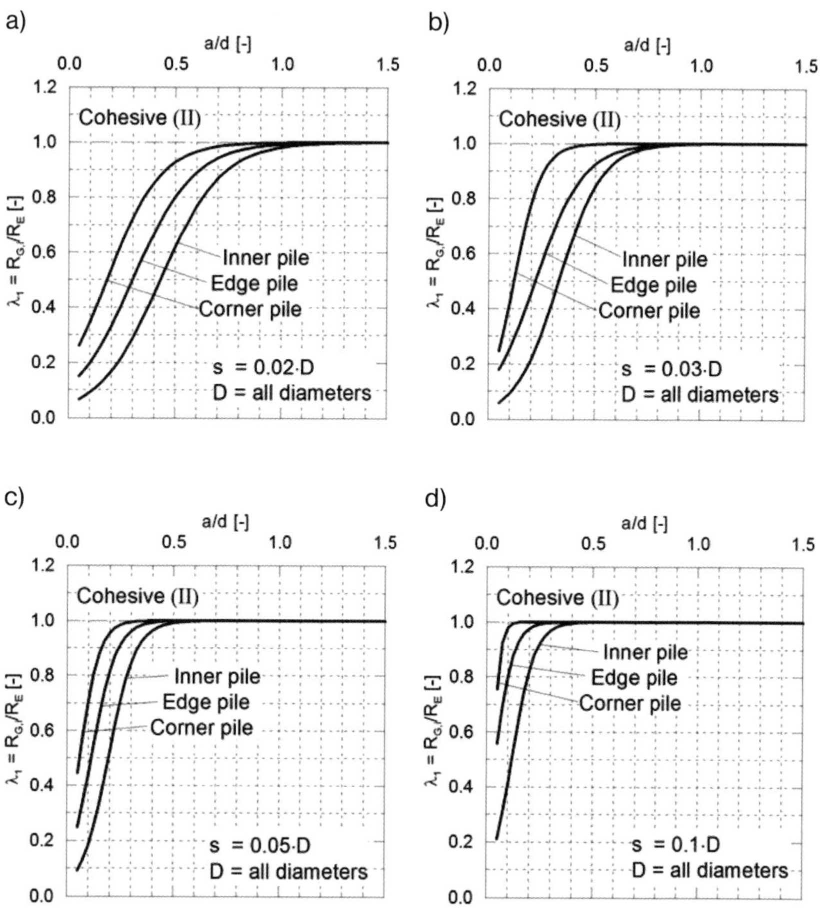

Figure 8.9 Nomograms for consideration of the group effect of bored pile groups in cohesive soils (II) for E_s = 15–30 MN/m² and a relative settlement of a) s = 0,02 · D, b) s = 0,03 · D, c) s = 0,05 · D, d) s = 0,1 · D

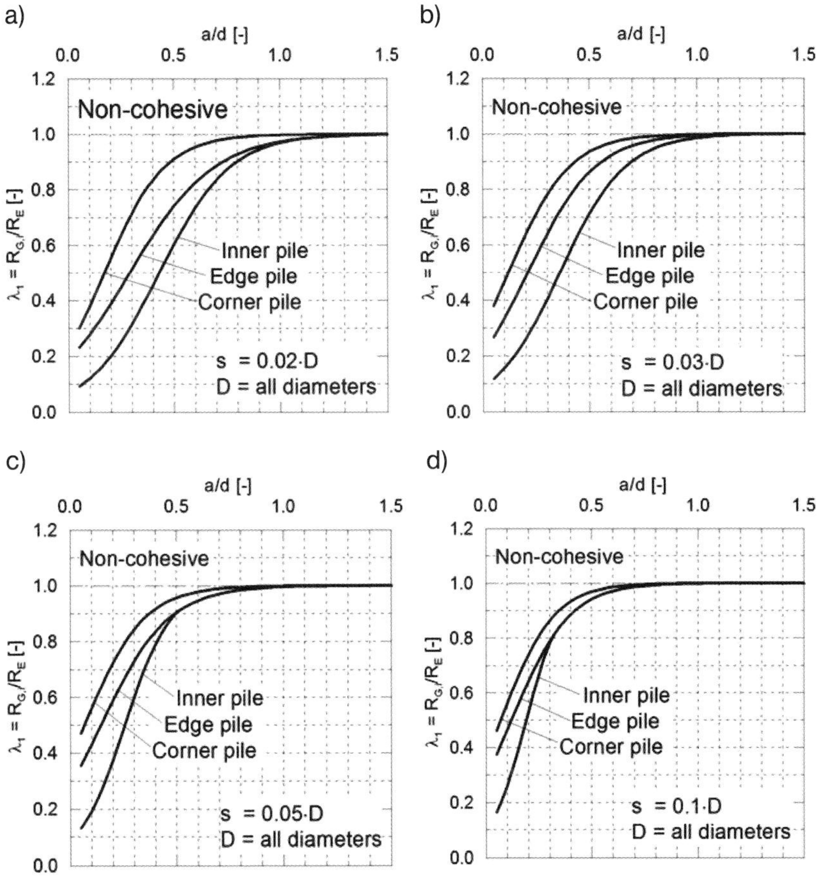

Figure 8.10 Nomograms for consideration of the group effect of bored pile groups in non-cohesive soils for $E_s \geq 25$ MN/m² and a relative settlement of a) s = 0,02 · D, b) s = 0,03 · D, c) s = 0,05 · D, d) s = 0,1 · D

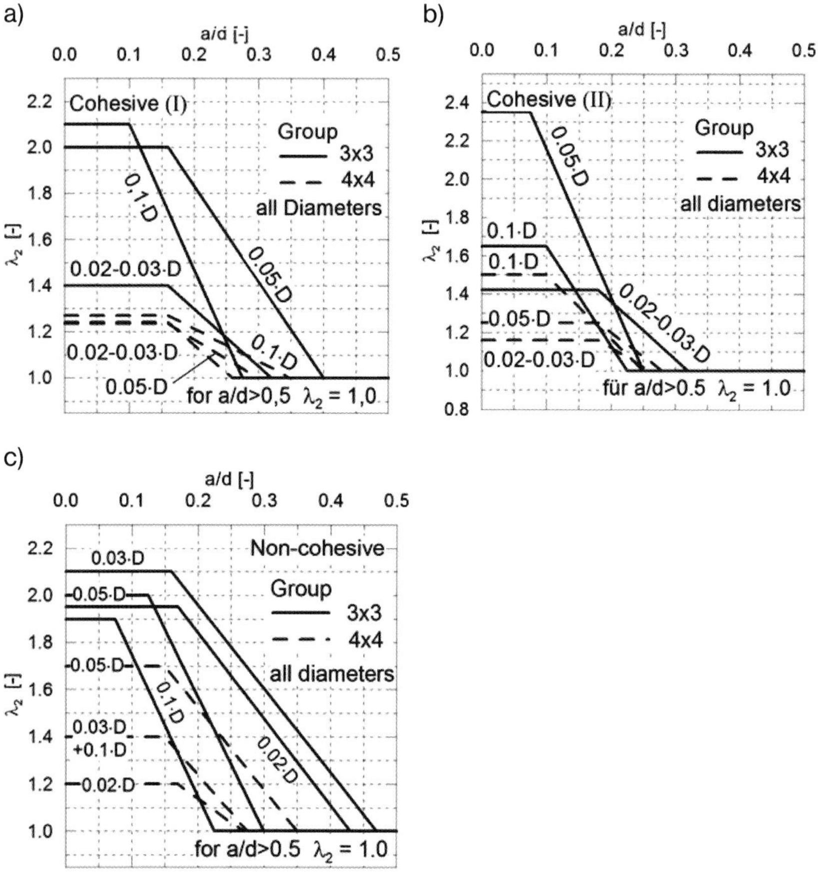

Figure 8.11 Nomograms of the influence of the group size on the group effect for a) cohesive soils (I), E_s = 5–15 MN/m², b) cohesive soils (II), E_s = 15–30 MN/m², c) non-cohesive soils, $E_s \geq$ 25 MN/m²

8.2.1.4 Displacement pile groups

(1) To date, little consolidated knowledge is available concerning the load-bearing capacity of displacement pile groups. Some initial guidance is provided below. Project-specific applications should be confirmed by a geotechnical expert.

(2) A group factor $G_R \geq 1{,}0$ can be adopted for full displacement pile groups in non-cohesive soils for the ultimate limit state. Group factors up to $G_R = 1{,}50$ can be used for favourable ratios of pile centre distances to-embedment depths of a/d = 0,3–0,7.

(3) In the serviceability limit state the group factor for full displacement pile groups in non-cohesive soils and with unfavourable pile centre distances-embedment depth ratios should be adopted as $G_R < 1,0$. Values of $a/d \leq 0,5$ can be classified as unfavourable. The group factor can also be around $G_R \geq 1,0$ for favourable pile centre distances-embedment depth ratios.

(4) A group factor $G_R \leq 1,0$ should be adopted for full displacement pile groups in cohesive soils. For small pile centre distances-embedment depth ratios, it could be adopted at around $a/d = 0,1$ where $G_R = 0,7$. Cohesive soils with undrained cohesion greater than $c_u \geq 100$ kN/m² can display more favourable load-bearing capacity if the group factor is assumed to $G_R = 1,0$.

(5) The possibility of a reduction in pile capacity as a result of increases in porewater pressure initiated by the installation process must be taken into consideration for displacement pile groups in cohesive soils. This effect can result in a temporary group factor $G_R = 0,4$. Under circumstances this effect can continue for weeks.

(6) As a general rule for paragraphs (1) to (4) above, the group factor trends towards $G_R = 1,0$ for increasing pile centre distance-embedment depth ratios.

(7) Partial displacement pile groups can display a smaller group factor than full displacement piles.

(8) Overall, it can be assumed that displacement pile groups normally display settlement behaviour deviating from that of bored pile groups. Because of pile group preloading, e.g. due to the driving energy expended on driven displacement piles and the stressing effects between the piles, smaller settlements than for bored pile groups should be anticipated, at least close to the serviceability limit state.

8.2.1.5 Micropile groups

(1) To date, little consolidated knowledge is available concerning on the load-bearing capacity of groups of grouted micropiles. Based on the tensile tests carried out on micropile groups described in [117], group factors for micropile groups should not be adopted more favourable than those for bored pile groups after 8.2.1.3. The adopted group factors should be confirmed by a geotechnical expert for the specific application.

(2) It should also be noted that the failure state of micropiles is generally reached at substantially smaller settlements than for bored piles.

8.2.1.6 Layered ground

(1) If a pile group is founded in ground where several layers are classified as load-bearing after 8.2.1.2 (3), the share of the pile resistance activated in the respective strata may be estimated and procedure can follow (2) to (5).

(2) If one stratum displays a substantially higher load-bearing capacity than the other strata and it can be anticipated that the governing share of the pile resistance is activated in this stratum, the thickness of this stratum only shall be adopted to form the pile centre distance-embedment depth ratio a/d. The nomograms corresponding to the soil type in this stratum shall be used.

(3) If the stratum with the highest capacity is considerably thinner than the other strata, such that it is not anticipated that the governing share of the pile resistance is activated in this stratum, the nomograms for the soil type of the thickest stratum shall be used. However, the thicknesses of all load-bearing strata shall be added for assessment of the embedment length in the load-bearing stratum.

(4) If the strata thicknesses and soil parameters are such that it is anticipated that approximately equal pile resistance shares are activated in each stratum, the nomograms for the soil type of each individual stratum shall be adopted, and the mean of the derived group factors be formed. The thicknesses of all load-bearing strata shall be adopted as the embedment length in the load-bearing stratum.

(5) Annex B11 includes an example.

8.2.2 Tension pile groups

(1) Analysis of uplift of a tension pile group is performed for the ultimate limit state. No resistances occur; instead, the resisting variables are treated as actions, see 8.1.2.

(2) Analysis of sufficient safety against pull-out of tension piles in the GEO-2 limit state may be performed in analogy to the procedure for single piles, see Sections 5 and 6.

(3) An example analysis for a tension pile group can be found in Annex B12.

8.2.3 Laterally loaded groups

(1) In pile groups in which all piles receive practically equal horizontal head deflections, the individual piles participate to differing degrees in accepting the action H_G on the pile group. In double-symmetrical groups of equal piles the distribution H_i of the action on piles of the group may be calculated using:

$$\frac{H_i}{H_G} = \frac{\alpha_i}{\Sigma\alpha_i} \tag{8.9}$$

whereby:

$$\alpha_i = \alpha_L \cdot \alpha_Q \tag{8.10}$$

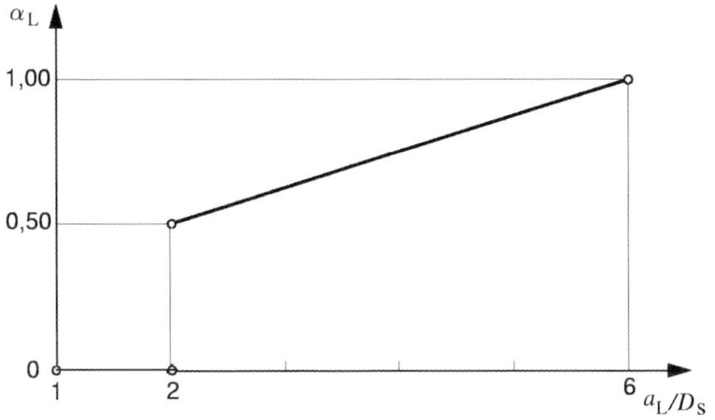

Figure 8.12 Reduction factor α_L for the ratio a_L/D_s of pile centre distances a_L in the direction of the force to the pile shaft diameter D_s

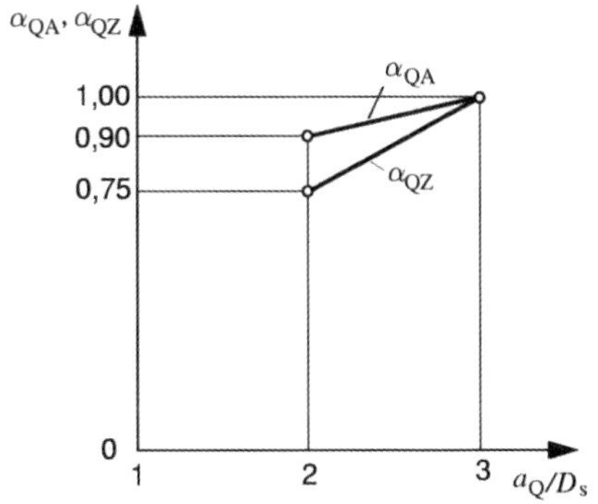

Figure 8.13 Reduction factors α_{QA} and α_{QZ} for the ratio a_Q/D_s of pile centres a_Q lateral to the direction of the force to the pile shaft diameter D_s; for $a_Q/D_s < 2$ the same conditions apply as for a continuous wall, see e.g. DIN 4085

The factors α_L and α_Q depend on the pile centre distances a_L in the direction of force, the distances a_Q lateral to the direction of force and the position of the pile within the pile group. They shall be taken from Figures 8.12 and 8.13 and be adopted within the pile group as shown in Figure 8.14. The procedure recommended here is based on investigations by [69], [129] and [130].

(2) When determining the action effects and deformations based on subgrade reaction moduli, the reduction factors α_i for a pile in the group correspond to the subgrade reaction moduli reductions given in (3) and (4) below.

(3) For a linear with depth z increasing subgrade reaction modulus (applicable as an approximation for bored piles in normally consolidated cohesive soil and

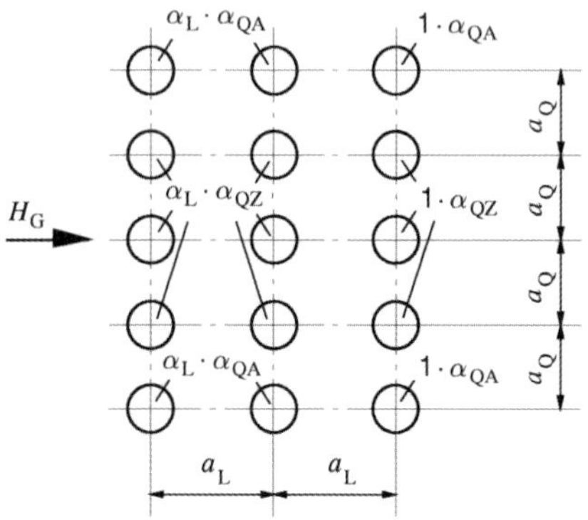

Figure 8.14 Reduction factors α_i as a function of the position of the pile within the group and to the direction of the action H_G

in non-cohesive soil):

$$k_{s,k}(z) = k_{hE,k} \cdot z/D_s \tag{8.11}$$

and the elastic length L_E of the individual pile:

$$L_E = \left(\frac{E \cdot I}{k_{hE,k}} \right)^{0,2} \tag{8.12}$$

applies for $L/L_E \geq 4$:

$$k_{hi,k} = \alpha_i^{1,67} \cdot k_{hE,k} \tag{8.13}$$

and for $L/L_E \leq 2$:

$$k_{hi,k} = \alpha_i \cdot k_{hE,k} \; . \tag{8.14}$$

For values $4 > L/L_E > 2$, linear interpolation is allowed. Where:

$E \cdot I$ flexural stiffness of pile;
$k_{hE,k}$ the characteristic value of the modulus of subgrade reaction of the individual pile at the depth $z = D_s$;
$k_{hi,k}$ the characteristic value of the modulus of subgrade reaction of pile n in the group at the depth $z = D_s$;
L the length of the pile.

(4) For subgrade reaction moduli that are constant with depth (as the upper bound for piles in overconsolidated cohesive soil):

$$k_{s,k}(z) = k_{s,k} = \text{const.} \tag{8.15}$$

and the elastic length L_E and the modulus of subgrade reaction of the individual pile $k_{sE,k}$:

$$L_E = \left(\frac{E \cdot I}{k_{sE,k} \cdot D_s} \right)^{0,25} \tag{8.16}$$

applies for $L/L_E \geq 4$:

$$k_{si,k} = \alpha_i^{1,33} \cdot k_{sE,k} \tag{8.17}$$

and for $L/L_E \leq 2$:

$$k_{si,k} = \alpha_i \cdot k_{sE,k} \tag{8.18}$$

Linear interpolation is allowed for values $4 > L/L_E > 2$.

(5) Equations (8.9) to (8.18) apply equally to piles with hinged connections to and part or full restraint in the pile cap.

(6) The axial stiffness of the piles (pile resistance/settlement of the pile head) can have a substantial influence on the design related to bending. The action effects should therefore be determined using upper and lower bound values of the subgrade reaction moduli.

(7) In pile groups with irregularly distributed piles the α_i values may be determined using Figures 8.12 and 8.13 accordingly.

(8) In pile groups with varying flexural pile stiffnesses, the distribution H_i of the total action H_G acting on the individual piles (i) of the pile group may in approxation be determined using the α values shown in Figures 8.12 and 8.13, and using Equations (8.19) and (8.20).

$$\frac{H_i}{H_G} = \frac{C_i}{\Sigma C_i} \tag{8.19}$$

$$C_i = H_0 / y_0 \tag{8.20}$$

where:

H_0 arbitrary value of a horizontal action on the head (unit load) of the individual pile;

y_0 the associated pile head deflection.

The C_i values shall be calculated using the subgrade reaction moduli according to Equations (8.13) to (8.14) or (8.17) to (8.18), taking the degree of restraint or the deformation conditions at the pile head into consideration.

(9) An example calculation for the distribution of the subgrade reaction moduli in pile groups loaded laterally to the pile axis is provided in Annex B13.

8.3 Bearing Capacity Analyses

8.3.1 Compression pile groups

8.3.1.1 External capacity

(1) Bearing capacity analyses are required for the GEO-2 limit state of the whole pile group and of the individual piles . The individual pile is analysed in analogy to 6.3.1.

(2) For analysis of sufficient safety against failure of a compression pile group in the GEO-2 limit state, the limit state conditions:

$$F_{c,d} \leq \sum R_{c,d,i} \quad \text{or} \tag{8.21a}$$

$$F_{c,d,G} \leq \sum R_{c,d,G} \tag{8.21b}$$

must be fulfilled. Where $\Sigma R_{c,d,i}$ is the group resistance resulting from the position-dependent individual pile resistances for $s_{ult} = s_g = 0,1 \cdot D$, e.g. after 8.2.1. For this consideration, the group settlement is related to the settlement s_{ult} of a single pile. In contrast, $R_{c,d,G}$ is the group resistance after (5) and (6) for a pile group model representing a large, equivalent pile. Though normally selecting the limit values $q_{b,k}$ and $q_{s,k,j}$ for the tip resistance and skin friction of a single pile, they are initially adopted for the group, independent of settlements. The design value of the effects $F_{c,d}$ is given by:

$$F_{c,d} = F_{G,k} \cdot \gamma_G + F_{Q,rep} \cdot \gamma_Q \,. \tag{8.22}$$

(3) The design values of the resistances $R_{c,d,i}$ are given by:

$$R_{c,d,i} = R_{c,k,i} / \gamma_t \tag{8.23}$$

where γ_t is the partial factor to DIN 1054:2010-12, Table A2.3 (Annex A3.2), depending on whether pile resistances are based on empirical data or acquired from pile load tests. $R_{c,k,i}$ is given by (4).

(4) The group effect shall be taken into consideration when adopting the characteristic pile resistances $R_{c,k,i}$. The resistance R_E of a comparable single pile with the same settlement, factorised by the group factor $G_{R,i}$, can be adopted for the characteristic resistance of the group pile. The group factor $G_{R,i}$ is a measure of the group effect and can be determined after 8.2.1.3.

$$R_{c,k,i} = R_E \cdot G_{R,i} \tag{8.24}$$

(5) As an alternative to the settlement-dependent ($s_{ult} = s_g = 0,1 \cdot D$) design procedure after Eq. (8.21a), in conjunction with (4), analysis of the capacity of a pile group may, in consistency with EC 7 Eq. (8.21b), also be performed in approximation and independent of settlements after (2) and (3) for a large,

single pile:

$$R_{c,k,G} = q_{b,k} \cdot \sum A_{b,i} + \sum q_{s,k,j} \cdot A^*_{s,j} \qquad (8.25)$$

where:

$R_{c,k,G}$ the characteristic resistance of the whole pile group in the GEO-2 limit state, determined for the model of the pile group as a large, equivalent single pile;

$q_{b,k}$ the characteristic value of the pile tip resistance in the failure state for the single pile;

$A_{b,i}$ the nominal value of the pile base areas of the single piles i in accordance with Figure 8.15;

$q_{s,k,j}$ the characteristic value of the pile skin friction in the failure state of the single pile in stratum j, relative to the shaft surface $A^*_{s,j}$ of the equivalent single pile;

$A^*_{s,j}$ the nominal value of the circumference area of the pile group, using the equivalent single pile as model of the pile group, see Figure 8.15.

(6) The design values of the group resistance $R_{c,d,G}$ are given by:

$$R_{c,d,G} = R_{c,k,G} / \gamma_t \qquad (8.26).$$

The partial factor γ_t depends on whether pile resistances are based on empirical data or are acquired from pile load tests. It is selected after Annex A3.2 or the EC 7-1 Handbook [44], Table A2.3.

(7) The load-bearing and settlement behaviour of the pile group in the GEO-2 limit state cannot be unequivocally defined. The analyses after (2) to (6), which yield the largest utilisation of the pile group in the ultimate limit state, may be adopted as the governing analyses, assuming the serviceability of the pile group after 8.4.1 is adhered to.

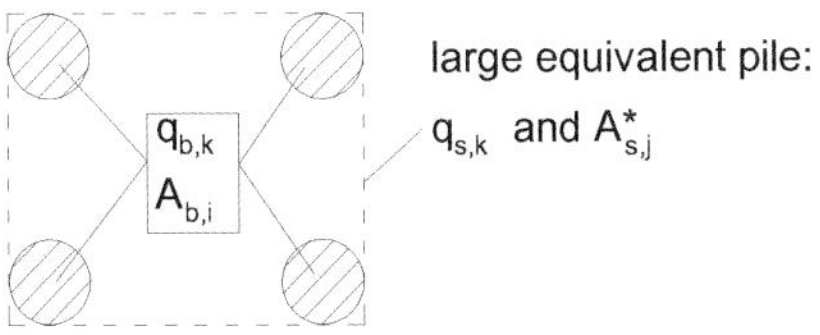

Figure 8.15 Example for applying the resisting components of a pile group as a large, equivalent pile, plan view

8.3.1.2 Structural analyses of the pile capping slab

(1) In axially loaded pile groups, the pile cap or the superstructure shall be checked for sufficient dimensioning to compensate for the variable resistance-

settlement behaviour of the individual group piles and to correspondingly redistribute the loads. To determine the characteristic effects as the basis for designing the pile capping slab or the structure, the piles in the structural analysis model may be replaced by equivalent springs. The spring stiffness $c_{p,k,i}$ is given after 8.4.1 by:

$$c_{p,k,i} = \frac{R_{k,i}\,(SLS)}{s_k}. \tag{8.27}$$

The structural analysis model using equivalent spring stiffnesses after Eq. (8.27), may only be applied to characteristic action combinations, which are also used as the basis for determining the equivalent stiffnesses. Because the equivalent spring stiffnesses postulate linear resistance-settlement behaviour, new equivalent spring stiffnesses shall be determined for other characteristic or representative action combinations.

(2) Also see 8.3.3.

(3) An example calculation of the external capacity of a pile group and the characteristic effects of the pile capping slab is included in Annex B11.

8.3.2 Tension pile groups

8.3.2.1 Introduction

(1) Always two limit state cases shall be investigated when analysing the stability of foundations or other structures anchored in the ground by tension piles:

a) Sufficient safety agaist uplift must be demonstrated for the UPL limit state for an assumed soil block formed of the piles and the enclosed surrounding ground.
b) Assuming that each pile acts as a single pile, sufficient safety against pull-out must be demonstrated after 8.3.2.3 for the GEO-2 limit state.

Also see the note on 8.1.2 (1).

(2) See 8.1.2 (6) and (7) when taking shear forces T_k in the superstructure into consideration, and particularly for the bottom of an excavation anchored by piles and subject to uplift forces.

(3) An example analysis for a tension pile group is included in Annex B12.

8.3.2.2 Analysis of the attached soil block in the UPL limit state

(1) In order to achieve sufficient safety against uplift of a foundation anchored by tension piles and subject to tensile forces, the following condition must be verified for the UPL limit state:

$$G_{dst,k} \cdot \gamma_{G,dst} + Q_{dst,rep} \cdot \gamma_{Q,dst} \leq G_{stb,k} \cdot \gamma_{G,stb} + G_{E,k} \cdot \gamma_{G,stb} \tag{8.28}$$

where:

$G_{dst,k}$ the characteristic value of permanent, destabilising, vertical actions;
$\gamma_{G,dst}$ the partial factor for destabilising, permanent actions in the UPL limit state after Annex A2 or the EC 7-1 Handbook [44], Table A2.1;
$Q_{dst,rep}$ the characteristic or representative value of variable, destabilising, vertical actions;
$\gamma_{G,dst}$ the partial factor for destabilising, permanent actions in the UPL limit state after Annex A2 or the EC 7-1 Handbook [44], Table A2.1;
$G_{stb,k}$ the lower characteristic value of stabilising, permanent, vertical actions imposed by the superstructure;
$\gamma_{G,dst}$ the partial factor for destabilising, permanent actions in the UPL limit state after Annex A2 or the EC 7-1 Handbook [44], Table A2.1;
$G_{E,k}$ the characteristic weight of the attached soil after 8.1.2 (4).

(2) When analysing the safety against uplift (UPL limit state), the action of shear forces T_k may be taken into consideration after the EC 7-1 Handbook [44]. The shear forces are treated as favourable permanent actions using:

$$G_{dst,k} \cdot \gamma_{G,dst} + Q_{dst,rep} \cdot \gamma_{Q,dst} \leq G_{stb,k} \cdot \gamma_{G,stb} + \left(G_{E,k} + T_k\right) \cdot \gamma_{G,stb} \quad (8.29)$$

T_k is the characteristic value of the shear resistance or the friction resistance that develops around a soil block in which a tension pile group acts, or in a soil-structure interface.

8.3.2.3 Analysis of the capacity of a single tension pile in the GEO-2 limit state

(1) This analysis may be performed in analogy to 6.3.1 for the pull-out of a single pile using the general equation:

$$F_{t,d} \leq R_{t,d} \quad (8.30)$$

(2) When determining the effect $F_{t,d}$ any favourably acting shear forces T_k may be taken into consideration as restraining actions after Eq. (8.2).

(3) The pile skin friction of closely spaced piles in the tension pile group can be influenced by a group effect, which normally leads to a reduction in the characteristic skin friction $q_{s,k}$ compared to an uninfluenced single pile, see e.g. [40]. However, analysis of safety against pull-out may, in approximation, be performed regardless of the group effect, as for an single pile. This is because the characteristic total tensile force cannot become smaller than the characteristic weight of the attached block of soil when a group effect exists.

8.3.3 Structural failure of group piles and pile cap structures

(1) The effects on group piles shall be analysed in analogy to 6.3.3, taking the group effect into consideration.

(2) If two or more actions on the pile group or the grillage virtually cancel each other with regard to effects on certain piles in the group, and the piles thus carry no, or only minor loads, these piles shall be structurally designed.

(3) When analysing the safety against structural failure of the pile capping structure or the superstructure, 8.3.1.2 must be observed.

8.4 Serviceability Analyses

8.4.1 Compression pile groups

(1) When determining the settlement in a pile group any possible increase in settlement as a result of a group effect must be taken into consideration. For this purpose, the factorised single pile settlement $s_{E,k}$ to which the group factor $G_{s,i}$ is applied can be adopted as the mean settlement of the pile group resulting from an action imposed on all group piles by uniform distribution of the total action. The group factor $G_{s,i}$ is a measure of the group effect and can be determined after 8.2.1.2.

$$s_k = s_{E,k} \cdot G_{s,i} \ . \tag{8.31}$$

(2) If the deformations of the pile foundation are of importance for the structure as a whole, sufficient safety against loss of serviceability (SLS) must be demonstrated. The analysis is fulfilled if the design value of the actions F_c(SLS) is smaller than or at least equal to the sum of the design values of the pile resistances $R_{d,i}$(SLS). After DIN 1054 the design values hereby normally correspond to the characteristic values.

$$F_{c,d}\,(SLS) = F_{c,k}\,(SLS) \le \sum R_{d,i}\,(SLS) = \sum R_{k,i}\,(SLS)\,. \tag{8.32a}$$

For the analysis also the allowable settlements, assuming characteristic effects on the pile foundation in the serviceability limit state may be used as follows:

$$\text{zul}\ s_k \ge s_k \ . \tag{8.32b}$$

(3) Any possible increased settlement resulting from a group effect after (1) must be taken into consideration when adopting the characteristic pile resistances $R_{2,k,i}$. The characteristic resistance of a group pile is given by the resistance $R_{E,k}$ of a comparable single pile with the same settlement, factorised by the group factor $G_{R,i}$. The group factor $G_{R,i}$ is a measure of the group effect and can be determined after 8.2.1.3.

$$R_{k,i}\,(SLS) = R_{E,k} \cdot G_{R,i} \ . \tag{8.33}$$

(4) In analogy to 8.3.1.2, any possible effect imposed on the pile capping slab or the superstructure must be taken into consideration also for this effect condition in the serviceability limit state.

(5) Differential settlements between piles must be anticipated in pile groups with flexible pile capping slabs. Assuming no pile capping slab exists and all piles receive actions of the same magnitude, after [124] the following may be assumed for the serviceability limit state: The differential settlements within the pile group and relative to the mean settlement s_G of all piles are in the order of:

- corner piles $\Delta s = -0.13$ to $-0.23 \cdot s_G$;
- edge piles $\Delta s = -0.02$ to $-0.06 \cdot s_G$ and;
- inner piles $\Delta s = 0.09$ to $0.15 \cdot s_G$

The above figures are valid under condition that the piles are arranged regularly, the group size $n_G \leq 81$, pile length L = 9–24 m, the pile centre distances a = 3–6 · D and the pile diameter D = 0,3–1,50 m. If a pile capping slab exists, the differential settlements decrease with increasing stiffness of the pile capping slab. In turn, approximate, pile position-dependent spring constants can be derived from this information.

(6) An example of a serviceability analysis for a pile group with a rigid capping slab is contained in Annex B11.

8.4.2 Tension pile groups

(1) There are currently no reliable approximation approaches for the deformation behaviour of tension pile groups. For simple structures which are not or only marginally sensitive to deformations, it may be assumed that the serviceability limit state analysis is also covered if the conditions after 8.3.2 are adhered to.

(2) If the structure is sensitive to deformations, more reliable investigations are required, e.g. using the finite element method, see 8.5.

8.4.3 Laterally loaded pile groups

(1) In laterally loaded pile groups it must be demonstrated that the lateral load causes only tolerable displacements, rotations and tilting are compatible with the serviceability limit state of the pile group. It should be observed that the lateral load at the pile head level is not equally distributed amongst all piles in cases where the piles are subjected to equal head deflections, see 8.2.3 and Annex B13.

8.5 Higher Accuracy Pile Group Analyses

(1) If required by the structural boundary conditions, the design or the sensitivity of the structures involved, supplementary investigations of better accuracy should be performed, in addition to the simplified approaches listed in 8.1 to 8.4.

(2) Higher accuracy of investigations on axially and laterally loaded pile groups can be achieved through:

a) Numerical analyses with complete or partial modelling of the pile group, e.g. using the finite element method, whereby the analysis model should be suitably calibrated.
b) Pile load tests on smaller pile groups and comparison with load tests on single piles. This allows extrapolation of the behaviour of larger groups or calibration of numerical models.

9 Static Pile Load Tests

9.1 Introduction

(1) The fundamental requirements on pile load tests are set out in the EC 7-1 Handbook [44], DIN EN 1536, DIN EN 12699 and DIN EN 14199. This section comprises a compilation of recommendations for executing static pile load tests.

(2) Concerning the directions of forces differentiation is made between:

- along the pile axis (axial), see (9.2), and
- lateral to the pile axis, see (9.3).

(3) Micropile load testing is dealt with separately in 9.4, even when static, axial loads only are involved.

(4) Further principles to be considered in connection with pile load tests are contained in Section 5.

9.2 Static Axial Pile Load Tests

9.2.1 Installation of test piles

(1) Test piles shall be installed at locations where the ground conditions are representative for the site and characteristic test results are anticipated. The test pile locations should be specified on the basis of ground investigations.

(1a) If a single pile load test is carried out, it must normally be located where the most adverse ground conditions are believed to occur. If this is not possible, an allowance must be made when deriving the characteristic resistance. This allowance shall be specified by the geotechnical expert.

(2) If load tests are carried out on two or more piles, the test locations must be representative of the site and one of the test piles must be located where the most adverse ground conditions are believed to occur.

(3) If the ground conditions on a site are highly variable, several pile load tests shall be carried out, allowing the capacity of the respective piles in the areas displaying different ground properties to be reliably delineated.

(4) Test piles should correspond as closely as possible to the working piles. The same pile type and installation method should be used as for the working piles.

(5) Test piles shall be constructed with the same, or similar, geometric dimensions as the working piles. If piles need to deviate from a 1 : 1 scale for technical reasons, it is permissible to reduce the diameter of bored piles. However, the ratio may not fall below 0,5 : 1. Only pile diameters $\geq 0{,}8$ m may be reduced and the reduced diameter of the test pile may not be $\leq 0{,}5$ m.

Recommendations on Piling – EA Pfähle.
Editors: German Geotechnical Society (*Deutsche Gesellschaft für Geotechnik e.V.*)

(6) If shaft resistance mobilisation needs to be reduced along the upper end of the pile shaft during the test (complete eradication is not technically possible), the following methods can be used:

- wrapping of the reinforcement cage with slip films or applying a slip coating, e.g. on prefabricated piles;
- core-drilling over the finished pile with an oversize casing and filling the annulus with a bentonite slurry, for example on prefabricated piles or small- to medium-diameter bored piles;
- ground loosening bores;
- sleeves or sacrificial linings in combination with lubricants, for example bentonite slurry or;
- double linings.

Note: For bored piles constructed with casings it should be noted that sacrificial linings always lead to a reduction of the pile cross-section and that sealing of the annulus formed during the extraction of the casing against the rising concrete can be problematic.

9.2.2 Test planning

9.2.2.1 General notes

(1) Under compressive loads, pile load testing provides a resistance-settlement curve for the entire pile resistance R and the load-dependent displacements, and a resistance-heave curve for tensile loads, together with time-settlement curves or time-heave curves. Suitably instrumented, the components pile base resistance R_b and pile shaft resistance R_s can be identified and their respective dependence on the displacements can be determined. Figure 9.1 shows an example of a pile's resistance-settlement curve.

(2) Additional piles can be constructed for pile load tests or working piles can be used. In both cases, pile testing should be performed in a timely manner to allow any modifications to the foundation design resulting from the tests to be implemented.

(3) Where working piles are employed for pile load testing, the influence of any preloading applied during the test shall be taken into consideration for the assessment of the pile capacity and displacement behaviour within the structure.

(4) If pile load tests are planned to identify pile base and pile shaft resistances using compressive and tensile load tests, these test shall be carried out on different piles.

Note: If the direction of loading is reversed, the maximum skin friction values can decrease heavily, see 5.9 and 9.2.2.5.

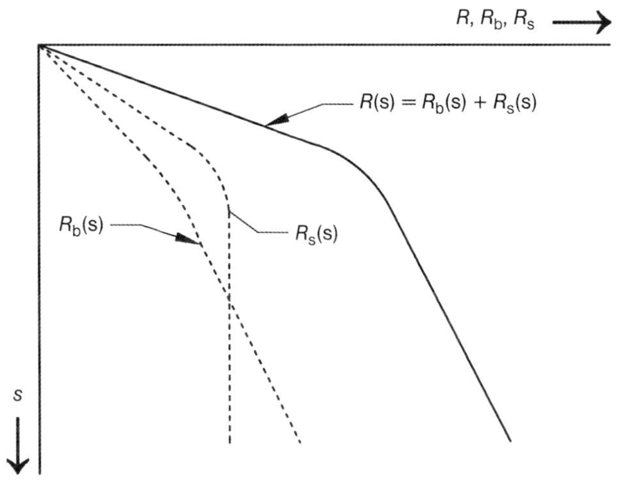

Figure 9.1 Pile resistance-settlement curve (RSC)

(5) Tensile load tests performed to investigate piles normally under compressive loads are permissible (e.g. for micropiles, Section 9.4.2.1), because this method yields conservative skin friction values. Displacements under tensile loads are greater than under compressive loads.

(6) If both axial and lateral pile load tests are performed, axial load testing should be carried out first. If the sequence is reversed the values and the development of skin friction can be influenced, together with respective displacements for their mobilisation.

(7) Pile load tests on displacement piles may not be performed immediately following installation. In non-cohesive soils, the load should not be applied less than three days after installation, and not less than three weeks in cohesive soils.

Note: By monitoring the pore water pressure it is possible to identify the time at which the excess pressure caused by the installation has dissipated.

(8) Load testing on cast-in-place concrete piles may only begin when the concrete has reached the strength necessary to accept the test load. Compressive strength testing can be carried out on additional test cubes for evidence that the pile concrete has developed the required strength.

9.2.2.2 Number of test piles

(1) At least one test pile should be planned for each pile type and each geotechnical uniform ground situation. The number of test piles shall be specified in consultation with the geotechnical expert.

Note: According to the EC 7-1 Handbook [44] a larger number of load tests lead to lower correlation factors.

9.2.2.3 Test load

(1) The test loads P_p shall normally selected sufficiently that the ultimate limit state (GEO-2 analysis) can be achieved, thus fulfilling one of the following criteria:

a) Limit settlement criterion:

$$R_{c;m} = R(s_g) \text{ or } R_{t;m} = R(s_{sg,t}) \quad (9.1)$$

where

$s_g = 0{,}1 \cdot D$, with D to be taken from Figure 9.2.

b) Creep behaviour criterion:

$$R_{c;m} = R(k_s) \text{ or } R_{t;m} = R(k_s). \quad (9.2)$$

Whereby k_s as shown in Figure 9.3, represents an individually specified creep ratio, used to define the time-dependent displacement of the pile head under constant pile load.

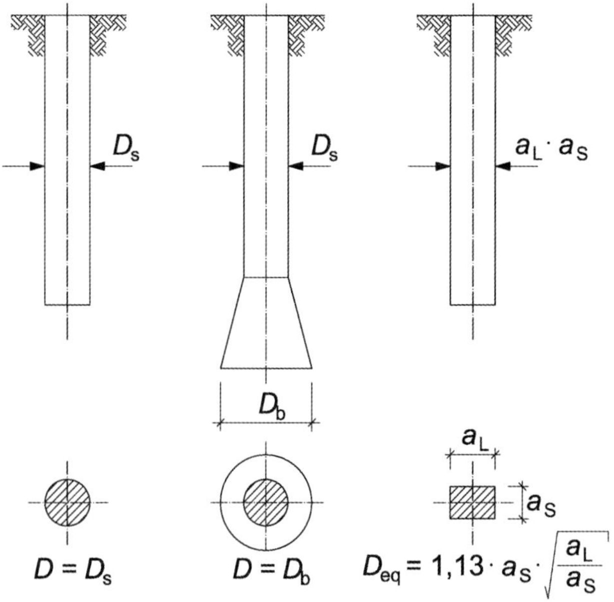

Figure 9.2 Definition of the pile diameter D governing the limit settlement s_g

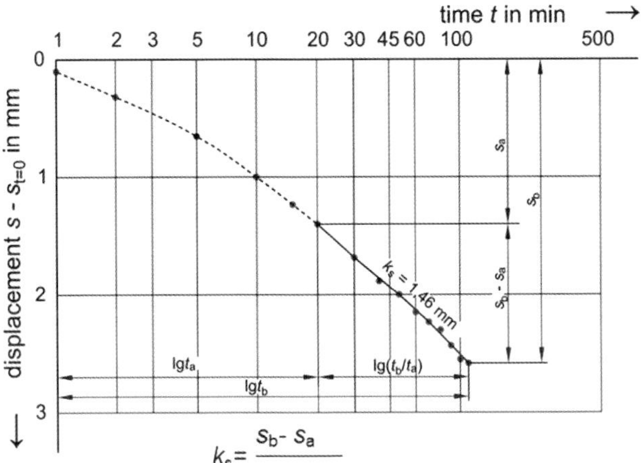

$$k_s = \frac{s_b - s_a}{\lg (t_b/t_a)}$$

Figure 9.3 Definition of the creep ratio k_s

(2) When estimating the test load with the aid of empirical data after 5.4, it should be taken into consideration that even if the 50% quantile values are adopted, a higher resistance than indicated by the empirical data is anticipated for half of all pile load tests. Selecting a suitably large test load is therefore recommended.

Note: The creep ratio k_s used to determine the pile resistances in the ultimate limit state normally has a magnitude of $k_s \approx 2$ mm, but shall be specified in agreement with the geotechnical expert.

(3) If the load test is only performed to verify the adopted characteristic pile resistance, the EC 7-1 Handbook [44] stipulates the following, minimum test loads:

for compression: $P_p = F_{c,d} \cdot \gamma_t \cdot \xi_1$ (9.3a)

for tension: $P_p = F_{t,d} \cdot \gamma_{s,t} \cdot \xi_1$ (9.3b)

where:

$F_{c,d}$ design value of the compressive pile load;

$F_{t,d}$ design value of the tensile pile load;

γ_t partial safety factor for the total resistance of a pile after Annex A3.2;

$\gamma_{s,t}$ partial safety factor for the tension pile resistance after Annex A3.2;

ξ_1 correlation factor after Annex A4.1 for pile resistances measured in static load tests, relative to the mean.

(4) The EC 7-1 Handbook [44] also stipulates that a model factor η_M shall be considered for grouted tension pile systems:

$$P_p = F_{t,d} \cdot \gamma_{s,t} \cdot \xi_1 \cdot \eta_M \tag{9.4}$$

(5) The test pile shall be designed for the chosen test load.

9.2.2.4 Principles for the instrumentation

(1) Test pile instrumentation must be compatible with the test objectives and the required quality of results, and be specified by, or in agreement with, the geotechnical expert. Three requirement levels are differentiated:

a) Basic requirements
Monitoring of the pile resistance: the displacement of the pile head, the applied load and the time are measured. It is not necessary to separately monitor the pile base capacity and the pile skin friction.

b) Enhanced requirements
Monitoring of the pile base resistance and the pile shaft resistance: the displacement of the pile head, the applied load, the pile base capacity q_b and the time are monitored. The pile base resistance R_b is calculated from q_b. The pile shaft resistance R_s is derived from the difference between the total pile resistance and the pile base resistance.

Special instrumentation is required at the pile base.

c) High requirements
Monitoring of the pile base resistance and the distribution of skin friction over the length of the pile: the displacement of the pile head, the applied load, the pile strain in various sections along the pile length, the pile base capacity q_b and the time shall be monitored.

Special instrumentation is required at the pile base and along the pile shaft.

(2) Enhanced or high requirements apply, for example:

- if the skin friction can fall to a lower residual value after exceeding a peak value;
- in ground with complex stratification;
- if high negative skin friction is anticipated;
- if the test pile heads are considerably higher for testing than those of the working piles;
- if special demands are placed on limiting deformations;
- if measures for reducing skin friction are implemented, see 9.2.1(6).

9.2.2.5 Special load situations

(1) If considerable pulsating or alternating loads or creep are possible in the pile serviceability limit state, these load situations should be simulated as far as

possible in the load test, also see 5.9 and Section 13. In such cases, the requirements on the testing process shall be specified by the geotechnical expert or the geotechnical designer. Notes on procedures and examples are given in [42], [53], [118] and [135].

(2) It is not necessary to simultaneously simulate loading parallel and square to the pile axis.

9.2.3 Loading systems

9.2.3.1 Introduction

(1) The load shall be introduced centrally and axially by means of hydraulic jack(s). Anchored beams, load frames or kentledge (dead loads) serve as reaction systems for compressive load tests, see Figure 9.4.

(2) Anchor bars embedded in the test pile are used for tensile load tests. The tensile load is applied by means of hollow piston jacks acting against a reaction beam.

(3) Optionally, the test pile itself, or parts thereof, can be utilised as the reaction system for hydraulic jacks concreted into the pile base or the pile shaft, see 9.2.3.4.

9.2.3.2 Reaction systems

(1) The elements of an anchored reaction system are the steel girder or load frame, anchoring elements and connections. The elements of a non-anchored reaction system are load distribution beams, support structures and surcharge masses (kentledge).

(2) The steel components shall be designed to DIN 18800 to remain in the elastic state ("elastic-elastic method"). The partial safety factor for actions may be adopted at 1.1. The reaction system elements shall be designed compliant with the EC 7-1 Handbook [44] for the transient design situation BS-T partial factors (actions and resistances).

(3) The distances of the kentledge (dead load) supports or anchoring elements for the reaction system (tension pile or grouted anchor) from the test pile shall be as shown in Figure 9.4. For bored piles and displacement piles the clear distance of the test pile to a kentledge support (Figure 9.4a) or to vertical anchor piles (Figure 9.4b) should be at least $a = 2{,}5 \cdot D$ or 2,5 m, whichever the larger, whereby the governing pile diameter D is as shown in Figure 9.4. The minimum distances shall be increased as necessary for load tests on raking piles, for special load situations or for more stringent test requirements.

(4) Smaller distances between the reaction beam supports and the test pile may be adopted for tensile loads tests, if structural measures are taken or the

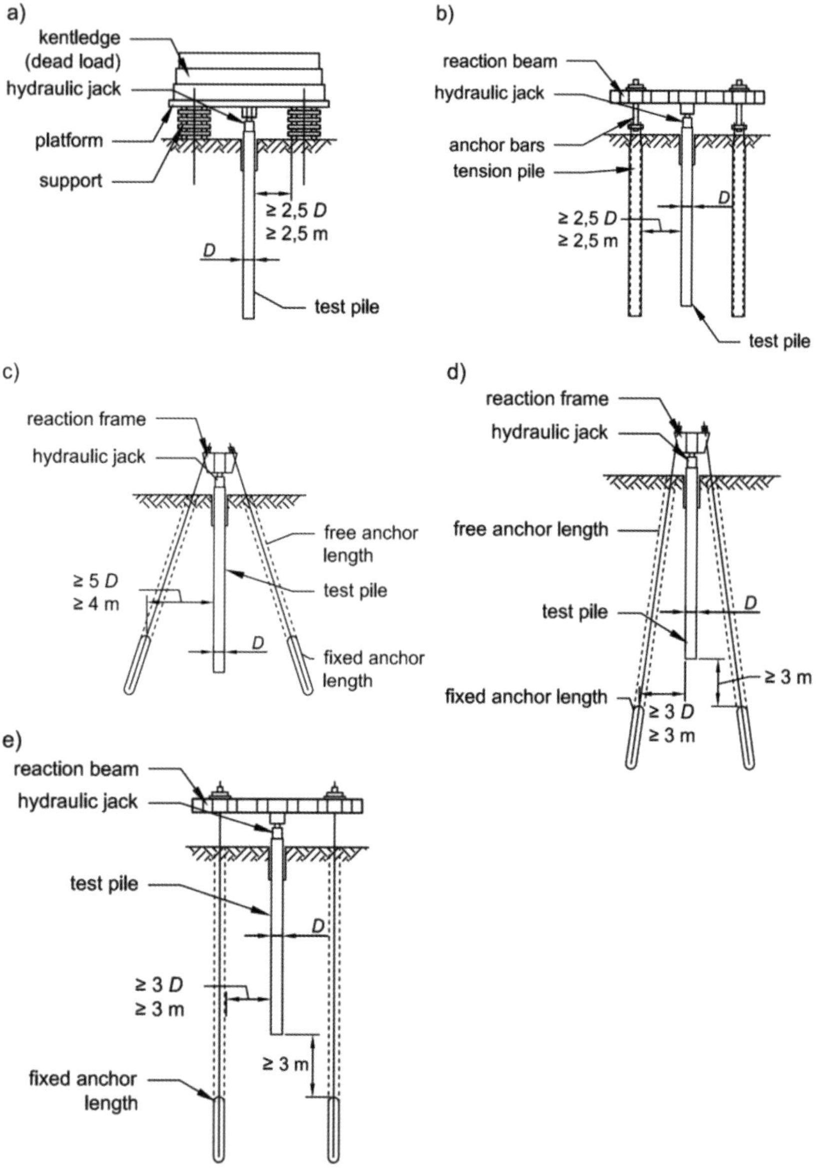

Figure 9.4 Minimum dimensions between load apparatus and test pile for compressive load tests for a) supports for kentledge (dead loads), b) tension piles, c) radial, inclined grouted anchors with shallow fixed anchor lengths, d) radial, inclined ground anchors with deep fixed anchor lengths, e) ground anchors parallel to the test pile and with deep fixed anchor lengths

sequence of ground layers ensures that the support does not substantially influence the performance of the test pile. In particular, any bracing of the pile against parts of the reaction system shall be prevented.

(5) If ground anchors are used to hold a reaction frame or beam as shown in Figures 9.4c) to e), the fixed anchor lengths must be clearly defined. In the free anchor length any load transfer must be ruled out by the use of sleeves or by flushing.

(6) Before testing begins, ground anchors should be either prestressed by at least 1,1 times the maximum individual share of the test load, to compensate for unequal initial deformations and equalise deformation behaviour, or care must be taken to ensure that unequal lengths can be compensated for during testing.

(7) The elastic elongations of long reaction anchors can become so large that the cylinder stroke is no longer sufficient. In these cases use of considerably larger anchor steel cross-sections is recommended in order to stresses and, consequently, elongations of the anchors.

(8) Reaction piles and anchors should be arranged symmetrically around the test pile. If they are arranged asymmetrically, the resulting influences must be taken into consideration for the load application and compensated for if necessary.

9.2.3.3 Hydraulic jacks

(1) The test load should always be applied using hydraulic jacks, even where a dead load is available as reaction mass. The maximum possible force and the necessary piston stroke should be compatible with the scheduled test load, and the anticipated displacements of the pile and of the reaction system.

(2) The elastic deformations of the reaction system can get to considerable magnitude and should be assessed before testing begins. This is especially the case if grouted anchors with high-strength steel tension tendons are used. The initial deformations of the loading system must also be taken into consideration when specifying the necessary piston stroke.

(3) If several hydraulic jacks are used to apply the necessary test load, they should be the same make and model and be supplied by a common supply from one hydraulic unit. Each hydraulic jack should be provided with a shut-off valve.

(4) Spherical bearings on the jack pistons have proven useful in compensating for tilting.

(5) If the piston stroke is insufficient, the pistons must be shimmed using distance pieces. It is important that the hydraulic jacks act in symmetry and guarantee axially symmetrical load transmission during all test phases.

Note: It has proven useful to work with at least three or four hydraulic jacks in linear or cruciform arrangement, respectively.

9.2.3.4 Embedded hydraulic jacks

(1) In bored piles the pile load can also be applied using hydraulic jacks or flat jacks, which are embedded in the concrete at the pile base or in the pile shaft. In this case, the pile itself, or parts thereof, serve as the reaction system and the load is applied from below or in two opposite directions, see Figure 9.5.

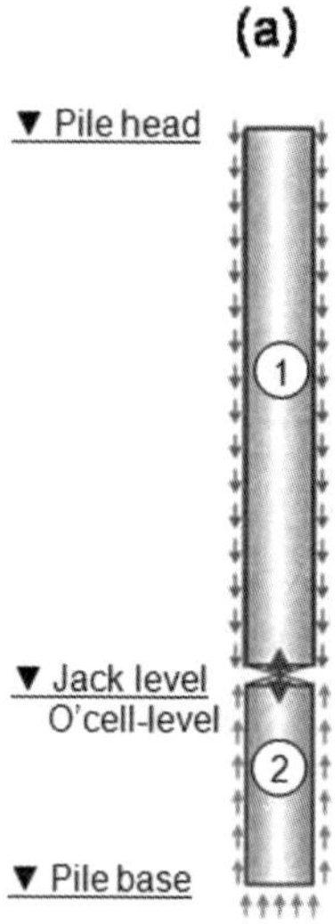

Figure 9.5 Schematic of pile division into segments when executing pile load tests using the Osterberg method, for single-level tests (a) and multi-level tests (b), after [101]

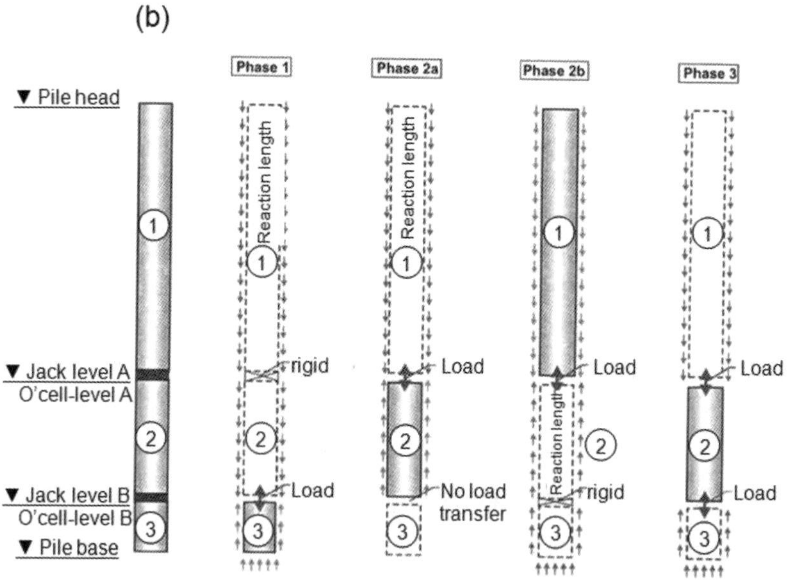

(2) This type of test requires complex pile instrumentation for force and displacement measurement, and is suitable for high test requirements or confined site conditions, where other reaction systems cannot be implemented.

(3) When executing load tests using this method, division of the pile shaft into two segments with an intermediate load cell is possible for a single-level testing as shown in Figure 9.5a, or in three or four segments with two or more intermediate embedded jacks for multi-level testing as shown in Figure 9.5b.

(4) The maximum possible load in any test phase is limited by the pile segment with the least resistance. Piles with base resistance greater than skin friction are therefore unsuitable for this test method.

(5) Figure 9.5a shows pile loading in two opposing directions (bidirectional). Tensile resistance is activated in the upper segment, compressive resistance in the lower segment.

(6) In the multi-level test (Figure 9.5b), different load conditions can be generated within the pile by activating or deactivating individual embedded jacks. For example, this primarily allows the base capacity (Phase 1) or the skin friction in segment 1 or 2 (Phase 2a and 2b) to be examined.

(7) For the verification of the stiffness of the reinforced concrete unconfined compression tests can be performed by simultaneously loading a pile section between two embedded jacks, for example for segment 2 in phase 3, as shown in Figure 9.5b.

(8) If the embedded jack is placed in the pile shaft, measures must be taken to ensure that the concreting pipe can pass through the jack and the concrete can flow freely. Sufficiently large cut-outs with thread-in aids for the concreting pipe should be left in the load distribution slabs. The lower face of the jack must be formed such that any contaminated concrete at the head of the rising concrete column cannot be trapped below the embedded jack.

(9) The test does not directly yield the resistance-settlement curve of a comparable pile loaded at its head. It must be derived from the individual segment test results and the individual phases, where applicable. The forces measured for equal segment displacements shall be added for this purpose.

(10) Guidance and reports can be found in [24], [101], [107], [108], [109] and [147].

9.2.3.5 Pile head

(1) The pile head must be carefully prepared to allow the load and displacement monitoring instruments to be placed centrally, flush and perpendicular to the pile head. If the pile's cross-sectional area is insufficient to accept the hydraulic jack(s) and the instrumentation, the pile head shall be enlarged.

(2) Directly loaded pile heads and pile head enlargements must be designed to for the tensile splitting forces developing during loading. Use of a tubular steel sleeve has proven useful.

(3) Care must be taken that no forces are transferred to the ground by a pile head enlargement.

9.2.4 Instrumentation and monitoring

9.2.4.1 Displacement measurements

(1) The vertical displacement of the pile head must be measured using redundant systems: mechanically using precision gauges or electrical displacement transducers, and in addition optically, e.g. by means of precise levelling.

(2) A reference system is used for mechanical measurement. It consists of reference beams, arranged on either one or both sides of the test pile and supported under statically defined conditions approximately 2,5 · D, but at least 2,5 m, from the pile, see Figure 9.6. During the test, the reference system shall be protected from actions resulting from the test load, vibrations and temperature differentials.

Note: Timber reference beams are less sensitive to temperature changes than steel beams.

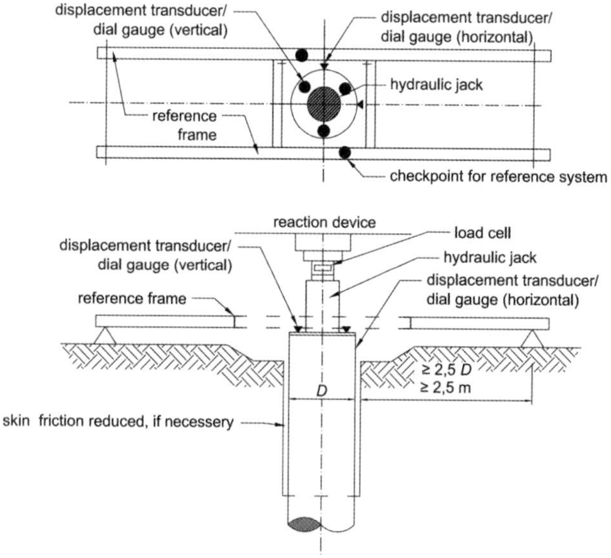

Figure 9.6 Reference system for displacement measurements at the pile head (example)

(3) Precise levelling reference points must be established outside the zone of influence of the entire test apparatus.

(4) The range of the monitoring instruments must be such that they do not need to be moved during the test.

(5) The measurement uncertainties (deviation of the displayed from the true values) for the pile head displacements may not exceed ±0,2 mm. Transducers with a 0,01 mm scale division are required for this.

(6) At least two, preferably three, transducers arranged symmetrically around the pile shall be used to measure axial displacements. Glass plates should be fixed beneath the sensors on the pile head to reduce friction.

(7) Two sensors arranged perpendicular to each other shall be installed to check pile head movements square to the pile axis. These measurements should always be made during load tests on raking or slender (flexible) piles.

Note: The measurements primarily serve to confirm that the test load is transmitted centrally and axially into the pile, and thus allow a timely reaction in case of disorders or eccentric loading.

(8) In displacement-controlled load tests (constant rate penetration tests) or for cyclic loads, electrical, continuously monitoring instruments shall be used for displacement measurement.

9.2.4.2 Load measurement at the pile head

(1) Load cells must always be used, the hydraulic pressure shall be recorded for control purpose only.

(2) Load cells shall correspond at least to precision class 1.

(3) All load cells must be calibrated, whereby the calibration may not be older than twelve months.

9.2.4.3 Pile base resistance

(1) The pile base resistance mobilised as a result of settlement is recorded by means of hydraulic or electric load cells installed in the pile base, see Figure 9.7.

(2) Care must be taken to ensure that the entire load is transmitted through the monitoring instrument. Where possible, the load cell should cover the entire area of the pile base. Undesirable load transfer around the cell can be prevented by using soft, foam rubber inserts, for example. The foam rubber inserts shall formed such that they are not completely compressed by the weight of the fresh concrete.

(3) Before installing the load cells the base of the excavation shall be cleared of loose or softened material and prepared with a mortar bed.

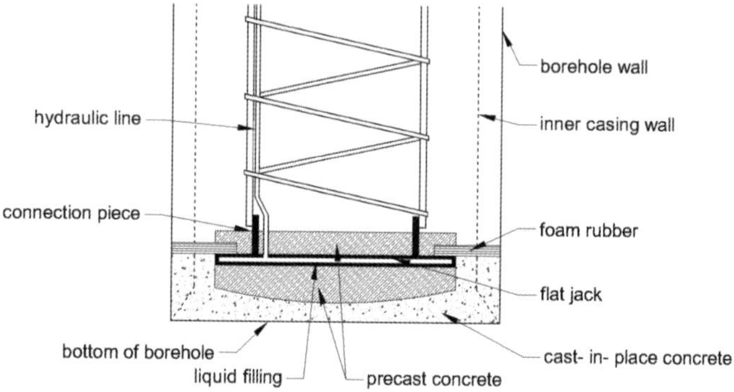

Figure 9.7 Schematic arrangement of a pile base load cell

(4) A conical or domed load cell underside has proven useful in providing a form-fitting support in the mortar bed.

9.2.4.3 Pile shaft resistance

(1) The pile shaft resistance can only be recorded indirectly. It can either be determined as a whole by subtracting the activated pile base resistance from the applied load (Figure 9.8a), or in sections via the longitudinal pile strain, adopting the Young's modulus and the cross-sectional area.

(2) The axial strains can be measured as follows:

a) Using extensometers as shown in Figure 9.8b): the strain of predefined pile sections between the pile head and the anchorage point of the respective extensometer is measured. By suitably staggering the measuring levels, the distribution of the changes in length are acquired by subtraction and, derived from this, the normal force distribution along the pile is calculated. Depending on the type, normally not more than 6 extensometer bars can be bundled in a multiple extensometer. The extensometer head should be placed in a recess in the pile head. Electrical transducers should be used. When carrying out the test and evaluating the results, it should be noted that the change-in-length distributions are acquired by subtraction and from relatively long measurement sections. When arranging the transducers and the reference system it should be noted that eccentricities and unintentional bending of the pile head could falsify the results. The extensometers should therefore be installed as close as possible in the pile centre.

b) Using electrical concrete strain sensors (strain gauges or vibrating wire gauges) as shown in Figure 9.8c: these can measure locally at predefined pile depths over lengths of a few centimetres or integrated over longer pile sections. The transducers are fixed to the reinforcement cage such that they cannot change orientation during installation and concreting, and the strain

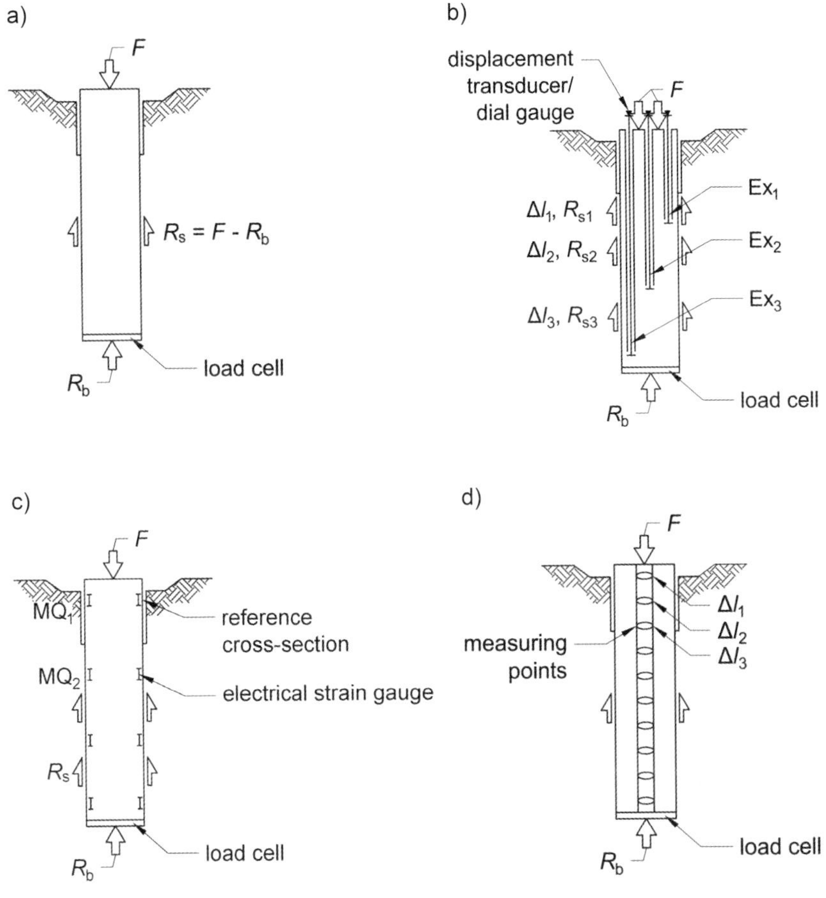

Figure 9.8 Monitoring of pile shaft resistance a) measuring the base resistance by a load cell, b) measuring the base resistance by a load cell and longitudinal deformations using extensometers, c) measuring the base resistance by a load cell and longitudinal deformations using local or integrated strain transducers, d) measuring the base resistance by a load cell and continuous pile deformation measurement by means of a sliding micrometer

in the concrete is reliably transmitted, i.e. without slip. A configuration including two or three symmetrically arranged strain transducers per level and a number of measurement levels normally allows the distribution of longitudinal strains to be determined with sufficient precision. A prerequisite for reliably monitoring length changes using strain transducers is their complete embedment in the pile material and their placement at locations representative for the development of strains of the pile.

c) Using sliding micrometers as shown in Figure 9.8d: the pile strain is continuously measured in subsequent sections by means of a mobile probe in a tube with predefined measuring marks. The pile head with the loading apparatus is equipped such that the measuring tube can be arranged as centrally as possible and remains accessible. Because monitoring requires manual work within the heavily loaded structure, special safety measures must be taken to protect personnel.

(3) If longitudinal strains are measured locally or in predefined pile sections, it is necessary to interpolate those for the intermediate regions. This requires "engineering judgement" of results and experience in the assessment and evaluation.

9.2.4.5 Special instrumentation for tests with embedded hydraulic jacks

(1) When performing pile load tests where the hydraulic jacks are embedded in the concrete at the pile base or within the pile shaft, the absolute and relative displacements of individual pile sections must be monitored using extensometer systems and electrical displacement transducers.

(2) In addition, measurements by means of concrete strain transducers can be taken at selected pile depths to determine the pile shaft resistances along individual pile sections.

9.2.4.6 Pile cross-sectional area and deformation properties

(1) Precise knowledge of the pile cross-section and of the true deformation properties of the pile material are required to determine the variation with depth of the axial pile forces and the effective skin friction, from the longitudinal pile strains. In practice and for cast-in-place concrete piles, these values cannot be recorded with sufficient accuracy.

(2) Instead, with the aid of a reference measurement level near the level of load introduction, where uniform stress distribution across the pile cross-section can be assumed and the skin friction resistance mobilised in the test is still negligible, the strain signal can be "calibrated" in terms of the acting load. Assuming the constant pile cross-section along this length, it is possible to directly correlate the measured strain signal to the effective longitudinal force, without directly determining the Young's modulus,.

9.2.4.7 Protection of monitoring instruments

(1) The monitoring instruments at the pile head must be protected against weather influences and other disturbances. Encasing the entire test apparatus in a housing is recommended for tests involving enhanced or high requirements. Shielding of monitoring instruments and of the reference system is sufficient for short-term tests or tests involving lesser requirements.

(2) Fittings, cables and the monitoring instruments themselves, both inside and outside of the pile, must be marked and protected from damage during all installation phases. In particular, this includes sufficient insulation of electrical transducers and their cables against water and fresh concrete ingress, as well as mechanical protection against damage during concreting, cut-off of piles and during the phase leading up to the commencement of the test.

9.2.5 Testing procedure

9.2.5.1 Load steps and loading rates

(1) The load should be applied in at least two cycles. The maximum load during the first cycle should correspond to the characteristic value of the effect on the pile. The maximum load during the final cycle corresponds to the test load.

(2) The test load should be applied in at least eight load steps. It has proven useful to begin loading with a small preload step to settle the loading apparatus and set up the displacement monitoring instruments.

Note 1: When the characteristic pile resistance is approached or for creep $k_s > 0,5$ mm, it can be necessary to reduce the load steps to allow more detailed study.

Note 2: See Figure 9.3 for definition of k_s.

(3) The load should be increased slowly and carefully, in particular where the load is predominantly transferred by skin friction or at the higher load steps. Jolts and vibrations shall be avoided.

(4) The duration for increasing the load from one step to the next should be at least 3 minutes, see Figure 9.9.

Note: If the load is increased too quickly, long-term creep deformations in the pile-soil system can result. The loading rate should therefore be kept deliberately low at the higher load steps.

(5) For basic test requirements, the load shall be kept constant during the individual load steps for at least the observation periods given in Figure 9.9.

(6) For enhanced and high test requirements and until the characteristic pile effect $F_{c,k}$ or $F_{t,k}$ is reached the load shall be maintained during the individual load steps at least until the increase of displacements has fallen to 0,1 mm every 20 minutes.

In the subsequent load steps the load may be increased once the displacement rate has decreased to below 0.1 mm every 5 minutes. If possible, readings should be taken at equal intervals to allow the creep behaviour to be correctly identified.

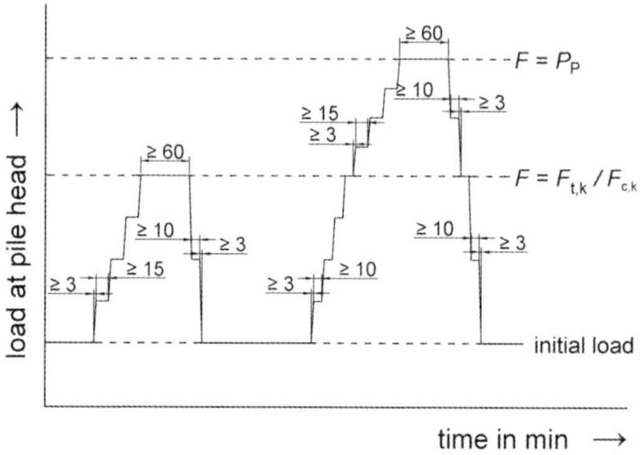

Figure 9.9 Recommended load steps

Note: In terms of the characteristic pile effect $F_{c,k}$ or $F_{t,k}$ and the test load P_p, it has proven useful to maintain the load for a longer period than until the dissipation of movements defined above, in order to record any creep redistributions in the ground and assess long-term behaviour.

(7) An intermediate unload step shall be applied once the characteristic pile effect $F_{c,k}$ or $F_{t,k}$ has been reached. The readings should be made at the same intervals as for the load steps; however, it is only necessary to wait for the movements to cease at complete unloading or unloading to the initial load step. The same criteria apply for reloading and further loading as for unloading or initial loading, see Figure 9.9.

(8) Care must be taken to ensure that the load remains constant for the duration of a load step. For basic test requirements this is achieved by hand pumping or manual control of the hydraulic pump, while for enhanced and high requirements the load shall be maintained by automatic electrical or hydraulic means.

(9) If sufficient time is not available to allow movements to dissipate, for loads exceeding the characteristic pile effect a displacement-controlled test with constant rate of displacements can be used to determine the ultimate load (constant rate of penetration test, CRP test). A CRP test can also be necessary in cohesive soils with high plasticity for example.

(10) The penetration rate used for the CRP test shall be selected to between 0,25 mm/min and 1,25 mm/min in cohesive soils and to between 0,75 mm/min and 2,5 mm/min in non-cohesive soils, or be adapted to the capacity of the hydraulic unit.

(11) Where possible, the CRP test should be automatically controlled, and the data be automatically recorded.

(12) If the CRP test is performed using a number of different penetration rates ("jump test"), a soil viscosity index can be determined. Using the viscosity index, in turn, the load-displacement curve can be adapted to a standardised penetration rate [72].

9.2.5.2 Monitoring intervals

(1) During each load step the axial displacements at the pile head should be recorded at least after the following observation periods: 0, 2, 5, 10, 20, 40, 60, 80, 100 min., etc., after reaching the respective load.

(2) The intervals take account of the plotting of the recorded displacements on semi-logarithmic scale and also allow the determination of creep k_s.

Note: If long observation periods are envisaged once displacement have ceased (e.g. overnight), readings and load corrections in intervals of several hours are sufficient.

(3) Larger intervals are sufficient to record the horizontal movements of the pile head, e.g. at the beginning and end of a load step.

(4) Automatic data acquisition should be used for CRP tests. The data should be recorded at least every 60 seconds, depending on the penetration rate.

(5) Sliding micrometer measurements are carried out at the end of each load step.

(6) Before the beginning, and at the end of the loading test, the elevation of the pile head and of the reference system shall be checked by a datum measurement and a final measurement. These controls are carried out e.g. by precision levelling, with reference to a reference point. For enhanced and high test requirements the control measurements should additionally be performed after movements have ceased following every load step.

(7) If large ground settlement or heave can occur during the test, e.g. when using dead loads as reaction systems, the displacements of the pile and of the reference system must be checked once every load step.

(8) The actual pile load is to be recorded, and adapted if necessary, in the same intervals as the settlement or heave of the pile head.

(9) Displacement of the pile head should have ceased before control measurements are taken of the pile and of the reference system, and for non-automatic recording of the base capacity and changes of lengths along the pile axis.

(10) Fixed intervals should be adhered to for automatic recording and storage of load, deflection, stress and strain data.

9.2.5.3 Records

(1) Field reports in tabled format shall be compiled to contain all measurement results obtained at the monitoring intervals and all observations made during the test. Among others, they shall contain:

- hydraulic pressures and load cell readings;
- displacement measurements at the pile head;
- electrical stress and strain readings, if not recorded electronically;
- control measurements at the pile head and of the reference system;
- weather conditions (temperature, wind, cloud cover);
- people present during the test;
- special test conditions, e.g. readjustments;
- reasons for deviations from planned test sequence.

(2) The records shall be signed and attached to the pile test report, see 9.2.7.2.

(3) If automatic data recording and storage is used, the data should be visualised during the test.

(4) The load and displacement data should be plotted as a resistance-displacement curve already during the load testing process. For higher load steps, and in particular for load tests on working piles, the development of pile displacements and the associated creep k_s with time should also be plotted to allow the testing procedure to be adapted where necessary.

(5) The semi-logarithmic graph of the displacement with time as shown in Figure 9.11 allows conclusions about creep behaviour and the anticipated characteristic pile capacity. Where tests are carried out on working piles this allows potential overloading beyond the characteristic capacity to be recognised at an early stage.

(6) To end tests under enhanced and high requirements, the pile shall be completely unloaded after displacements have ceased at the initial load step and the measurement shall continue until the displacements have completely ceased.

9.2.6 Evaluation

(1) In tests with basic requirements it is sufficient to plot the resistance-settlement or heave curves as shown in Figure 9.10.

(2) For enhanced and high test requirements the time-displacement curves shall also be drawn on a semi-logarithmic scale, the creep be determined and then be presented with the corresponding loads as shown in Figure 9.11.

(3) Where the pile base and pile shaft resistances are monitored separately, the corresponding components $R_b(s)$ and $R_s(s)$ shall be presented in a single diagram, see Figure 9.12.

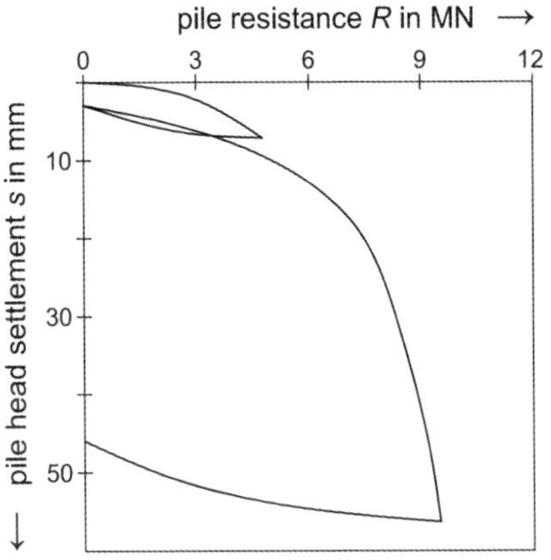

Figure 9.10 Plot of resistance-settlement curve

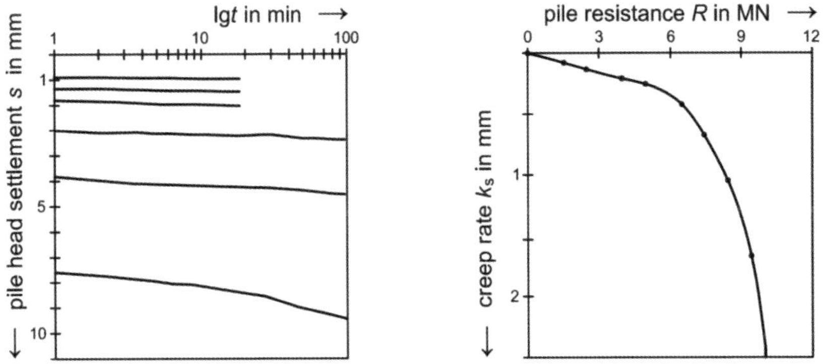

$$k_s = \frac{s_b - s_a}{\lg(t_b/t_a)} = \frac{\Delta s}{\Delta \lg t}$$

Figure 9.11 Plot of time-displacement and creep

(4) For high test requirements in addition to the resistance-settlement or resistance-heave curves, the measured strains along the pile axis and the derived longitudinal force and skin friction shall be plotted against the depth, see Figure 9.13. The cross-section used for evaluation and the Young's modulus of the pile material shall also be given.

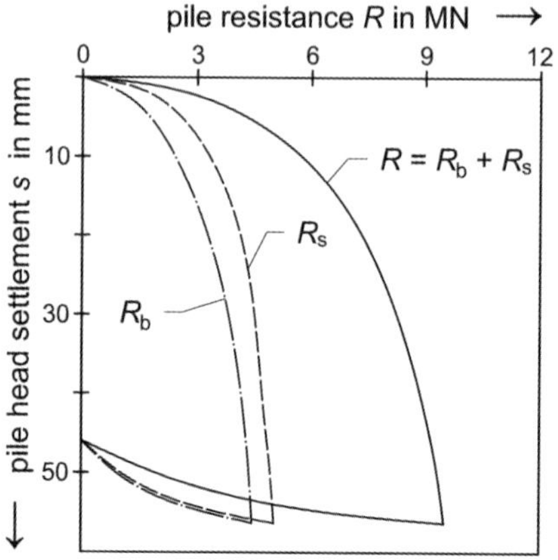

Figure 9.12 Separate plot of base and shaft resistance

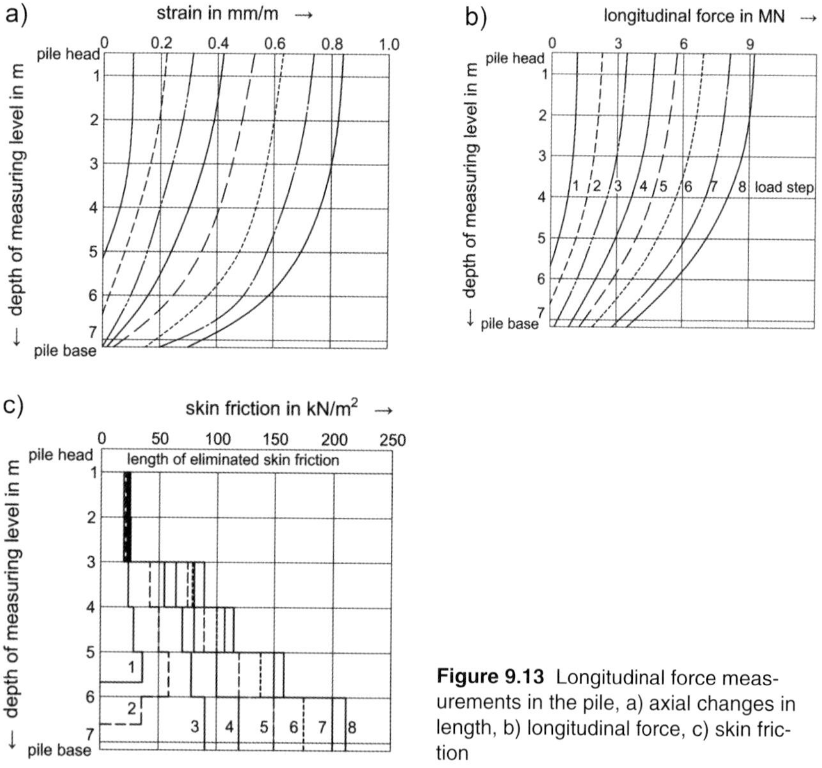

Figure 9.13 Longitudinal force measurements in the pile, a) axial changes in length, b) longitudinal force, c) skin friction

9.2.7 Documentation and reports

9.2.7.1 Introduction

(1) The complete pile test report shall document the load test in two separate parts.

a) Part 1: Test report including all details of pile installation, the test, test results, presentation of test results and their evaluations.
b) Part 2: Interpretative report including the test results and their interpretation and resulting recommendations for the design and the execution.

(2) The test report (part 1) is normally compiled by the company installing the piles and/or executing the load test, the interpretative report (part 2) by the geotechnical designer or the geotechnical expert.

Note: For basic test requirements the interpretative report can also be compiled by the pile manufacturer and/or the company carrying out the test.

9.2.7.2 Test report

(1) The report should contain the following information:

1. General information
   - time and place of the test, site maps;
   - special local conditions;
   - involved personnel;
   - description of the ground and the groundwater conditions; as appropriate, abstracts from the ground investigation report (geotechnical investigation report) shall be included.
2. Information on the pile system
   - pile designation;
   - pile type;
   - dimensions, section, material quality, reinforcement;
   - execution method, execution records;
   - pile material strengths at the time of load testing.
3. Test installations
   - configuration of testing installations ;
   - reaction system (drawings, details, information on anchors or anchor piles);
   - loading system (hydraulic jacks with configuration and controls, maintenance of constant load);
   - monitoring instruments at the pile head (pile head preparation, load cells, reference system, type and configuration of displacement transducers, type of control measurements, reference points, calibration certificates);
   - monitoring instruments at the pile base (system, installation details, dimensions, calibration certificates);

   - monitoring instruments in the pile shaft (type, configuration of measurement levels, calibration certificates);
   - method of data acquisition.
4. Test execution
   - planned and actual loading regime;
   - reason for deviations from the planned test sequence;
   - testing records (copies of original records);
   - saved data storage media in cases of electronic data acquisition;
   - observations during the test execution (pile, reaction system, reference system, loading system, environment, weather conditions);
   - photographic documentation.
5. Test evaluation
   - resistance-displacement curves for $R(s)_{m,i}$, where appropriate including $R_b(s)_{m,i}$ and $R_s(s)_{m,i}$;
   - pile head displacements as a function of load duration;
   - changes of lengths along the pile axis;
   - information on the assumed or true cross-sectional dimensions of cast-in-place concrete piles as used for the evaluation;
   - determination of the effective Young's modulus; if expedient, reports of laboratory investigations of pile sections;
   - longitudinal forces along the pile;
   - tested or measured values of skin friction with depth and of the pile shaft resistance.
   - configuration of test installations;

9.2.7.3 Interpretative report

(1) The pile load test interpretative report shall be based on the geotechnical investigation report and the test report. It should include the following information:

- evaluation of the load test results;
- recommendations on pile type, depth and diameter;
- design specifications for the foundation to be executed;
- information on possible group effects (where several piles are loaded);
- recommendations for the execution;
- recommendations for any necessary additional investigations or long-term monitoring.

9.3 Static Lateral Load Test

9.3.1 Introduction

(1) These Recommendations apply to piles designed for loads square to the pile axis, not to piles subjected to lateral pressures as a result of the movement of soft, cohesive soils, see 4.5.

(2) Static lateral load tests are normally carried out to determine the lateral resistance and deformations in the serviceability limit state. In long, slender piles the failure state cannot normally be achieved under loads lateral to the pile axis; this is only possible in short, rigid piles.

9.3.2 Installation of test piles

(1) Test piles shall be installed at locations where the ground conditions are representative for the site and characteristic test results are anticipated. The test pile locations should be specified on the basis of ground investigation results.

(2) If a single pile load test is carried out, it should be located where the most adverse ground conditions are believed to occur. If this is not possible, an allowance must be made when deriving the characteristic resistance. This allowance shall be specified by the geotechnical expert.

(3) If load tests are carried out on two or more piles, the test locations must be representative of the pile site and one of the test piles must be located where the most adverse ground conditions are believed to occur.

(3a) If the ground conditions on a site are highly variable, several pile load tests shall be carried out, allowing the capacity of the respective piles in the areas displaying different ground properties to be reliably delineated.

(4) Test piles should correspond as closely as possible to the structural piles. The same pile type, installation method and dimensions should be used as for the structural piles.

9.3.3 Test planning

9.3.3.1 General notes

(1) Resistance-deflection curves and time-deflection curves are determined from a pile load test for the pile resistance and the load-dependent displacement.

(2) Additional test piles can be constructed for pile load tests or working piles can be used. In both cases, pile testing should be performed in a timely manner to allow any modifications to the foundation design resulting from the tests to be implemented.

(3) Where working piles are employed the influence of preloading on the performance within the structure should be considered.

(4) In addition, it shall be considered that the flexural stiffness of such working piles – as far as reinforced concrete piles are concerned – can depend on the axial load, which is normally not yet present at the time of the test.

(5) In terms of crack formation, a steel stress f_{yk} = 200 MN/m² should not be exceeded when testing reinforced concrete working piles. As long as the bending tensile stress does not exceed:

$$f_{ctm} = 0,3 \cdot f_{ck}^{(2/3)} \tag{9.7}$$

where

f_{ctm} tensile strength of the concrete

empiricism indicates that the bending resistance of the whole concrete section is retained (no cracks of the concrete, Condition I) and the flexural stiffness therefore remains the same as in the later structure, where the transition to Condition II (cracked concrete in the tension zone) is generally prevented by an axial, compressive load., The influence of reinforcement on flexural stiffness should be examined and taken into consideration separately if necessary [130].

(6) If pile load tests are performed on the same pile both axially and lateral to the pile, the axial load test should be performed first. If the sequence is reversed the magnitude of the skin friction can be influenced.

(7) It is not necessary to simultaneously simulate loading parallel and lateral to the pile axis.

(8) Realistic simulation of the support conditions of the pile head in the superstructure can be realised only with great difficulty for lateral load testing. It is usual to test piles restrained in the ground and with free head.

(9) The concrete of concrete piles must have achieved at least 80% of its nominal strength at the time of the load test, so that Condition I (no cracks) is retained for as long as possible in the zone of the maximum bending moment,. For proof of the actual concrete strength compressive strength tests can be carried out on additional test cubes.

(10) Pile load tests on displacement piles may not be performed immediately following installation. In non-cohesive soils, the load should be applied not less than three days after installation, and not less than three weeks in cohesive soils.

Note: By monitoring pore water pressures it is possible to determine the time at which the excess pressure caused by the installation has dissipated.

9.3.3.2 Number of test piles

(1) At least one test pile should be planned for each pile type and each geotechnical uniform ground situation. The number of test piles shall be specified in consultation with the geotechnical expert.

9.3.3.3 Test load

(1) The load to determine the lateral resistance and deflections of long, flexible, slender piles shall be as follows:

$$P_p = F_{tr,d} \quad (9.8)$$

where:

$F_{tr,d}$ design value of the lateral load on a pile.

(2) The load to determine the characteristic lateral resistance of short, rigid piles shall be as follows:

$$P_p = F_{tr,d} \cdot \gamma_{R,e} \quad (9.9)$$

where:

$F_{tr,d}$ design value of the lateral load on a pile;
$\gamma_{R,e}$ partial factor for passive earth pressure.

9.3.3.4 Ground investigations

(1) The soil strata in the immediate vicinity of the test pile must be investigated and their elasticity moduli E_S be determined to enable evaluation of the test results and adoption of the subgrade reaction moduli from the test for the design of piles in comparable ground conditions.

9.3.3.5 Principles for the instrumentation

(1) Test pile instrumentation must be compatible with the test objectives and the required quality of results, and be specified by, or in agreement with, the geotechnical expert.

(2) For load tests lateral to the pile axis only one level of requirements is formulated, namely "monitoring of the pile resistance lateral to the pile axis and of the deflection curve". The deflection of the pile head, the applied load, the incremental change in the pile inclination and the time are to be measured for these purposes.

(3) The magnitude of the modulus of subgrade reaction along the length of the pile can be determined from the deflection curve.

9.3.3.6 Load situations

(1) Because of the non-linear behaviour of the pile-ground system, the serviceability limit state should be modelled as closely as possible in the test. If certain load changes in a structure are anticipated at defined intervals, e.g. repeated pulsating and alternating loads as a result of braking loads, or tempera-

ture changes of a bridge superstructure, the respective load sequences should be simulated in the test. The increase in the deflection s_h at the pile head and the change in the deflection curve as a result of creep deformations in cohesive soils under permanent loads, should also be determined in the test, see [38].

(2) The pile head normally receives a combined loading by a horizontal load and a bending moment. The moment can amplify the action of the horizontal load, e.g. a load acting above the pile head for an isolated pile or row of piles, or it can reduce it, e.g. by a frame effect in pile groups with rigid capping slab, see Figure 9.14 and [6].

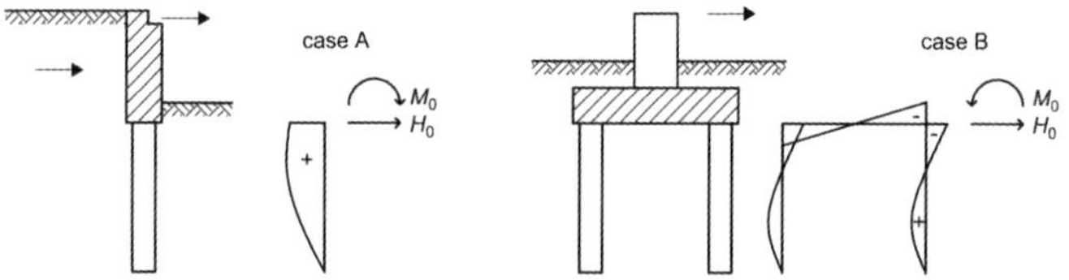

Figure 9.14 Types of horizontal loads on structural piles, after [6]

(3) In most cases, transferring a bending moment to a test pile makes test execution considerably more difficult and expensive. Instead, the test load lateral to the pile axis may be selected such that more or less the same ground reactions are mobilised as in the serviceability limit state. Where pile heads will later be restrained in the superstructure, but are load tested with a free head, the deflection curve prevalent in the restraint case must be taken into consideration to determine the test load [6].

(4) If the final load cannot be defined with certainty at the time of the load test, the test load should be large enough to allow realistic calculation of expected deformations during the future service conditions.

(5) For cyclic lateral loads, the deflection of the pile head increases with the number of load repetitions. In soil-pile systems in which no pore water pressure accumulations occur, these displacements approach a final value. To derive a relationship between the accumulated pile displacement and the number of load repetitions it is necessary to apply a sufficiently large number of loads to the test pile (see calculations and examples in Annex D).

9.3.4 Loading systems

(1) The load shall be applied centrally using hydraulic jacks. Where possible, two neighbouring piles should be loaded against each other, allowing special reaction systems to be dispensed with, see Figure 9.15 and [130].

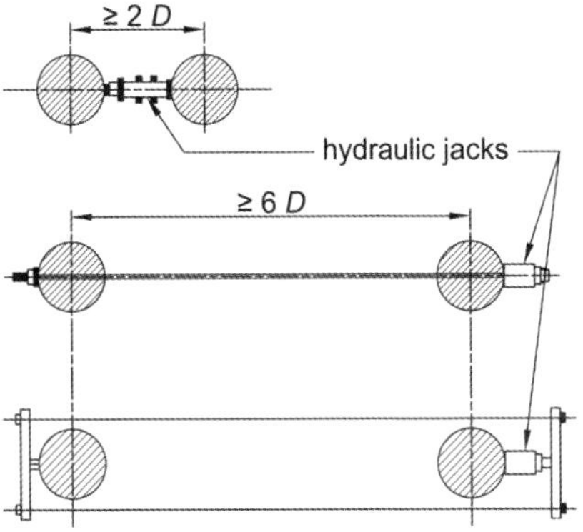

Figure 9.15 Various loading apparatus for pile load tests lateral to the pile axis, after [130]

(2) A device for automatically maintaining the load constant should be installed in the hydraulic system.

(3) If the piles should be pulled together by a tensile member their axial distance shall be at least 6 · D; if they are pushed apart it shall be at least 2 · D. In the latter case buckling of the compression member must be prevented.

(4) The steel components shall be designed to DIN EN 1993-1-3 to remain in the elastic state ("elastic-elastic method"). The partial factor for actions may be adopted at 1.1. The elements of the reaction system shall be designed compliant with the EC 7-1 Handbook [44] using the BS-T safety factors.

(5) The pile head must be carefully prepared to allow the load and deflection monitoring instruments to be placed centrally, flush and perpendicular to the pile head.

(6) Directly loaded pile heads must be designed to accept the forces imposed during loading. Use of a steel sleeve for protection of the pile head has proven useful.

9.3.5 Instrumentation and monitoring

9.3.5.1 Deflection measurement at the pile head

(1) The horizontal deflection of the pile head must be monitored using redundant systems: mechanically using precision gauges or electrical displacement transducers, and in addition optically, e.g. by means of electro-optical distance measurement.

(2) A reference system is used for mechanical measurements. It consists of measuring beams arranged on either one or both sides of the pile and supported

under statically defined conditions at 1,5 · D, perpendicular to the loading direction, see Figure 9.16a. During the test, the reference system shall be protected from actions resulting from the test load, vibrations and temperature differentials.

(3) Reference points shall be established outside the zone of influence of the entire testing system.

(4) The measuring range of the monitoring instruments must be such that they do not need to be moved during the test.

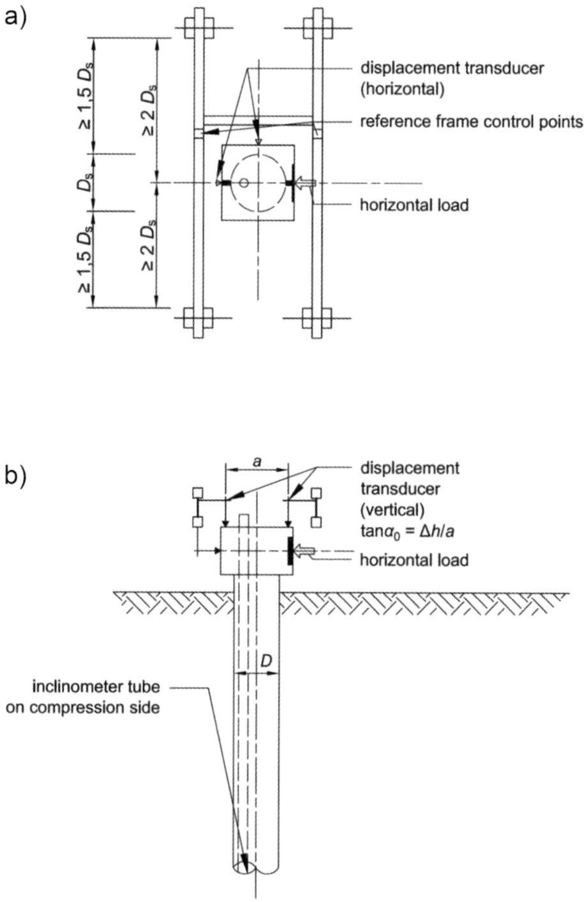

Figure 9.16 Example of logging instruments for pile load tests lateral to the pile axis

(5) The measurement uncertainties (deviations of the displayed values from the true values) for pile head displacement may not exceed ±0,2 mm. Displacement transducers with a 0,01 mm scale division are required for this.

(6) The transducers for measuring the horizontal displacement shall be arranged perpendicular to each other. Glass plates should be fixed beneath the sensors on the pile head to reduce friction, see Figure 9.16b.

(7) For cyclic load testing, continuously monitoring electrical instruments shall be used for deflection measurement.

9.3.5.2 Monitoring of the deflection curve

(1) The deflection curve is determined using inclinometer or deflectometer measurement along the entire length of the pile shaft.

(2) Monitoring can be by stationary installed measuring links or mobile using measurement probes.

(3) Where monitoring makes use of a mobile probe, the pipe shall be arranged such that it remains accessible. Because monitoring requires manual work within the heavily loaded structure, special safety measures must be implemented to protect personnel.

9.3.5.3 Load measurement at the pile head

(1) Load cells shall always be used; the hydraulic pressure shall be registered for control purposes only.

(2) Load cells must correspond at least to precision class 1.

(3) All load cells must be calibrated, whereby the calibration may not be older than twelve months.

9.3.5.4 Protection of monitoring instruments

(1) The monitoring instruments at the pile head must be protected against weather influences and other disturbances to prevent the results being invalidated.

(2) Embedded parts must be protected from damage during all installation phases.

9.3.6 Testing procedure

9.3.6.1 Load steps and loading rates

(1) The test load should be applied in six steps to determine the characteristic lateral resistance of long, flexible, slender piles. Unloading should be carried out in three steps to determine permanent displacement (see Figure 9.17).

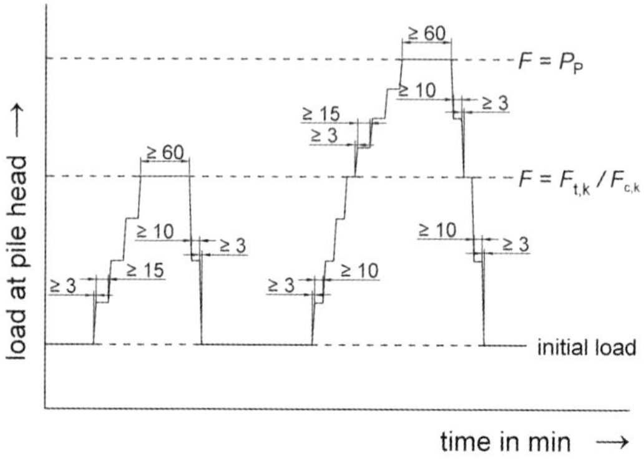

Figure 9.17 Schematic of the minimum requirements on the loading programme for a long, slender pile

(2) For the determination of the characteristic lateral resistance of short, rigid piles, the test load should, in analogy to axially loaded piles, be applied in two load cycles up to the proposed test load, see Figure 9.18.

(3) If transition to Condition II (cracked concrete in the tension zone) is anticipated, intermediate steps should be introduced in the respective range of the loading.

(4) During initial loading or unloading the load may only be increased or reduced if the change in the horizontal pile head displacement has fallen to below a value of 0,1 mm in 5 min. The minimum duration of a load step and the duration of the load increase can be taken from Figures 9.17 and 9.18.

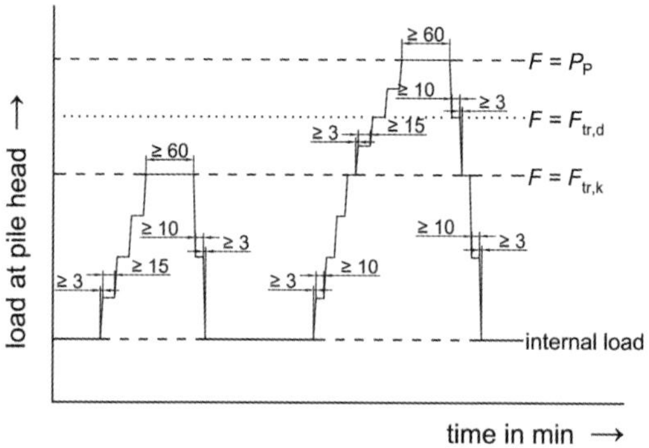

Figure 9.18 Schematic of the minimum requirements on the loading programme for a short, rigid pile

9.3.6.2 Monitoring intervals

(1) During each load step the horizontal displacements at the pile head should be recorded at least after the following observation periods: 0, 2, 5, 10, 20, 40, 60, 80, 100 min., etc., after reaching the respective load.

(2) The intervals take account of the plotting of the recorded displacements on semi-logarithmic scale and also allow the determination of creep k_s .

Note: If long observation periods are envisaged once displacements have ceased (e.g. overnight), readings and load corrections in intervals of several hours are sufficient.

(3) The deflection curve is measured at the end of each load step.

(4) Before the beginning, and at the end of the load test, the position of the pile head and of the reference system shall be checked by a datum measurement and a final measurement.

(5) The actual pile load shall be recorded, and adapted if necessary, in the same intervals as the displacement of the pile head.

(6) Displacements of the pile should have ceased before control measurements are taken of the pile head and of the reference system.

(7) Fixed intervals should be adhered to for automatic recording and storage of load, displacements and the deflection curve.

9.3.6.3 Records

(1) Field reports in tabled format shall be compiled to contain all measurement results obtained at the monitoring intervals and all observations made during the test. Among others, they shall contain:

- hydraulic pressures and load cell readings;
- displacement measurements at the pile head;
- control measurements at the pile head and of the reference system;
- weather conditions (temperature, wind, cloud cover);
- people present during the test;
- special test conditions, e.g. readjustments;
- reasons for deviations from planned test sequence.

(2) The records shall be signed and attached to the pile test report.

(3) If automatic data recording and storage is used, the data should be visualised during the test.

(4) Already during the time of testing, in addition to registering the measured values of displacements and loads, the load-displacement curve should be plotted.

9.3.7 Evaluation

(1) The measured pile head displacements are plotted as resistance-deflection curves and the deflection curves are presented for each load step, see Figure 9.19.

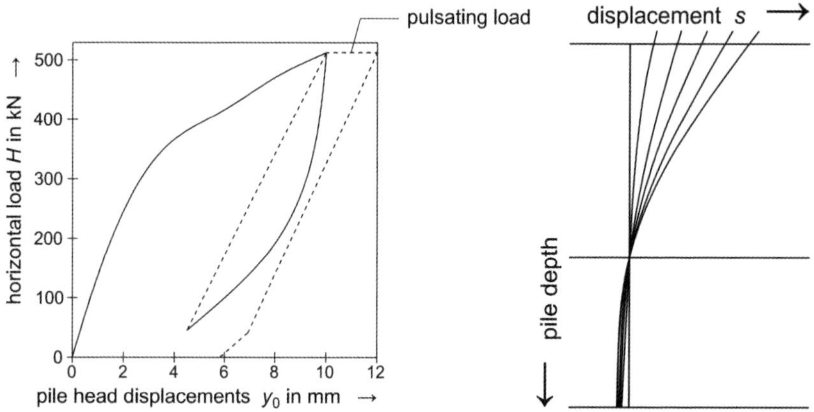

Figure 9.19 Pile head displacements and deflection curves for a test pile

(2) Figure 9.20 shows deflection curves for piles of varying stiffness.

9.3.8 Documentation and reports

9.3.8.1 Introduction

(1) The complete pile test report shall document the load test in two separate parts.

a) Part 1: Test report including all details of pile installation, the test, test results, presentation of test results and their evaluations.
b) Part 2: Interpretative report including the test results and their interpretation and resulting recommendations for the design and the execution.

(2) The test report (part 1) is normally compiled by the company installing the piles and/or executing the load test, the interpretative report (part 2) by the geotechnical designer or the geotechnical expert.

9.3.8.2 Test report

(1) The report should contain the following information:

1. General information
   - time and place of test, site maps;
   - special local conditions;

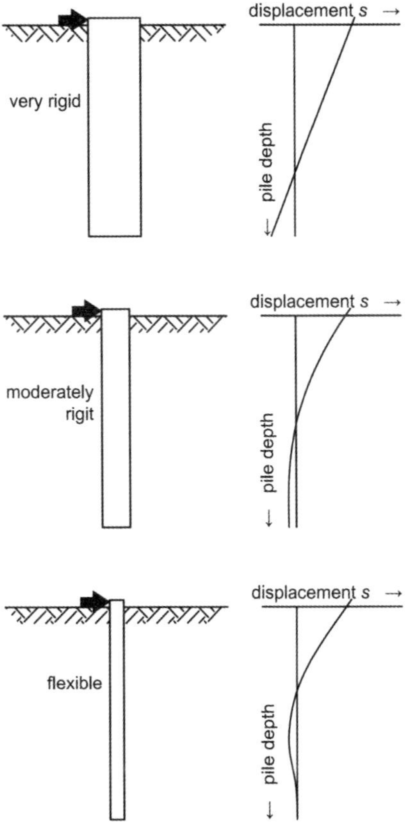

Figure 9.20 Deflection curves for piles of varying stiffness

   - involved personnel;
   - description of the ground and the groundwater conditions; as appropriate, abstracts from the ground investigation report (geotechnical investigation report) shall be included.
2. Information on the pile system
   - pile designation;
   - pile type;
   - dimensions, section, material quality, reinforcement;
   - execution method, execution records;
   - pile material strengths at the time of load testing.
3. Test installations
   - configuration of testing installations;
   - reaction system (drawings, details);

- loading system (hydraulic jacks with configuration and controls, maintenance of constant load);
- monitoring instruments at the pile head (pile head preparation, load cells, reference system, type and configuration of displacement transducers, type of control measurements, reference points, calibration certificates);
- monitoring instruments for measurement of the deflection curve (type, configuration and calibration certificates);
- method of data acquisition.

4. Test execution
   - planned and actual loading regime;
   - reason for deviations from the planned testing procedure;
   - testing records (copies of original records);
   - saved data storage media in cases of electronic data acquisition;
   - observations during the testing (pile, reaction system, reference system, loading system, environment, weather conditions);
   - photographic documentation.
5. Test results
   - resistance-displacement curves;
   - pile head displacements as a function of load duration;
   - pile deflection curve for each load step.

9.3.8.2 Interpretative report

(1) The interpretative report shall be based on the geotechnical investigation report and the test report. It should include the following information:

- evaluation of load test results with indication of subgrade reaction moduli;
- recommendations for pile execution;
- recommendations for any necessary additional investigations or long-term monitoring.

9.4 Static Axial Load Tests on Micropiles (Composite Piles)

9.4.1 Installation of test piles

(1) Test piles shall be installed applying the following principles:

a) Test piles shall be installed at locations where the most adverse ground conditions are anticipated. If the ground conditions on a site are highly variable, additional pile load tests shall be carried out, allowing reliable assessment of capacities in the areas of different ground conditions. The test pile locations should be specified on the basis of the ground investigation results.
b) Test piles should correspond as closely as possible to the working piles. The same pile type, execution method and dimensions should be used as for the working piles.

(2) Where micropiles are used for anchoring, the test piles shall be constructed with a confined load transfer zone, in accordance with the design boundary conditions. Load transfer above the specified transfer zone shall be eliminated by structural measures.

(3) If shaft resistance mobilisation needs to be reduced along the upper end of the pile shaft during the test (complete eradication is not technically possible), the following methods can be used, e.g. as shown in Figure 9.21:

- flushing (not permitted for working piles);
- smooth sleeve tube sealed with foam material at its bottom end. It should be noted that this detail does not conform to the approval certificates for micropile construction. If the pile shall be used as a working pile, the corrosion protection must comply with approval certificate;
- sleeves in combination with lubricants, for example bentonite slurry or grease.

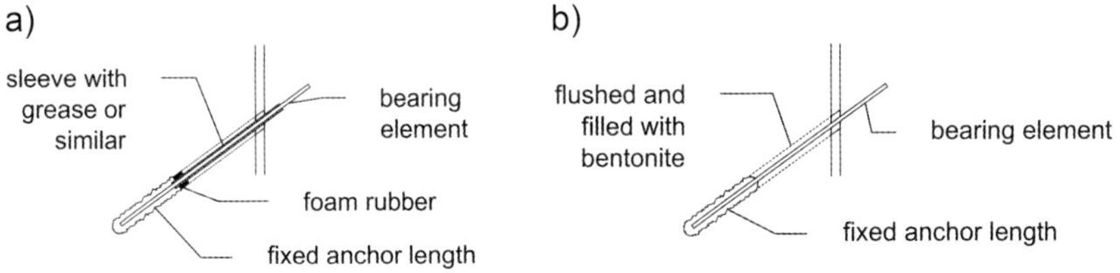

Figure 9.21 Reduction of load transmission in the upper length of micropiles

9.4.2 Test planning

9.4.2.1 General notes

(1) Results of pile tests are resistance-settlement curves for compressive load testing and resistance-heave curves for tensile testing, respectively, for the entire pile resistance R and the load-depending displacements together with the corresponding time-settlement and time-heave curves.

(2) Additional piles can be constructed for pile load tests or working piles can be used. In both cases, pile testing should be performed in a timely manner to allow any modifications to the foundation design resulting from the tests to be implemented..

(3) Where working piles are employed for pile load testing, the influence of any preloading applied during the test shall be taken into consideration for the assessment of the pile capacity and displacement behaviour within the structure.

(4) Tensile load testing to investigate piles designed to carry compressive loads are permissible, because this method returns a conservative skin friction.

Displacements under tensile loads are greater than under compressive loads of the same absolute magnitude.

(5) For compressive load testing structural measures are necessary to ensure that no unscheduled pile head deflection occurs and premature buckling failure is prevented.

(6) The load testing may only begin when the cement grout/fine-grained concrete has gained the strength necessary for the test loads. Evidence of actual compressive strength can e.g. be provided by compressive strength testing on additional test specimens.

(7) If piles intended for load testing are subjected to loads greater than permissible from internal pile capacity design, e.g. if failure of the grout-ground interface should be reached, it can be necessary to use a stronger steel bearing member.

(8) In cases where a stronger steel bearing member cannot be installed, tests can be carried out on piles with shorter force transmission length in order to achieve failure at the ground-grout interface.

(9) If failure has occurred in piles with a shorter force transmission length, a directly proportional increase of the pull-out resistance may not be automatically assumed when the force transmission length is increased. The increase is sub-proportional due to the strain in the bearing member.

9.4.2.2 Number of test piles

(1) DIN 1054:2010-12 requires load tests on micropiles:

a) on at least 3% of the projected number of piles;
b) but on at least $n \geq 2$ piles.

9.4.2.3 Test load

(1) The test loads used in pile tests shall normally high enough that the ultimate limit state (GEO-2 analysis) can be achieved, thus fulfilling the creep criterion:

$$R_{c,m} \text{ or } R_{t,m} = R(k_s) \tag{9.10}$$

where k_s represents a creep value, which is to be individually specified and which is used to define the time-dependent displacement of the pile head under constant pile load (see Figure 9.3).

Note: For determining the pile resistances in the ultimate limit state creep in the order of $k_s \approx 2$ mm is frequently used; it shall however be specified in agreement with the geotechnical expert.

(2) When estimating the test load with the aid of empirical data (see 5.4), it should be taken into consideration that even if the 50% quantile values are

adopted, a higher resistance than indicated by the empirical data is anticipated for half of all pile load tests. Selecting a suitably larger test load is therefore recommended.

(3) If the load test is only carried out to verify the adopted characteristic pile resistances, the EC 7-1 Handbook [44], 7.5.2.1 A (5), stipulates the following, minimum test loads:

for compression: $P_p = F_{c,d} \cdot \gamma_t \cdot \xi_1$ (9.11)

for tension: $P_p = F_{t,d} \cdot \gamma_{s,t} \cdot \xi_1 \cdot \eta_M$ (9.12)

where:

$F_{c,d}$: design value of the compressive pile load;
$F_{t,d}$: design value of the tensile pile load;
γ_t: partial factor for the total resistance of a pile after Annex A3.2;
$\gamma_{s,t}$: partial factor for the tension pile resistance after Annex A3.2;
η_M: model factor from the EC 7-1 Handbook [44], 7.6.3.2 A (3c);
ξ_1: correlation factor for pile resistances measured in static load tests, relative to the mean, after Annex A4.1.

(4) The test pile shall be designed for the adopted test load.

9.4.2.4 Principles for the instrumentation

(1) Test pile instrumentation must be compatible with the test objectives and the required quality of results, and be specified by, or in agreement with, the geotechnical expert. Two requirement levels are differentiated for micropiles:

a) Basic requirements
Monitoring of the pile resistance: the displacement of the pile head, the applied load and the time are measured.
c) High requirements
Monitoring of the pile resistance and pile shaft resistance: the displacements of the pile head, the applied load, the pile strain in various sections distributed along the pile length and the time are measured. Special instrumentation of the pile base and along the pile shaft is required.

9.4.2.5 Special loading situations

(1) If considerable pulsating or alternating loads or creep are possible in the pile's serviceability limit state, these load situations should be simulated as far as possible in the load test, also see Section 13. In such cases, the requirements on the testing process shall be specified by the geotechnical expert or the designer. Notes on procedures and examples are given in [42] and [135].

(2) Testing of pile groups shall be carried out if characteristic tensile pile loads $F_{t;k}$ are greater than 700 kN and distances of force transfer zone are less than 1.5 m. In this case the load test shall be by simultaneous testing of three neighbouring piles.

9.4.3 Loading systems

9.4.3.1 Reaction systems

(1) For tensile loads the reaction system consists of a testing girder and the support structure, see Figure 9.22a. For compressive loads the reaction system consists of a reaction-beam or a load frame, anchoring elements and connections, see Figure 9.22b and c.

(2) The steel components shall be designed to DIN 18800 using the elastic-elastic (E-E) method. The partial factor for actions may be adopted at 1.1. The elements of the reaction system shall be designed compliant with the EC 7-1 Handbook [44] using the BS-T partial factors (actions and resistances).

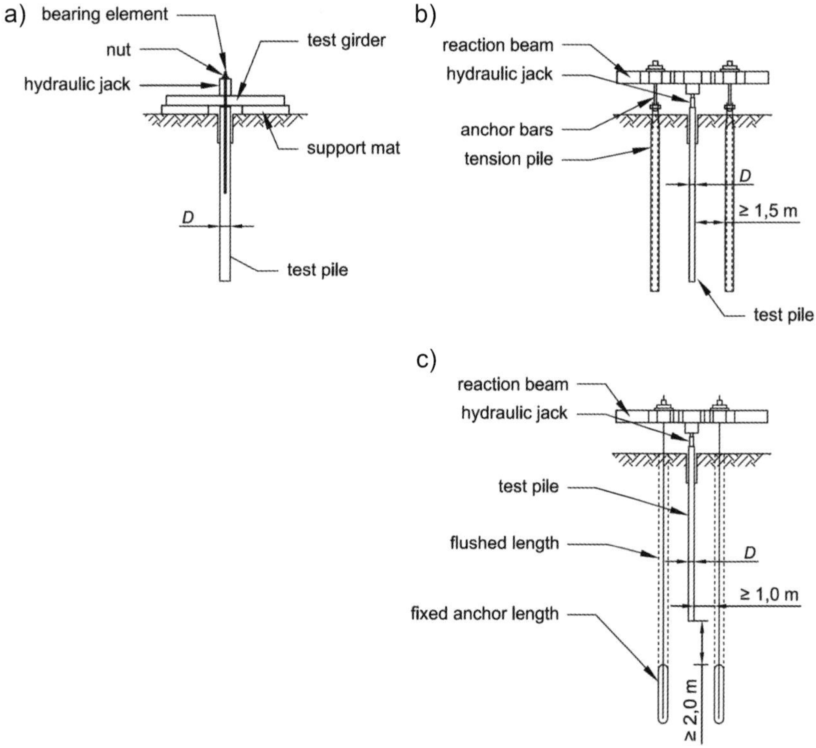

Figure 9.22 Micropile load tests: minimum spacing between the loading system and the test pile a) tension pile, reaction system formed by test girder on support mats b) compression pile, reaction beam anchored by parallel tension piles with load transfer along their full lengths c) compression pile, reaction beam anchored by parallel ground anchors with deep fixed anchor lengths

(3) For micropile load tests the clear distance between the test pile and vertical anchor piles or kentledge supports should be at least 1.5 m, see Figure 9.22b.

(4) In tensile load tests it shall be prevented that the test pile braces itself against any part of the reaction system.

(5) If grouted anchors are used as reaction elements as shown in Figure 9.22c the anchor load transmission zones must be clearly defined. Along the free anchor length any load transmission must be ruled out by using sleeves or by flushing.

(6) Of long reaction anchors the elastic elongation can become so large that the cylinder stroke is no longer sufficient. In these cases it is recommended to considerably increase the anchor steel cross-section in order to reduce the elongation.

(7) Reaction piles and anchors should be arranged symmetrically around the test pile. If arranged asymmetrically, any resulting influence must be taken into consideration during the loading and be compensated if necessary.

9.4.3.2 Hydraulic jacks

(1) The load shall be applied both centrally and axially. Hollow piston hydraulic cylinders and spherical bearings have proven useful.

(2) The test load should always be applied using hydraulic jacks, even where kentledge or a dead load is available as reaction mass. The maximum possible force and the necessary piston stroke should be compatible with the scheduled test load, and the anticipated displacements of the pile and of the reaction system.

9.4.3.3 Pile head

(1) For compression pile tests the pile head shall be prepared to allow the loading and displacement monitoring instruments to be placed centrally, flat and perpendicular to the pile head, see Figure 9.23.

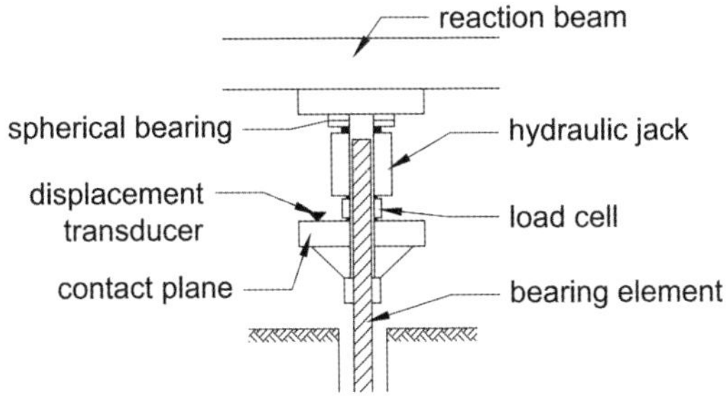

Figure 9.23 Compressive load testing of micropiles, example of pile head configuration

9.4.4 Instrumentation and monitoring

9.4.4.1 Displacement measurement

(1) The axial displacement of the pile head is measured mechanically using precision gauges or electrical displacement transducers. Displacements of vertical piles can also be recorded optically using precision levelling.

(2) The reference system consists of measuring beams arranged on either one or both sides of the pile and anchored under structurally defined conditions at least 1.5 · D. During the test, the reference system shall be protected from actions resulting from the test load, vibrations and temperature differentials.

(3) Precise levelling reference points must be established outside the zone of influence of all testing installations.

(4) The range of the monitoring instruments must be such that they do not need to be replaced during the test.

(5) The measurement uncertainty (deviation of the displayed value from the true value) for pile head displacements may not exceed ±0,2 mm. Transducers with a 0,01 mm scale division are required.

(6) Glass plates should be fixed beneath the sensors on the pile head to reduce friction.

(7) Where cyclic load tests are performed, continuously logging, electrical instruments shall be used to measure displacements.

9.4.4.2 Load measurement at the pile head

(1) Load cells must always be used; the hydraulic pressure shall be recorded for control purposes only.

(2) Load cells must correspond at least to precision class 1.

(3) All load cells must be calibrated, whereby the calibration may not be older than twelve months.

9.4.4.3 Pile shaft resistance

(1) In individual cases, e.g. for piles that are designed to transfer their load to the ground along a limited transfer zone, it can be necessary to determine the pile shaft resistance with depth when load testing micropiles.

(2) The pile shaft resistance can only be determined indirectly by monitoring the longitudinal pile strain and consideration of the Young's modulus and the cross-section at the respective measurement level.

(3) The longitudinal strains can be logged using electrical strain transducers (strain gauges). The strains can be measured over lengths of only a few centimetres at specified pile depths. They shall be fixed to the micropile's bearing

member such that their position does not change during installation and grouting. A configuration including two strain transducers per measurement level (on opposing sides) and several measurement levels over the length of the pile normally allow sufficient accurate assessment of the distribution of longitudinal pile strains.

(4) If longitudinal strains are measured locally or along predefined pile sections, interpolation are necessary for the lengths between. This requires "engineering" judgement and experience.

9.4.4.4 Protection of monitoring instruments

(1) The monitoring instruments at the pile head must be protected against weather influences and other disturbances. For short-term tests it is sufficient to shield the monitoring instruments and the reference system.

(2) If the pile shaft resistance is measured over the pile depth, the embedments, cables and the monitoring instruments themselves, both inside and outside of the pile shall be marked and be protected from damage during all installation phases. In particular, this includes sufficient insulation of electrical transducers and their cables against water ingress and fresh concrete, as well as mechanical protection against damage during concrete placement.

9.4.5 Testing procedure

9.4.5.1 Introduction

(1) When load testing micropiles, differentiation shall be made between piles designed to transfer the loads into the ground:

- along their entire length (System A) or;
- only along limited transfer zone (System B).

9.4.5.2 Load steps and loading rates for System A

(1) The load shall be applied in at least two cycles. The maximum load during the first cycle should correspond to the characteristic value of the effect on the pile. The maximum load during the final cycle corresponds to the test load.

(2) The test load shall be applied in at least eight load steps. It has proven useful to begin loading with a small initial load step to settle the loading system and set up the displacement monitoring instruments.

Note: When nearing the characteristic pile resistance or at a creep ratio $k_s > 0{,}5$ mm, the load steps should be reduced, if appropriate, to allow more detailed study.

(3) The load should be increased slowly and carefully. Jolts and vibrations should be avoided.

(4) The load increase from one step to the next should take at least 3 minutes, see Figure 9.24.

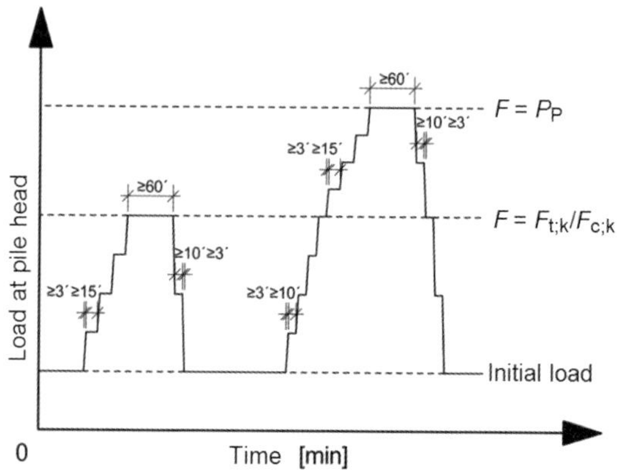

Figure 9.24 Recommended load steps for System A

Note: If the load is increased too quickly, long-term creep deformations in the pile-soil system can result. The loading rate should therefore be kept deliberately low at the higher load steps.

(5) The load shall be held constant during the individual load steps for at least the observation periods shown in Figure 9.24. The load may only be increased once the displacement rate has decreased to below 0,1 mm in 5 minutes. If possible, readings should therefore be made at equal intervals to allow the creep behaviour to be correctly identified.

Note: It has proven useful to maintain the loads at the characteristic pile effect $F_{c,k}$ or $F_{t,k}$ and the test load P_p for longer periods than until the decreasing of movements as defined above. This serves the recording of any creep redistributions in the ground and the assessment of long-term behaviour.

(6) An intermediate unloading sequence shall be applied once the characteristic pile effect $F_{c,k}$ or $F_{t,k}$ has been reached. The readings should be made at equal intervals; however, it is only necessary to wait for movements to cease at complete unloading or unloading to the initial load step. During reloading, and up to the load step prior to unloading, the same minimum observation periods shall be kept as for the unloading. During the further loading the same minimum observation periods apply as for the initial loading during the initial load cycle, see Figure 9.24

(7) Care shall be taken to ensure that the load is maintained constant for the duration of a load step. An automatic electrical or hydraulic load maintenance device should be used. If such means are not available this can also be achieved by hand pumping or manual control of the hydraulic pump.

9.4.5.3 Load steps for System B

(1) The test load should be applied in at least eight load steps and cycles, where the maximum load of each cycle should be increased by one load step (see Fig. 9.25). The load steps should be approximately equal. The maximum load during one of the cycles should correspond to the characteristic value of the effect on the pile. The maximum load during the final cycle corresponds to the test load.

(2) It has proven useful to begin loading with a small initial load step to settle the loading system and set up the displacement monitoring instruments.

Note: When nearing the characteristic pile resistance or at a creep ratio $k_s > 0{,}5$ mm, the load steps to allow more detailed study.

(3) The load should be increased slowly and carefully. Jolts and vibrations should be avoided.

(4) The load increase from one step to the next should take at least 3 minutes, see Figure 9.25.

Note: If the load is increased too quickly, long-term creep deformations in the pile-soil system can result. The loading rate should therefore be kept deliberately low at the higher load steps.

(5) The load shall be held constant during the maximal load of each individual cycle for at least the observation periods shown in Figure 9.25. The load may only be increased once the displacement rate has decreased to below 0,1 mm in 5 minutes. If possible, readings should therefore be made at equal intervals to allow the creep behaviour to be correctly identified.

Note: It has proven useful to maintain the loads at the characteristic pile effect $F_{c,k}$ or $F_{t,k}$ and the test load P_p for longer periods than until the decreasing of movements as defined above. This serves the recording of any creep redistributions in the ground and the assessment of long-term behaviour.

(6) An intermediate unloading sequence shall be applied once the peak load step of a cycle has been reached the first time. The readings should be made at equal intervals; however, it is only necessary to wait for movements to cease at complete unloading or unloading to the initial loading step. During reloading, and up to the load step prior to unloading and further loading, the same minimum observation periods apply as for unloading (Figure 9.25).

(7) Care shall be taken to ensure that the load is maintained constant for the duration of a load step. An automatic electrical or hydraulic load maintenance device should be used. If such means are not available this can also be achieved by hand pumping or manual control of the hydraulic pump.

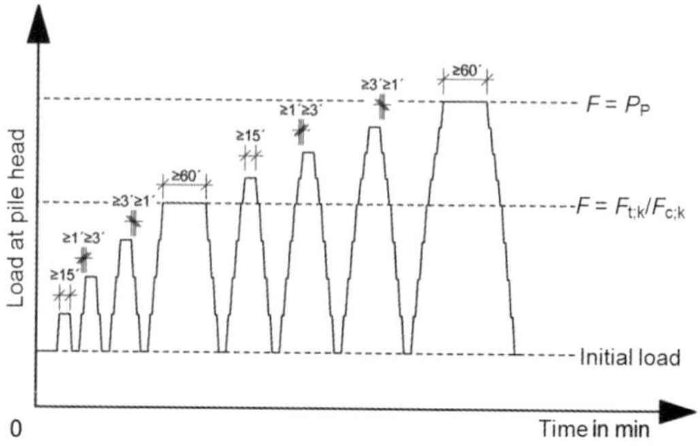

Figure 9.25 Recommended load steps for System B

9.4.5.4 Monitoring intervals

(1) During each load step the axial displacement at the pile head should be recorded at least after the following observation periods: 0, 2, 5, 10, 20, 40, 60, 80, 100 min., etc., after reaching the respective load.

(2) The intervals take account of the plotting of the recorded displacements on semi-logarithmic scale and also allow the determination of creep k_s.

Note: If long observation periods are envisaged once displacement have ceased (e.g. overnight), readings and load corrections in intervals of several hours are sufficient.

(3) Displacements of the pile head should have ceased before control measurements are taken of the pile and of the reference system, and for non-automatic recording of the changes in length along the pile axis. Fixed intervals should be adhered to for automatic recording and storage of load, deflection, stress and strain data.

(4) Before the beginning, and at the end of the loading test, the elevation of the pile head and of the reference system shall be checked by a datum measurement and a final measurement. These controls are carried out e.g. by precision levelling, with reference to a reference point. The actual pile load shall be recorded, and adapted if necessary, in the same intervals as the settlement or heave of the pile head.

9.4.5.5 Records

(1) Field reports in tabled format shall be compiled to contain all measurement obtained at the monitoring intervals and all observations made during the test are logged in tables in field logs in line with the logging intervals. Among others, they shall contain:

- hydraulic pressures and load cell readings;
- displacement measurements at the pile head;
- electrical stress and strain readings, if not recorded electronically;
- control measurements at the pile head and of reference system;
- weather conditions (temperature, wind, cloud cover);
- people present during the test;
- special test conditions, e.g. readjustments;
- reasons for deviations from planned test sequence.

(2) The records shall be signed and attached to the pile test report, see 9.4.7.2.

(3) If automatic data recording and storage is used, the data should be visualised during the test.

(4) The load and displacement data should be plotted as a resistance-displacement curve already during the load testing process. In addition to the load steps and the displacements, the development of pile displacements with time, together with the associated creep k_s, shall be recorded in order to allow the adaptation of the testing procedure where necessary.

(5) The semi-logarithmic graph of deflection with time allows conclusions about creep behaviour and the anticipated characteristic pile capacity to be drawn. Where tests are carried out on structural piles this allows any violation of the characteristic capacity to be recognised at an early stage.

(6) To end the test after displacement have ceased, the pile is completely relaxed as far as the initial load step and the data logged until the displacement have ceased.

9.4.6 Evaluation

(1) The displacement measured at the pile head are plotted either as resistance-settlement or resistance-heave curves as shown in Figure 9.10 or, for micropiles designed to transfer their loads along a limited transfer zones into the ground (System B), as shown in Figure 9.26.

(2) For each peak load step per load cycle also the time-displacement curves shall be plotted on semi-logarithmic scale, and the creep k_s be determined and visualised together with the corresponding loads as shown in Figure 9.11.

(3) When monitoring the pile shaft resistance over the pile depth, in addition to the resistance-displacement curves, the strains measured along the pile axis and the derived longitudinal force and skin friction shall be plotted in relation to depth and load, see Figure 9.12. The cross-section used for evaluation and, where necessary, the determined Young's modulus of the pile material shall also be given.

(4) For micropiles designed to transfer their loads into the ground along limited transfer zones (System B), the test evaluation shall demonstrate that the

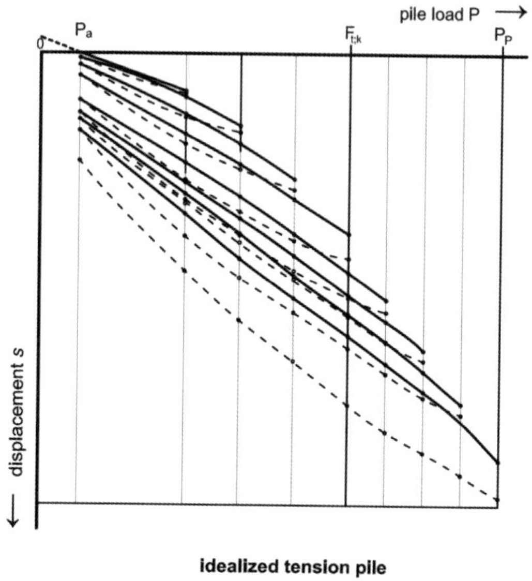

Figure 9.26 Resistance-displacement curve for a micropile with limited force transfer zone (System B) – example shown for a tension pile

confinement of the force transfer zone was actually achieved. The confinement of the load transfer zone achieved during pile installation can be determined via the measured displacements. To this end, the displacements at the end of each loading cycle are divided into an elastic and a plastic component. If the confinement of the force transfer zone is achieved by flushing, the pile above the transfer zone theoretically consists of the steel bearing member only. In this case, the actual confinement of the load transfer zone can be checked by determining the numerical length of the flushed region. The numerical elastic length of the steel bearing member L_{app} until to the near end of the load transfer zone is given by:

$$L_{app} = (A_t \cdot E_t \cdot \Delta s)/\Delta P \tag{9.13}$$

where:

A_t cross-section of the steel bearing member;
E_t Young's modulus of steel bearing member;
ΔP test load minus initial load;
Δs displacement s_p of the steel bearing member measured at anchoring point, for test load P_P minus the displacement s_a after unloading to the initial load P_a.

9.4.7 Documentation and reports

9.4.7.1 Introduction

(1) The load test shall be documented in two separate partial reports.

a) Part 1: Test report including all details of pile installation, the test, test results, presentation of test results and their evaluations.
b) Part 2: Interpretative report including the test results and their interpretation and resulting recommendations for the design and the execution.

(2) The test report (part 1) is normally compiled by the company installing the piles and/or executing the load test, the interpretative report (part 2) by the geotechnical designer or the geotechnical expert.

9.4.7.2 Test report

(1) The report should contain the following information:

1. General information
   - time and place of the test, site maps;
   - special local conditions;
   - involved personnel;
   - description of the ground and the groundwater conditions; as appropriate, abstracts from the ground investigation report (geotechnical investigation report) shall be included.
2. Information on the pile system
   - pile designation;
   - pile type;
   - dimensions, diameter of tension member, material quality;
   - execution records.
3. Test installations
   - configuration of testing installations;
   - reaction system (drawings, details, information on anchors or anchor piles);
   - loading system (hydraulic jacks with configuration and controls, maintenance of constant load);
   - monitoring instruments at the pile head (load cells, reference system, type and configuration of displacement transducers, type of control measurements, reference points, calibration certificates);
   - monitoring instruments in the pile shaft (type, configuration of measurement points, calibration certificates);
   - method of data acquisition.
4. Information on test execution
   - planned and actual loading regime;
   - reason for deviations from the planned testing procedure;
   - testing records (copies of original records);
   - saved data storage media in cases of electronic data acquisition;

– observations during the test execution (pile, reaction system, reference system, loading system, environment, weather conditions);
– photographic documentation.

5. Test results
   – resistance-displacement curves for R(s) as shown in Figure 9.10 and Figure 9.26;
   – pile head displacements as a function of load duration as shown in Figure 9.10;
   – time-displacement curves and determination of creep k_s for the respective peak steps of each loading cycle as shown in Figure 9.11;
   – for piles with limited load transfer zone and high test requirements: changes in length along the pile axis, skin friction and pile shaft resistance as shown in Figure 9.13.

9.4.7.3 Interpretative report

(1) The report shall be based on the geotechnical investigation report and the test report. It should include the following information:

a) the valuation of load test results;
b) recommendations on pile type, depth and diameter;
c) design specifications for the foundation to be executed;
d) information on possible group effects (where several piles are loaded);
e) recommendations for the execution;
f) recommendations for any necessary additional investigations or long-term monitoring.

Note: For micropiles designed to transfer their load into the ground along limited load transfer zones (System B), the evaluation of the load test results is based on the limits of the transfer zone which were actually achieved during the test pile execution.

10 Dynamic pile load tests

10.1 Introduction

(1) As described in 5.3, the EC 7-1 Handbook [44] stipulates that, under certain conditions, axial compression pile resistances may also be determined solely on the basis of dynamic pile load tests. In addition to dynamic pile load tests the EC 7-1 Handbook [44] also mentions impact tests.

(2) Dynamic pile load testing refers to methods using time-dependent recording of forces and movements at the pile head during an impact pulse (< 1 s). Dynamic pile load tests in the context of these Recommendations are methods in which the load is applied to the pile head by either braking (impact) or accelerating a mass and where the static resistance is determined by evaluating measured pile head movements.

(3) The result of a dynamic pile load test is the measured value of the pile resistance in the ultimate limit state $R_{c,m,i}$ (ULS) and, depending on the method used, the resistance-settlement curve.

(4) Dynamic load tests and impact tests can be performed on all types of piles. The different methods were originally developed to test prefabricated driven piles and were subsequently refined by improved monitoring techniques, modelling and analysis methods. Compared to static load tests, dynamic load testing requires less time and is normally less complex in terms of the test configuration, such that several piles can often be tested in a single day. For piles for coastal or offshore applications dynamic load tests are often the only method of testing or confirming the assumed capacities, because the execution of static load tests need an extremely high technically preparation effort and thus cannot be seen as feasible.

(5) For the application of impact test simple or improved driving formulae are used for analysis that allow the axial pile resistance to be calculated from data recorded during the driving process. Empirical data of the ground and the pile and the driving systems are adopted. Also wave equation analysis is used to evaluate impact tests.

(6) The EC 7-1 Handbook [44] stipulates that driving formulas should only be used in exceptional cases, because dynamic pile load testing provides a more precise method for determining the pile resistance at limited additional expense. The methods summarised under the term impact tests are therefore not dealt with any further in these Recommendations.

10.2 Range of Application and General Conditions

(1) The principal field of application of dynamic pile load tests comprises piles with a pile capacity predominantly resulting from their embedment in non-cohesive soils.

Recommendations on Piling – EA Pfähle.
Editors: German Geotechnical Society (*Deutsche Gesellschaft für Geotechnik e.V.*)

(2) The general limits of the method are given by the theoretical principles of one-dimensional wave propagation, see 10.3 and [138].

(3) See 5.3 (5) for details of limitations in conjunction with cohesive soils.

(4) The methods should only be adopted for use with foundations with the shape of bars or cylinders. Where planar foundation structures such as diaphragm wall panels or pile walls are used, two-dimensional wave propagation occurs. This is not dealt with by these methods.

(5) When testing sheet piles in a wall, skin friction in the ground and interlock friction can not be differentiated. The capacity can only be tested on free-standing piles.

(6) Dynamic pile load tests on piles in soils sensitive to dynamic actions, for example poorly graded, saturated, loose fine sands or normally consolidated, cohesive soils can lead to implausible results, such that dynamic load test in these soils are only possible with limitations.

(7) Because of the short time dynamic action the measurement or the test result does not provide a time-dependent settlement component for constant load. Any statements on serviceability are therefore only valid to a limited degree, unless the dynamic pile load test takes time-dependent effects into due consideration when calibrated against comparable static pile load test results, see 10.6.

(8) The magnitude of the load used in dynamic pile load testing shall be selected such that all resistances on the pile shaft and the pile base can be fully activated. If smaller loads are applied, the ultimate bearing capacity of the pile cannot be determined.

(9) The stresses in the pile resulting from the dynamic action may not exceed the allowable tensile and compressive stresses of the pile material. This fact shall be observed especially for cast-in-place concrete piles.

10.3 Theoretical Principles

(1) The effects on a pile embedded in the ground resulting from short-term, dynamic actions, e.g. such as those from a driving impact, represent a wave propagation problem in the elastic half-space. Physically, the local action in the half-space leads to a disturbance emitted from the source and transmitted to the continuum by propagating waves. In this process, the applied mechanical energy is radiated into the continuum and is converted to a different form of energy. Generally, the mechanical energy applied by an impact pulse is radiated in hemispherical form. As a result of the large differences between the ground soil and the pile material regarding stiffness and density, the major part of the applied wave remains within the pile, with the consequence that, in approximation, one-dimensional wave propagation can be assumed.

(2) The reduction of the pile dynamics to a simple, one-dimensional wave propagation problem allows analytical solutions for the partial differential equation for a free, undamped pile, and numerical solutions for a pile embedded in the ground. Generally, the attenuation in the amplitudes of the applied waves, reflected at the base of the pile, is utilised to assess the energy radiated into the ground and thus the ground resistances, in order to determine the pile capacity. A comprehensive description of the theoretical principles and the definitions of the underlying physical variables is given in [138].

(3) Mechanical variables, e.g. strains and accelerations are recorded at the pile head, see 10.4.5.3, and are converted to velocities, displacements and forces, which are the basis for the calculation of the bearing capacity.

(4) The force F(t) is calculated from the measured strain ε(t) using Eq. (10.1).

$$F(t) = \varepsilon(t) \cdot E_{dyn} \cdot A \tag{10.1}$$

where:

E_{dyn} Young's modulus at the measurement level,
A cross-section at the measurement level.

(5) The velocity is determined from the measured accelerations a(t) by integration and the displacement of the pile head by further integration.

(6) The energy E(t) applied to the pile is given as the time integral over the force and velocity using Eq. (10.2).

$$E(t) = \int F(t) \cdot v(t)\, dt \tag{10.2}$$

(7) The impact passes through the pile as a mechanical compression wave. The wave is reflected at the pile base as a result of stiffness and density differentials. Normally, the stiffness of the pile base support is lower than that of the pile material, such that the wave returns to the pile head as a tensile wave.

(8) The velocity c with which the wave passes through the pile is, in theory, a material constant. This velocity is related to Young's modulus E_{dyn} and the density ρ of the material by Eq. (10.3).

$$c = \sqrt{\frac{E_{dyn}}{\rho}} \quad [m/s] \tag{10.3}$$

(9) The time until an applied impact wave has travelled from the pile head to the pile base and back again is, see Eq. (10.4):

$$T = \frac{2 \cdot L}{c} \tag{10.4}$$

where:

L pile length, see Figure 10.1.

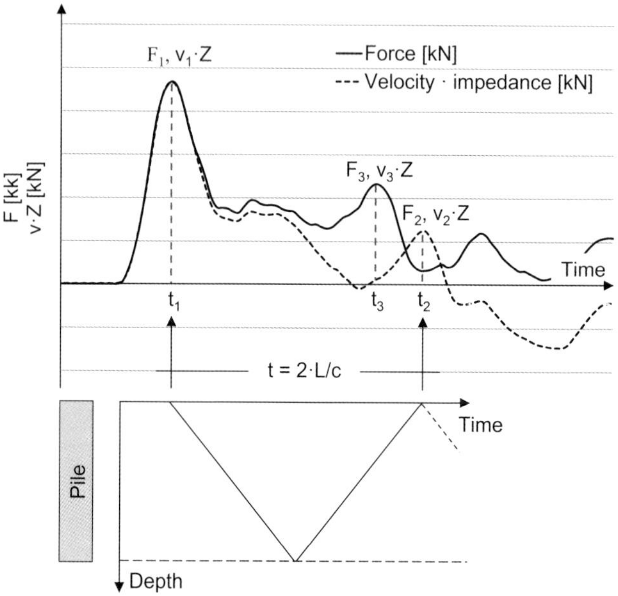

Figure 10.1 Example of pile head force and pile head velocity time histories

(10) From the dynamic equilibrium condition for the pile head follows, that the particle velocity of the mass particles v(t) at the pile head is proportional to the applied impact force, see Eq. (10.5):

$$F(t) = v(t) \cdot Z\,. \tag{10.5}$$

(11) The proportionality factor Z is known as the impedance (dynamic stiffness) and is given by the cross-sectional area, the stiffness and the density of the pile using:

$$Z = A \cdot c \cdot \rho = A \cdot E_{dyn}/c = A \cdot \sqrt{E_{dyn} \cdot \rho} \tag{10.6}$$

(12) Influences of the ground reduce with depth the pile force and the particle velocity. While the pile head force is in equilibrium with the contact force imposed by the falling weight, the velocity reduces. The magnitude of the difference gives an indication of the ground resistance, and especially the skin friction.

(13) Figure 10.1 shows the force F(t) as a continuous line and the velocity multiplied by the impedance $Z \cdot v(t)$ as a dashed line. When measurement begins the pile behaves similar to an infinitely long bar, i.e. force and velocity are proportional or, after multiplying the velocity by the impedance, congruent. At the time of the maximum value t_1 the force is $F(t_1) = F_1$, i.e. equal to

$Z \cdot v(t_1) = Z \cdot v_1$. The reflections caused by ground resistances then lead to a reduction of the velocity at the pile head. At time t_3 the difference between the force and the velocity is at its largest. At time t_2 the reflection from the base of the pile reaches the pile head or the measurement level.

10.4 Description of Testing Methods, Test Planning and Execution

10.4.1 Evaluation methods and type of load testing

(1) These Recommendations deal with what are referred to as the "direct methods" and the "advanced methods", also known as "complete modelling methods" (or "signal matching"). The following clauses give descriptions of the methodological principles and serve for selecting the method for the respective construction project. A detailed and more extensive description of the testing methods can be found in 10.4.5 and 10.4.6.

(2) The direct methods are the CASE and TNO methods, based on damping factors, and the rapid load test (RLT - statnamic test), based on the unloading point method.

(3) The complete modelling methods include the CAPWAP and the TNOWAVE methods.

(4) In terms of the type of pile loading, the direct methods CASE and TNO are no different to the complete modelling methods CAPWAP and TNOWAVE. All these methods base on an impact pulse applied for a relatively short period, which therefore generates a relatively short wave. In the rapid load test the impact pulse is applied for a relatively long period and generates a longer wave.

(5) The complete modelling methods (3) should be preferentially adopted to determine the pile resistance.

(6) The complete modelling methods allow the division of pile resistances into skin friction and end bearing capacity. Thus they enable for a plausibility test against the ground model. Analysis comprises a computed resistance-settlement curve which, however, does not contain time-dependent settlement components. Because of the modelling complexity and the time involved, on-site analysis is in general not possible.

(7) As result of the relatively short wavelengths, tensile stresses can be caused in the pile from the reflection at the base. As a consequence, unreinforced piles or pile-like foundation structures can be damaged. Adherence to allowable pile material stresses should be demonstrated, if necessary, see 10.2 (9).

(8) The direct methods permit on-site evaluations.

(9) The direct methods base on damping factors to be assessed as function of the type of ground or to be calibrated against static load tests. A calibration

leads to improved conclusions about the pile resistance. Homogeneous pile material and a constant cross-section are assumed for evaluation purposes. The loading system requires comparatively small expense. For driven piles the existing driving rig can normally be used. For cast-in-place concrete and bored piles loading systems employing free-fall masses are standard practice. The CASE and the TNO methods allow monitoring and evaluation during driving.

(10) The rapid load method, one of the direct methods, is also used to determine the measured pile resistance. It does not allow the division of the pile resistance into skin friction and end bearing capacity. The analysis yields a resistance-settlement curve, computed directly on-site. Because of the special loading system, availability should be clarified well in advance. Tensile stresses can be suppressed as a consequence of the relatively long wavelength and the resulting superpositioning of the wave components [138]. The method is therefore also suitable for use with unreinforced piles or pile-like foundation structures.

10.4.2 Number of load tests

(1) The number of dynamic pile loading tests shall be specified on a project- and ground-specific basis by the geotechnical designer or the geotechnical expert. It shall take account of the minimum 2 load tests per pile type and homogeneous ground situation required by the EC 7-1 Handbook [44] and the Annex A4.2. The validity of the results increases with the number of loading tests, which also pays off in the form of lower correlation factors ξ when converting the measured resistance into characteristic pile resistances, see Annex A4.2.

10.4.3 Ground investigations and pile installation documentation

(1) The ground investigations for pile foundations shall conform to the EC 7-2 Handbook [44] and Section 3.2. Homogeneous ground situations with similar stratification and ground properties shall be defined for the site. The dynamic pile tests shall be performed on piles at locations characterising the site's homogeneous situations.

(2) To facilitate the evaluation of dynamic pile tests the results of dynamic penetration or cone penetration tests should be taken into consideration together with the specific information contained in the pile construction records, concerning driving, concrete consumption, obstructions, founding depths, etc..

10.4.4 Time of testing and internal capacity

(1) Depending on the time of testing the following differences exist for displacement piles (prefabricated piles),

a) Tests are carried out during driving in order to determine changes in the penetration resistance with depth. The pile resistances determined at the end of driving represent a temporary value only. Any consolidation effect or relaxation at the pile base area can only be determined by restriking.
b) Tests are carried out immediately after installation to determine the pile capacity at the end of the driving. Direct methods are normally used (CASE, TNO). Any consolidation effect or relaxation at the pile base area can possibly be determined if an interruption of several hours was observed and tests accompanying the driving have been carried out.
c) Tests following the completed installation process and longer waiting times are preferentially recommended for determining the pile capacity, because the disturbances induced in the ground as result of the installation process have generally dissipated. The determined pile resistance is thus more realistic. For prefabricated driven piles any consolidation effect or relaxation at the pile base area can be determined if also tests have been carried out either accompanying the driving, see a), or tests shortly after installation, see b). As the pile resistances might have increased, the driving energy should be appropriately increased in this case

(2) Cast-in-place concrete piles should be tested not earlier than after approximately 7 days to take account of the hydration process of the pile concrete. Depending on the ground type an adhesion effect should be anticipated. Experience shows that the relevant ground disturbances have dissipated after approximately 10 days. Later testing normally yield higher pile resistances. Seven to forty days are common periods depending on concrete grade and necessary loading.

(3) With regard to prefabricated concrete and cast-in-place concrete piles, the calculated maximum compressive and tensile stresses in the pile should not exceed 80% of the concrete compressive or tensile strength during driving and testing, taking any prestressing into consideration.

(4) With regard to steel piles, the maximum stresses in the pile should not exceed 95% of the yield strength of the steel during driving and testing.

10.4.5 Dynamic load testing using the high-strain method

10.4.5.1 Brief description

(1) In high-strain dynamic load testing a pile is subjected to a very short impact pulse. The recording of accelerations and strains at the pile head allows the pile capacity to be determined using direct (CASE or TNO) and extended procedures (CAPWAP or TNOWAVE).

10.4.5.2 Loading system

(1) Driven piles are generally tested using the driving rig available on-site.

Notes: a) Because the hammer is usually only designed to overcome the static resistance to driving during the pile installation, it is possible that the piles can not be mobilised due to increased driving resistances after setup. If necessary, a larger hammer or suitable free-fall system should be used for testing.

b) Rapid-stroke piling hammers are not suitable for testing, because they cause continuous pile movements and do not allow individual strokes to be evaluated.

(2) Special loading systems are normally used for testing cast-in-place concrete piles. An example is shown in Figure 10.2. The systems consist of a free-fall mass, a guidance apparatus and the supporting structure. A sufficiently large and stable working platform is required. The lifting of the free-fall mass requires external lifting equipment, e.g. a mobile or a cable crane. The necessary mass m_{FG} is roughly calculated using:

$$P_P = (0{,}01 \text{ to } 0{,}02) \cdot R_{c,m} \quad (10.7)$$

where:

P_P weight of free-fall mass;
$R_{c,m}$ pile resistance as measured value;

Figure 10.2 Loading system for cast-in-place concrete piles
Note: in this case the pile was concreted 2,5 m high above the working platform and the loading system was placed on supports in order to ensure sufficient drop height.

and:

$$m_{FG} = P_P/g \tag{10.8}$$

where:

g gravitational acceleration.

10.4.5.3 Instrumentation

(1) It is virtually impossible to achieve a coaxial and perfect introduction of the impact load acting uniformly on the full area of the test pile. Because of the structural conditions at the pile head and of the configuration of the loading system, in particular of the free-fall mass, the guiding system, the pile cushion and the helmet, a uniform wave front can only be assumed at a distance of approximately 1,0 to 1,5 times pile diameter D or a from the pile head. The transducers required to measure strains and accelerations must therefore be installed at least at this distance from the pile head, see Figures 10.3 and 10.4. The pile must be exposed to the appropriate depth and any existing water must be pumped off.

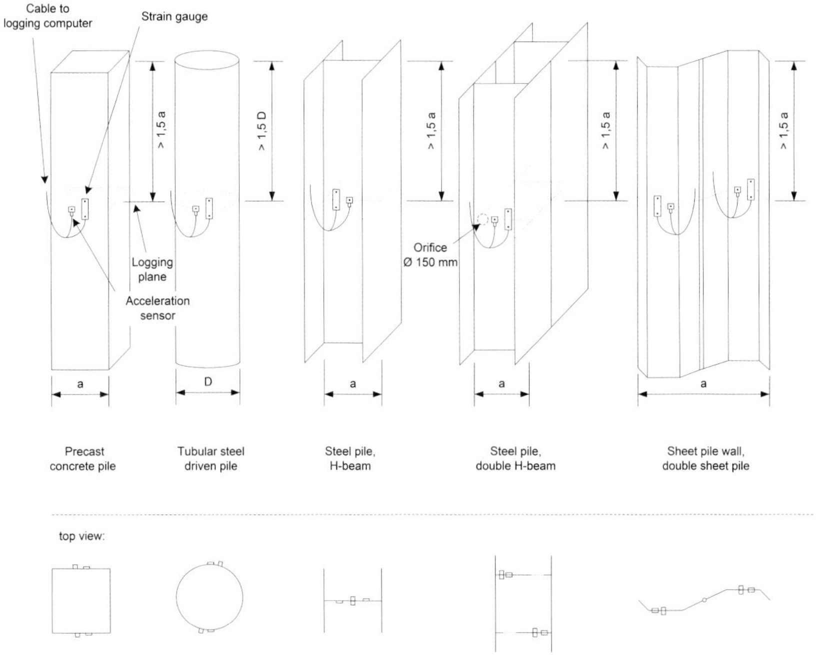

Figure 10.3 Schematic of logging configuration for driven piles

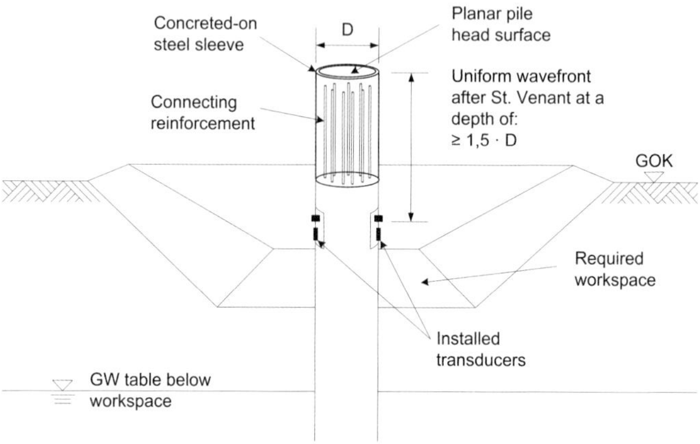

Figure 10.4 Schematic of monitoring configuration and preparation of a cast-in-place concrete pile for dynamic load testing

(2) To compensate for eccentricities the transducers shall be arranged at least on two opposing sides, that is, on a single axis. Arrangement of transducers in more axes increases evaluation precision, in particular if the action of the impact load might not be central. If central force impact is not possible arrangement of transducers in more axes is generally required.

(3) The following steps are necessary for driven piles (steel piles and prefabricated piles), also see Figure 10.3:

a) Preparation of the necessary workspace to a depth of at least 0,20 m below the measurement level.
b) Application of the transducers to the steel section or the surface of the prefabricated concrete as shown in Figure 10.3.
c) At double sections, the transducers shall be attached to the individual piles as the wave propagation is disturbed by the interlock. Where necessary, access holes should be cut into double sections consisting of H-beams as shown in Figure 10.3.
d) Connection of the transducers to the data acquisition system.

(4) The following steps are necessary for cast-in-place concrete piles, also see Figure 10.4:

a) Cut-off of piles at least down to the sound concrete.
b) Cleaning and preparation of the surface of the pile head and of the reinforcement.
c) Central placement and filling up with concrete of a steel casing to take the tensile shear stress resulting from the dynamic load and to protect the reinforcement. The casing should be of the same diameter as the pile and

must be axially centred. The casing length shall be selected such that its bottom end is not in the zone of the measurement level.

d) Preparation of the necessary workspace to a depth of at least 0,20 m below the measurement level.

e) Application of the transducers to flat, ground surfaces on the pile shaft. Windows must be cut into the steel if the measurement level is within the sleeved zone. The size of the windows shall be agreed with the institute carrying out the dynamic pile load tests. Dimensions around 0,25 m × 0,25 m are common.

f) Connection of the transducers to the data acquisition system.

(5) The transducer configuration for pile types not mentioned here, for example timber piles, shall be agreed with the institute carrying out the dynamic pile load testing.

(6) Watertight transducers shall be used for dynamic load tests where the transducers can get under water.

(7) Where tests accompany driving, suitably robust transducers and cables shall be employed.

(8) The permanent displacements shall be determined with accuracy to the millimetre using suitable monitoring instruments, for example precision levelling or a laser surveying system. In theory, these displacements correspond to the pile head displacements determined from the double integration of the measured acceleration. Comparison of values enables the control of the integration.

10.4.5.4 Performing the test

(1) In general, the test consists of several single strokes or series of strokes. The evaluation is based on one of these strokes.

(2) A sufficiently large displacement per stroke is necessary to activate the ultimate bearing capacity. Generally, displacements of a few millimetres are sufficient.

(3) An evaluation should be carried out on-site using one of the direct methods (CASE or TNO) in order to check that both the weight of the falling mass was correctly selected and the anticipated ultimate bearing capacity is plausible.

(4) The wave velocity c required for evaluation is in the one-dimensional wave theory a material constant, but is influenced in engineering terms by the material parameter tolerances and the composite pile/ground system. With regard to steel, the Young's modulus, the density and the wave velocity can be assumed as basically constant material parameters. They are substantially more variable for concrete, whereby the high strain rate (dynamic Young's modulus) must also be taken into consideration. In addition, the cross-section of cast-in-

place concrete piles can vary with depth depending on installation and ground parameters, and cannot always be precisely determined. The pile construction records and any integrity tests (Section 12) can provide information on anomalies. The basis for determining a site-specific wave velocity should therefore be established while performing the test.

Note: For example, the wave velocity in steel is $c = 5\,132$ m/s where $E_{dyn} = 210\,000$ MN/m² and $\rho = 7{,}85$ t/m³; the wave velocity in concrete is $c = 3\,200$ to $4\,000$ m/s where $E_{dyn} = 28\,000$ to $40\,000$ MN/m² and $\rho = 2{,}3$ to $2{,}5$ t/m³.

(5) With regard to concrete piles, and in particular cast-in-place concrete piles, these values are variable and must be determined as mean values over the length of the pile using the recorded data or the measured signal.

(6) The results of the dynamic load test are the force, determined from the strain, and the velocity, determined from the acceleration over time at the measurement level, see Figure 10.5. The velocity is multiplied by the impedance.

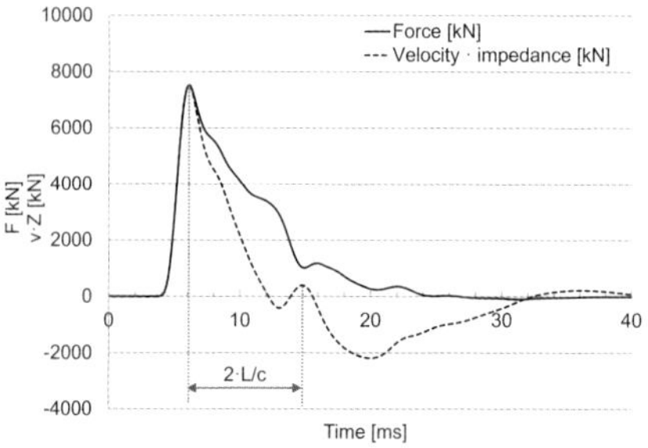

Figure 10.5 Force and velocity x impedance over time

(7) The data from the individual sensors shall be used to asses the quality of the measurements. The mean force and velocity graphs should remain congruent until the first maximum is reached, unless a separation is caused by skin friction or a change of the cross-section. In addition, the mean force and mean velocity graphs should approach zero and the displacement-time function calculated from double integration of the acceleration should horizontally approach the constant value of the permanent settlement.

(8) In the range from the first peak and the duration $t = 2 \cdot L/c$, i.e. the time the wave takes from the measurement level to the pile base and back again, the gap between the two curves gives an indication of the skin friction.

(9) The data recorded for all strokes are saved for further evaluation.

10.4.6 Dynamic load testing using the rapid load method

10.4.6.1 Brief description

(1) In a dynamic load test using the rapid load method, also known as the statnamic test, a pile is subjected to a long-duration impact such that the pile undergoes compressive loading during the entire test period. Using special evaluation methods the resistance-settlement curve can be determined via the initial system stiffness and the bearing capacity of the test or working piles, respectively.

(2) Detailed information on evaluation and interpretion of a rapid load test (RLT) can be taken from [41] and [49].

10.4.6.2 Testing types and timing

(1) Rapid load tests differentiate between the method using an accelerated reaction mass and the method using a free-falling weight and extended force transfer. The test can be regarded a rapid load test if the load duration T_f is greater than ten times the travel time of the applied strain wave from the pile head to the pile base.

$$T_f > 10 \cdot (L/c) \qquad (10.9)$$

where:

T_f duration of the loading;
L length of test pile;
c velocity of the strain wave in the test pile.

(2) The principle of the accelerated reaction mass method is based on activating a reaction mass acting on the pile by igniting a fuel in a pressure chamber. Igniting the fuel accelerates the reaction mass. An equal size reaction force acts axially on the pile.

(3) In the method using a free-fall mass and extended force transfer, the impact pulse of the free-fall mass impacting on the pile head is slowed by various damping elements, for example spring units, special rubber mats, or similar devices, see Figure 10.6.

Figure 10.6 Example of a spring unit for extending force introduction

10.4.6.3 Loading system

(1) When carrying out rapid load tests with accelerated reaction mass, special loading appliances are used, see the example in Figure 10.7a. These consist of the reaction mass, a pressure chamber, a cushioning apparatus for the reaction mass and the monitoring equipment required for testing.

Note: The availability of the loading system should be checked already during the test planning phase.

(2) Rapid load tests with a free-fall mass and extended force transfer (Figure 10.7b) require a similar test configuration to the high strain method. Additionally, devices for extending the time of force transfer, e.g. spring elements, are required (Figure 10.6). External lifting equipment is required for the erection of the loading system.

(3) The necessary mass for a loading system with an accelerated reaction mass or with a free-fall mass weight and spring unit m_{FR} is roughly calculated using:

$$P_P = (0{,}05 \text{ to } 0{,}10)\, R_{c,m} \tag{10.10}$$

and

$$m_{FR} = P_P/g \tag{10.11}$$

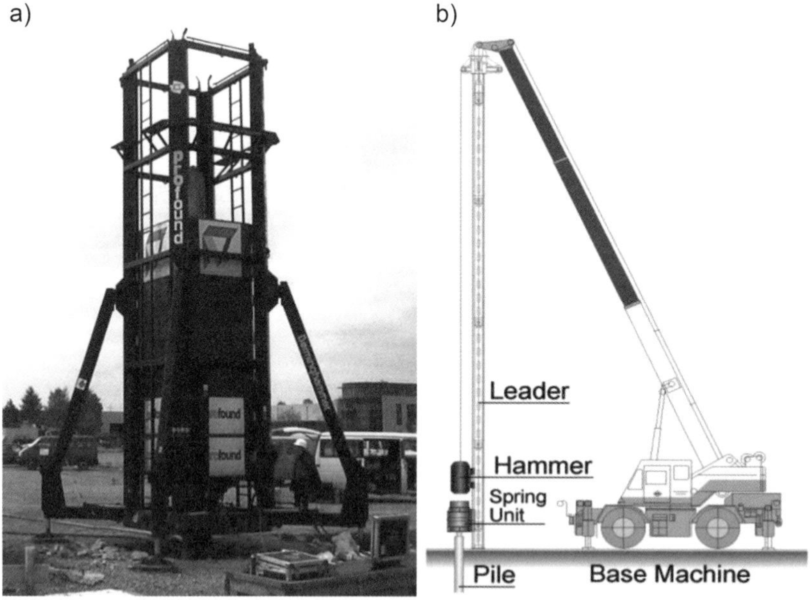

Figure 10.7 Loading apparatus with accelerated reaction mass and hydraulic cushioning system (a) or with free-falling weight and spring unit (b)

(4) During a test using an accelerated reaction mass the force acting on the pile can be varied by varying the amount of fuel used. In a test with a free-fall mass and extended force transfer the applied force is varied by adjusting the drop height.

10.4.6.4 Instrumentation

(1) Before the test the pile head must be level and undamaged, and the pile should protrude approximately 25 cm above the ground, see Figure 10.8.

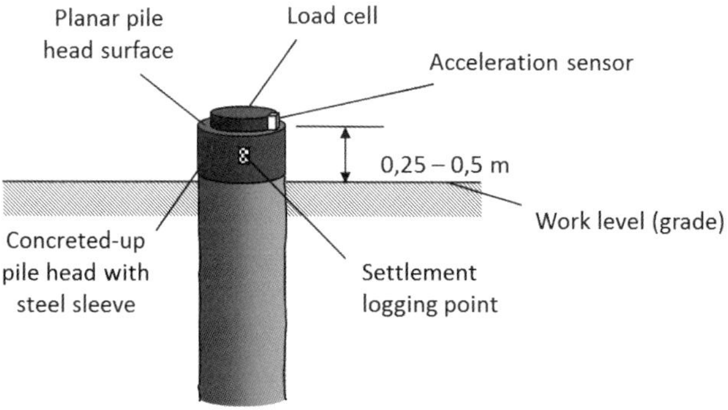

Figure 10.8 Schematic of measuring configuration and preparation of a cast-in-place concrete pile for rapid load testing

(2) The following steps are necessary when dealing with cast-in-place concrete piles:

a) Cut-off of piles at least down to sound concrete.
b) Cleaning and preparation of the surface of the pile head and of the reinforcement.
c) Central placement and filling-up with concrete of a steel casing to take the tensile shear stress resulting from the dynamic load and to protect the reinforcement. The casing should be the same diameter as the pile and must be axially centred.
d) Erection of the loading system: the pile head force shall be recorded by one or more calibrated load cells, placed between the pile head and the pressure chamber. Pile settlement is recorded below the pile head by an optical monitoring system. The optical monitoring system and the pile head must be kept at a sufficient distance from each other. Measurement of the pile head acceleration by an acceleration sensor serves for the evaluation using the unloading point method (10.5.3) and can also be used to check pile settlements.
e) Connection of the transducers to the data acquisition system.

(3) The following variables are recorded as a function of time during the test:

a) pile head force F(t);
b) pile head settlement s(t) and
c) pile head acceleration a(t).

10.4.6.5 Testing procedure

(1) In general, testing comprises a series of individual tests, each identifying a respective pile resistance. Taking the initial system stiffness into consideration, a resistance-settlement curve is derived.

(2) When carrying out a rapid load test with accelerated reaction mass, the reaction mass is activated by igniting the fuel in the pressure chamber. The ignition lifts the reaction mass off the pile head and generates a reaction force into the pile head. The reaction mass is prevented from impacting on the pile head again through gravity by various cushioning systems. Figure 10.9 shows an example loading system with accelerated reaction mass and gravel cushioning.

(3) When carrying out a rapid load test with free-fall mass and extended force transfer, the falling mass is placed directly above the pile head. An external lifting device raises the falling mass. The drop height should be chosen to suit the value of the pile load to be verified.

(4) The result of the test is a force-displacement curve based on the pile head force, pile head acceleration and pile head settlement measured over time, as well as the so-called unloading point. As a consequence, the pile's resistance-settlement curve can be determined, based on a hyperbolic approach, see Figure 10.10.

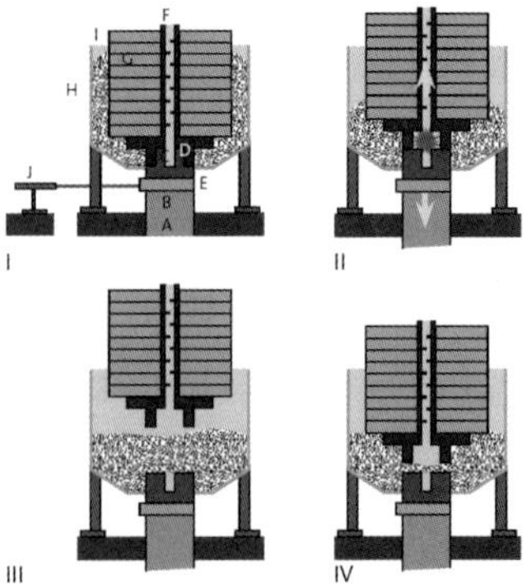

I – Initial condition, test configuration directly before test execution
II – Fuel ignition in combustion chamber followed by pile loading
III – Lateral gravel moves into center of cushioning device
IV – Soft cushioning of reaction mass by gravel

Figure 10.9 Rapid load testing with accelerated reaction mass and gravel cushioning

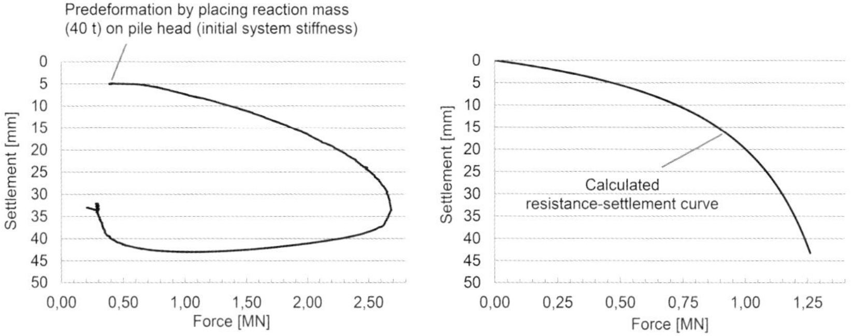

Figure 10.10 Recorded force-displacement curve with unloading point (left) and computed resistance-settlement curve (right)

(5) Once the test is completed the pile head should be examined for damage, in order to exclude any resulting influence on the evaluation. It is recommended to subject the test pile to integrity test before and after rapid load testing (see Section 12).

10.5 Evaluation and Interpretation of Dynamic Load Tests

10.5.1 Introduction

(1) Evaluation of dynamic pile load tests is based on the theoretical relationships outlined in 10.3. Depending on the testing method different evaluation methods are used for both direct and complete modelling methods,.

(2) The evaluation and interpretation of dynamic signal curves acquired from dynamic pile load tests is shown in Annexes C1, C2 and C3.

10.5.2 Direct methods using empirical damping values

10.5.2.1 Fundamentals

(1) The CASE and TNO methods are available for evaluation. After these methods the pile capacity for the static resistance R_{stat} is calculated from the total penetration resistance R_{tot} and the dynamic resistance component R_{dyn} resulting from the ground's inertial and damping forces.

$$R_{stat} = R_{tot} - R_{dyn} \quad [\text{kN}] \tag{10.12}$$

(2) The force and velocity profiles are determined from the measured strain and acceleration. The total penetration resistance is then given by the forces and velocities at the times t_1 and t_2, and the impedance Z, see Figure 10.1.

$$R_{tot} = \frac{1}{2}(F_1 + Z \cdot v_1) + \frac{1}{2}(F_2 - Z \cdot v_2) \quad [\text{kN}] \tag{10.13}$$

The difference in the times t_1 and t_2 is equal to the time required by the wave to travel through the pile from the head to the base and back again.

(3) The formula for the calculation of the dynamic resistance component contains empirical damping factors, e.g. derived from comparison with static load tests. These factors are primarily dependent on the ground type, the pile type, material and length and the stratigraphy.

Note: If damping factors cannot be reliably adopted, e.g. in cases of insufficient knowledge of the ground conditions or a lack of calibration against static load test data, substantial differences can result between the true pile capacities and those calculated using the CASE or TNO methods.

(4) According to the provisions and notations in the EC 7-1 Handbook [44] and Annex A4.2, the measured value of pile capacity from the dynamic pile load testing is, Eq. (10.14):

$$R_{c,m} = R_{stat} \quad [\mathrm{kN}] \tag{10.14}$$

10.5.2.2 CASE method

(1) The CASE method assumes that the dynamic resistance R_{dyn} is proportional to the penetration velocity of the pile base v_b. This means that, theoretically, the dynamic resistance is concentrated on this point. The proportionality factor is the product of the impedance Z and the damping factor J_c.

$$R_{dyn} = J_c \cdot Z \cdot v_b \quad [\mathrm{kN}] \tag{10.15}$$

For initial estimate of capacity and in relation to the type of ground, the damping factors in the ranges given in Table 10.1 can be adopted. Sensitivity studies are required for this purpose.

Table 10.1 Ranges of typical damping values for use with the CASE method

Soil	J_c [–]
Sand	0,05–0,20
Sandy silt	0,15–0,30
Silt	0,20–0,45
Silty clay	0,40–0,70
Clay	0,60–1,10

(2) Generally, the evaluation is based on the time t_1 at the point of the first peak in the force and velocity graph. The time t_2 is then:

$$t_2 = t_1 + 2L/c \tag{10.16}$$

This evaluation method is also known as the RSP method (Resistance Static taken at Peak 1).

(3) What is known as the RMX method (Resistance MaX) is better suited to piles with a high anticipated base resistance. The times t_1 and t_2 at the fixed distance of $2 \cdot L/c$ are evaluated for various sections of the graphs and the maximum resistance is determined from these. Because strong damping can be assumed for these types of piles, adopting the damping factors at the upper bound of the bandwidth is recommended. Here, too, sensitivity studies are required.

(4) Using:

$$v_b = v_1 + (F_1 - R_{tot})/Z \quad [m/s] \qquad (10.17)$$

and Eq. (10.13) and Eq. (10.15) the required value of R_{stat} results from Eq. (10.12).

(5) An example of evaluation using the CASE method is contained in Annex C1.

(6) The measured value required by the EC 7-1 Handbook is given by 10.5.2.1 (4), Eq. (10.14).

10.5.2.3 TNO method

(1) Using the TNO method the dynamic resistance is determined separately for the pile shaft $R_{s,dyn}$ and the pile base $R_{b,dyn}$

$$R_{dyn} = R_{s,dyn} + R_{b,dyn} \quad [kN] \qquad (10.18)$$

(2) The damping values are related to the pile shaft and the cross-sectional area of the pile base respectively. They are summarised in Table 10.2.

Table 10.2 Ranges of typical damping values – TNO formulas

Soil	C_s (TNO skin friction) [MN/m²/(m/s)]	C_b (TNO base resistance) [MN/m²/(m/s)]
Sand	0,002–0,010	0,4–2,0
Sandy silt	0,005–0,015	1,0–3,0
Silt	0,010–0,025	2,0–5,0
Silty clay	0,020–0,040	4,0–8,0
Clay	0,025–0,050	5,0–10,0

(3) The TNO method damping values are not comparable to those of the CASE method. The relationship to the respective cross-sectional areas and shaft surface areas requires that the pile/ground system damping properties are

differently considered. However, different damping influences can be taken into consideration for the shaft and the base.

(4) For the pile base:

$$R_{b,dyn} = v_b \cdot A \cdot C_b \quad [kN] \tag{10.19}$$

where:

C_b damping value for base resistance;
v_b velocity of pile base, Eq. (10.17).

(5) The dynamic resistance of the pile shaft is given by:

$$R_{s,dyn} = v_s \cdot O \cdot C_s \quad [kN] \qquad 10.20)$$

where:

C_s damping value for pile shaft;
v_s velocity at the pile shaft;
O portion of the shaft embedded in the ground.

(6) The governing velocity at the pile shaft is:

$$v_s = \frac{1}{2}(v_1 + F_1/Z) - \frac{1}{2}(F_3/Z - v_3) \quad [m/s] \tag{10.21}$$

The values F_3 and v_3 are taken from the graph at time t_3 ($t_3 < t_2$, within the time range 2L/c), representing the maximum difference between force and velocity, see Figure 10.1.

(7) The measured value required by the EC 7-1 Handbook is given by 10.5.2.1 (4), Eq. (10.14).

10.5.3 Direct method for evaluating a rapid load test using the unloading point method

(1) The evaluation of the force-displacement data recorded during rapid load testing uses the unloading point method (UPM) for non-cohesive soils and non-cohesive soils with small fines component. The Sheffield method (SHM) is recommended for use with cohesive soils. Refer to [41] for detailed description of both methods. These Recommendations only deal with the unloading point method.

(2) The UPM is based on the simplified assumption that the pile is regarded as a rigid body.

(3) The result of the evaluation is a resistance-settlement curve for the pile, which, however, does not contain time-dependent settlement components. A hyperbolic approach is adopted for the resistance-settlement curve. Basis for the assessment of the shape of the initial part of the graph are the settlements resulting from the placement of the reaction system or the free-fall mass on the

pile head and the initial system stiffness at times when the velocity is still small.

(4) The point of maximum settlement is defined as the unloading point. Because the direction of pile movement changes, the velocity is zero. Accordingly, it is assumed that at this time no velocity-dependent dynamic components are contained in the signal. The recorded associated force can thus be equated to the sum of the static ground resistance and the mass inertia of the pile.

(5) Taking the so-called "loading rate effect" in the signal [49] into consideration and using an empirical correction factor η dependent of the installation method, e.g. for sand: $\eta_{\text{driven pile}} = 0{,}94$, $\eta_{\text{bored pile}} = 1{,}06$, the results of (3) and (4) lead to the pile's resistance-settlement curve.

(6) An example of evaluation of a rapid-load test using the unloading point method is contained in Annex C3.

(7) The measured value required by the EC 7-1 Handbook is given by 10.5.2.1 (4), Eq. (10.14).

10.5.4 Extended method with complete modelling

(1) In principle, evaluations using the extended complete modelling method represent a form of system identification or the solution to an inverse problem, as described in [68]. More than one parameter set can exist for the solution. For this reason, the uniqueness of the solution must be checked using plausibility tests, in particular in terms of comparing the calculated parameters with the mechanical ground parameters derived from the ground investigations.

(2) The CAPWAP method and the TNOWAVE method are regarded as extended methods. The following information refers to the CAPWAP method; the TNOWAVE evaluation is similar.

(3) The methods are based on simulating real behaviour of a pile in the ground in a computer model. The pile is discretised into individual elements, to which pile properties are assigned. Linear elastic behaviour is assumed. The length of the elements is specified by the individual software systems. If the element lengths are altered manually the laws of dynamics and numerics must be observed. The modifications must be substantiated.

(4) The ground is modelled by rheological elements with allocated individual start parameters. For the stress-deformation behaviour of the ground, the static component is represented by a linear elastic-ideal plastic model and the dynamic component by a linear-viscous approach.

(5) Predominantly, the elements spring, damper and friction block are used, but elements for simulating a base gap are also possible. A pile-ground model is shown in Figure 10.11.

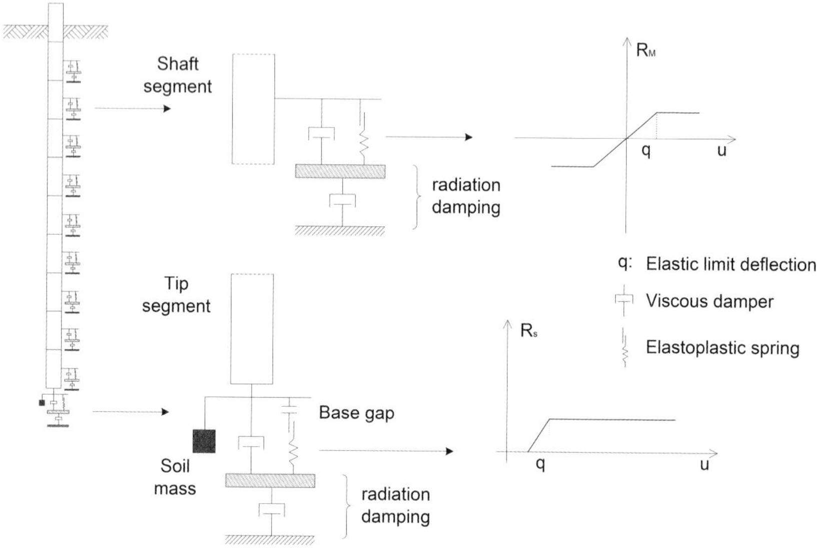

Figure 10.11 Pile-ground model for the extended complete modelling method

(6) On this system either the force-time history or the measured velocity-time history is superimposed as an action, and the system response is calculated as velocity-time history or force-time history. By varying the input parameters, predominantly the ground parameters, the calculated curve is adapted to fit the measured curve. An example of an imposed force-time profile is shown in Figure 10.12. Other methods for matching the recorded and the calculated time histories are also available.

(7) Result presentation comprises the force-time curves derived from recording and the product of velocity and impedance, see upper right graph in Figure 10.12. In addition, Figure 10.12 includes a comparison of the measured time history with those from the calculation. Here, it can be seen how well the model and parameter adaptations fit.

(8) Evaluation of the parameters determined following system identification gives the skin friction distribution over the pile length, the end bearing capacity and calculated resistance-settlement curves for the pile head and base. However, they do not contain any time-dependent settlement components.

(9) It is recommended to check the calculation results. To this end, the calculated ground parameters shall primarily be compared with the ground parameters given in the geotechnical report. In addition, sensitivity analyses, with varied calculation parameters can be carried out.

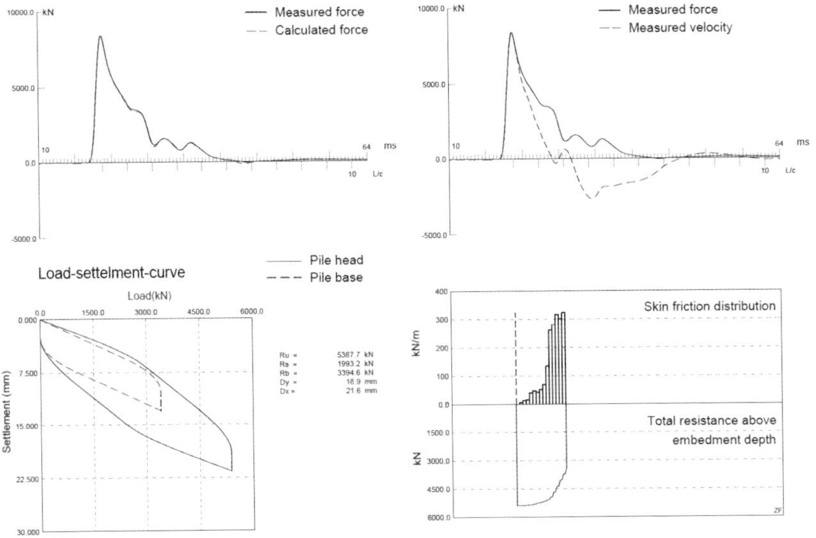

Figure 10.12 Results of CAPWAP evaluation

Note: In general, it cannot be assumed that the mathematically substantiated best curve fit represents the best solution to the problem in terms of soil mechanics. In addition, the quality of a curve fit should be assessed by alternately calculation using the measured force and the measured velocity as input variables. The validity can be checked by matching the curve over a time range greater than $4 \cdot L/c$.

(10) The extended complete modelling methods represent best engineering practice for evaluating dynamic pile load tests and should be preferred over the other methods outlined in Section 10. Moreover, they are also suitable for checking the empirical damping values used in the direct methods.

(11) An example of the evaluation of a dynamic pile load test using the extended complete modelling method is given in Annex C2.

(12) The measured resistance required by the EC 7-1 Handbook is given by 10.5.2.1 (4), Eq. (10.14).

10.6 Calibrating Dynamic Pile Load Tests

(1) The EC 7-1 Handbook [44], Recommendations on Piling 5.3 (2) and (3), and Annex A4.2 require that the results of the dynamic pile load tests are calibrated. Distinction is made between:

a) Calibration of dynamic evaluation methods against static pile loading test results from the same site.
b) Calibration of dynamic evaluation methods against static pile loading test results from a similar construction project.
c) Calibration of dynamic evaluation methods based on verifiable or general empirical data for pile resistances.

(2) The dynamic evaluation method can be calibrated against a static pile loading test on the same site. It is to be distinguished between the cases where two similar piles in similar ground are selected for testing or one single pile is subjected first to a dynamic and then a static pile loading test.

(3) When calibrating the dynamic pile load tests against the results of a static pile load test on the same site, the following must be ensured:

a) the limit state is achieved in both the static and the dynamic pile load tests;
b) a sufficiently long period between the two tests is adhered to when using the dynamic and the static pile loading test on one and the same pile;
c) where the static and the dynamic pile loading tests are applied to two different piles, the piles correspond as far as possible in terms of installation, geometry and ground conditions, and;
d) where necessary, the static pile load test evaluation results are made available for planning and evaluating of the dynamic pile load test.

(4) The EC 7-1 Handbook [44] and Annex A4.2 stipulate that the dynamic evaluation method can also be calibrated against static pile load test results from a similar construction project. A similar construction project means that the static pile load test adopted for the comparison was performed on a pile of the same type, of similar dimensions and in similar ground.

(5) The comparability of the dimensions must be ensured, especially in terms of the pile diameter. See 9.2.1 for additional notes. Similar ground is given if the same ground type, with similar compaction or strength, is prevalent both in the pile shaft area used for load transfer and in the pile base area.

(6) Calibrating dynamic evaluation methods against static pile load testing results from a similar construction project is often the rule, if no static pile load test is performed on the actual site.

(7) For prefabricated steel, reinforced concrete or prestressed concrete piles after 2.2.2, the correlation factors $\xi_{5,6}$ and $\Delta\xi$ may in simplification be adopted after 5.3 (4), if this is expressly agreed to by the geotechnical expert.

(8) For cast-in-place concrete piles, the results of calibration against the static pile loading tests shall in any case be presented, together with proof of the comparability.

(9) If the calibration of the dynamic evaluation method is based on verifiable or general empirical data for pile resistances, the damping coefficients given in

Tables 10.1 and 10.2 for the different ground and evaluation methods can provide an initial orientation.

(10) Calibration of dynamic pile load test evaluation methods against static pile loading test results, which is preferentially recommended, is performed as described below:

a) The resistance-settlement curve (RSC) of the static pile loading test forms the basis of the calibration, taking accuracy, tolerances and execution into consideration.
b) Reduce the static resistance-settlement curves to the initial deformation of the individual loading steps, i.e. remove the creep components occurring during the phases were the load was maintained constant, see Figure 10.13.
c) Analyse the dynamic pile loading test using a complete modelling method, aiming at the best possible fit between the recorded and the calculated curves during iteration.
d) Compare the resistance-settlement curve acquired from the dynamic pile load test using the complete modelling method to the reduced static pile loading test resistance-settlement curve adopted for calibration as mentioned in b).
e) If a static pile loading test on the same pile is used to calibrate the dynamic pile load test, the reloading branch of the reduced, static resistance-settlement curve mentioned in b) shall be adopted for the comparison.
f) If a static pile load test on a second, similar pile is performed to calibrate the dynamic pile load test, the initial loading branch of the reduced resistance-settlement curve as mentioned in b) shall be adopted for the comparison.

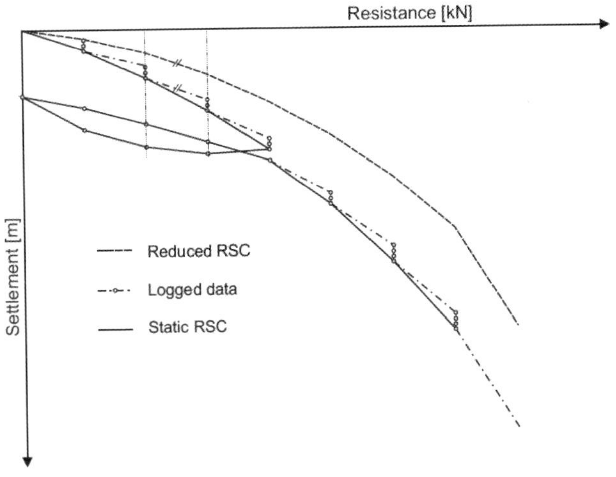

Figure 10.13 Static resistance-settlement curve (RSC) and resistance-settlement curve reduced by time-dependent components

g) If necessary, adapt the model parameters selected for use with the complete modelling method for the individual pile or ground elements, in order to facilitate matching the respective slopes of the resistance-settlement curves.
h) The model parameters determined in this way for the investigated pile system in the investigated soils should be transferred to the other, dynamically tested piles on the site as appropriate.

(11) In addition to the method for calibrating dynamic pile load tests against static pile loading tests described in (10), other calibration approaches for dynamic and static pile load test results may be employed, if their suitability can be verifiably demonstrated.

(12) Dynamic pile load tests using an extended complete modelling method can also be employed to determine tested pile skin friction data in the various ground strata.

10.7 Qualifications of Testing Institutes and Personnel

(1) The dynamic pile load tests shall be performed and evaluated by professional personnel, in possession of the necessary knowledge and experience.

(2) The necessary knowledge and experience of the professional personnel in executing and evaluating dynamic tests should be ensured by regular participation in professional training events presenting the actual state-of-the-techniques or even of the science.

10.8 Documentation and Reporting

(1) The test and its evaluation shall be completely and comprehensibly documented.

(2) The report must include information on the construction project, the place, time and reason for the test. The transducers used, the equipment and the names of the inspectors and evaluators shall also be documented.

(3) For the driving rig its type, manufacturer and the hammer or free-fall mass shall be listed. The choice shall be substantiated (e.g. free-fall mass/capacity ratio).

(4) Types and serial numbers of monitoring instruments and transducers, and all information on calibration as mentioned in 10.6 must be provided.

(5) For each tested pile designation, location, type, length, cross-section, construction company and time of installation shall be documented. Attaching driving or boring records of the tested piles to the report is recommended. During subsequent evaluation, information on the Young's modulus, impedance, unit weight and wave velocity shall be supplemented as necessary.

(6) The location and method of transducer installation shall be documented and photos be provided for proof where possible.

(7) All relevant monitoring parameters shall be documented for each test run. For driving, these include the drop height, driving energy/impact force and permanent pile penetration.

(8) Each set of raw data (recorded signals) shall be documented and stored for subsequent evaluations.

(9) The selection of all variable parameters (e.g. damping coefficients) shall be substantiated and all steps in the process be documented. As results of the direct CASE and TNO methods the total resistance R_{tot} and the ultimate load R_{stat} shall be provided. For a rapid load test the resistance-settlement curve calculated using the unloading point method shall be given. When using the complete modelling method, the comparison between the measured pile behaviour and the analysed model behaviour must be presented, and the calculated resistance-settlement curve and skin friction distribution be given.

(10) If static load tests are adopted for evaluation or calibration as described in Section 10.6, the method used shall be described and substantiated. This applies to both static pile loading tests on the same site and to using the results of comparable static pile loading tests.

(11) With regard to the provisions and notations adopted in the EC 7-1 Handbook [44] and Annex A4.2, the measured values of the pile capacity derived from the dynamic pile load test, $(R_{c,m})_{mitt}$ and $(R_{c,m})_{min}$, after Eq. (10.14) must be given. The use of the results of static loading tests for calibration after 10.6 when deriving $R_{c,m}$ shall be comprehensibly demonstrated.

(12) The report on the dynamic load testing or a rapid load test should include the following data:

Structure:	location	
	project designation	
	client	
	pile constuction company	
	type of loading test	
	consulting engineer	
Times:	pile installation	
	date/time of test	
Pile parameters:	pile type	
	pile number	
	pile length	[m]
	base of pile (reference to a Principle Datum)	[m]
	pile cross-section	[m^2]
	positions of transducers/sensors	[m]

Pile material:	Young's modulus	[MN/m^2]
	material density	[t/m^3]
	wave velocity c	[m/s]
Test execution:	monitoring instrument/transducer	
	weight of free-fall mass/hammer type	[t]
	drop height	[m]
	pile penetration	[mm/impact]
	driving energy (theoretical)	[kNm]
	driving energy (effective)	[kNm]
	max. impact force	[kN]
	max. velocity	[m/s]
	max. compressive stress	[MN/m^2]
	max. tensile stress	[MN/m^2]
Test evaluation:	evaluation agent/company	
	geotechnical documents used	
	pile construction record	
	CASE/TNO:	
	pile capacity	[kN]
	skin friction, end bearing capacity (TNO)	[kN]
	damping factor used	[–]
	rapid load test:	
	pile capacity	[kN]
	resistance-settlement curve	[–]
	correction factor	[–]
	CAPWAP/TNOWAVE:	
	pile capacity	[kN]
	skin friction, end bearing capacity	[kN]
	comparison of numerical and measured results resistance-settlement curve	[–]

(13) The report on the dynamic load testing shall indicate which calibration form was used for the evaluation, see 10.6 (1) a), b) or c). This calibration forms the basis for deriving characteristic pile resistances according to the EC 7-1 Handbook [44] or Annex A4.2 (specifying the correlation values $\xi_{5,6}$) by the geotechnical designer or the geotechnical expert, .

10.9 Testing Driving Rig Suitability

(1) The transducers used for dynamic load testing also enable analysis of driving impact and efficiencies. The following results are acquired:

- Characteristic impact-time diagrams for the hammer-pile-ground system;
- Hammer output, energy loss;
- Effective driving energy (single blow and complete driving process);
- Identification of anomalies during driving, e.g. premature ignition of diesel hammers, eccentricities;

- Behaviour of impact helmet and driving cushion (wear);
- Information on stresses in the pile;
- Hammer optimisation, optimal adaptation of hammer, driving rig, pile, anticipated capacity and calculation of driving time for planning and preparation of the works.

(2) The comparison of the applied driving energy to the measured energy transferred at the pile head allows the suitability of the driving system to be assessed, in particular the suitability of the hammer and the helmet, for each case respectively. If very large energy losses are noted, adjustments like technical and economical optimisation can achieve substantially improved results.

11 Quality Assurance during Pile Execution

11.1 Introduction

(1) With regard to quality assurance during execution, the requirements of the European execution standards issued for the individual pile types and the prestandards (DIN SPEC) published to facilitate their use in Germany shall be considered, see 1.1.

(2) The information of the following sections shall provide a practice-oriented supplement to the European execution standards and existing national technical approvals for special piles, as well as a broader basis for ensuring national quality standards.

(3) In the context of quality assurance during pile execution, particular emphasis is given on the requirements for ground investigation as stipulated in the EC 7-2 Handbook [45], see 3.2, and on the specific presuppositions for the works as stipulated in DIN EN 1536, DIN EN 12699 and DIN EN 14199.

Note: The needs and preconditions stipulated in DIN EN 1536, DIN EN 12699 and DIN EN 14199 should already be met during the planning phase, because they are of fundamental importance for selecting the construction methods, the plant and machinery, the tools and materials, and the quality assurance measures.

(4) The rules for the responsible monitoring and documentation of the pile execution by the contractor are given in the European execution standards. Where difficult boundary conditions prevail, independent monitoring by the geotechnical expert is recommended and should be specified on a case-by-case basis.

(5) The different pile systems and the most important construction methods and combinations of methods are described in Section 2. Quality-relevant aspects, to be observed during planning and execution, are outlined in the subsequent chapters.

11.2 Bored Piles

11.2.1 Principles

(1) It is immanent to the system that the formation of boreholes causes in any type of ground some stress release and some de-compaction in the immediate vicinity of the pile. Stress release and de-compaction shall be kept as small as possible by applying suitable support to the borehole walls and base, and by the selection and use of boring tools.

Note: Stress release and de-compaction can increase with time. Methods allowing expeditious execution should therefore be preferred and, the

Recommendations on Piling – EA Pfähle.
Editors: German Geotechnical Society (*Deutsche Gesellschaft für Geotechnik e.V.*)

last excavation step in the load-bearing strata may only be carried out if the pile can be completed during the same working day, DIN EN 1536:2010-12, 8.2.1.7 and 8.2.1.8.

(2) The support of the borehole serves “to prevent uncontrolled inflow of water and/or soil into the bore”, see DIN EN 1536:2010-12, 8.2.1.1. Common measures are:

- casings, normally installed in advance of excavation;
- soil-filled auger flights in boreholes produced using a continuous auger, or;
- support fluids.

If the conditions specified in DIN EN 1536:2010-12, 8.2.6 are met, boring can be carried out without support, either along the whole or sections of the bore.

(3) As a consequence of the type and stratification of the ground, the type of borehole support and the type and operation of the boring tools, the pile surface can become profiled, the pile cross-section can protrude outside the design cross-section and the pile base might not be strictly level. The outside diameter of the casing is regarded as the nominal pile diameter or, in unsupported, fluid- or earth-supported boreholes, the outside diameter of the boring tool. With regard to piles installed using diaphragm wall techniques (barettes), this should be applied to the separate excavation trenches.

(4) The construction of raking piles necessitates guidance for the excavation tools and when installing the reinforcement. These piles should therefore not be constructed without support or supported by fluids. Earth supported piles (CFA-piles) may be constructed only with inclinations up to 10:1 (DIN EN 1536).

(5) Structures similar to piles such as columns installed using jet grouting methods, grouted or not grouted vibro-displacement columns, caisson foundations or similar structures are not regarded as piles after 1.1. Whether, and to what extent the following principles should be applied shall be decided on a case-by-case basis.

11.2.2 Support to boreholes

11.2.2.1 Cased boreholes

(1) Rigid steel casings with appropriate connections for the individual casing segments are required for cased excavation. Casings must be designed for all stresses imposed during installation and withdrawal (torsion, compression, tension and bending stresses). The casings must be absolutely circular and the connections must be capable to withstand without deformation the high splitting forces building up during extraction.

(2) Casings are normally installed in advance of the excavation. Appropriate equipment capacities are necessary (torque and extraction force of an attached

casing oscillating or rotating machine or the power swivel (drilling head) of a rotary piling rig). Casing installation may only follow excavation in ground that remains stable for the duration of the excavation process.

(3) A cutting shoe with teeth or studs is normally necessary to be fitted to the bottom of the casing. The cutting shoe serves to reduce resistances when installing and withdrawing the casing. The teeth or studs project slightly relative to the casing and are often also fitted with cutting faces directing upwards. The projection causes a system immanent minor de-compaction in the immediately adjacent ground, even when the casing advance is sufficient. The projection and the type of the teeth shall therefore be selected to suit the ground strata to be excavated.

(4) Typical stripe or rib structures are produced on the surface of the pile when the casing is withdrawn, caused by the teeth at the cutting shoe: spiral structures occur when withdrawal is in one direction, zig-zag lines during oscillating withdrawal and vertical stripes during static withdrawal. These structures are typical for the system, see [59], for example.

(5) In ground in which the annulus caused by the projecting teeth can close again, displacements of ground material around the teeth can occur when the casing is withdrawn during concreting. The teeth cutting backwards during this phase "plough" the material to some extent into the pile surface. It can be useful to use a small tooth projection to minimize this ploughing effect.

(6) Ploughing effects with working ground material into the fresh concrete at the surface of the pile shaft is harmless in granular soils and, together with incrustations of ground, lead to a small increase in the pile cross-section and a certain increase of shaft friction. In cohesive ground, however, depressions or "blow-holes" filled with this material can occur at the surface. Normally these "blow-holes" are also harmless, provided that the minimum concrete cover to DIN EN 1992-1 (EC 2) is adhered to.

(7) With regard to cased boring using rotary methods with a rotary drilling head and Kelly bar, the lengths of the individual casing sections shall be selected such that, when a section is connected, the casing can be driven further as far as necessary to allow the proper operation of the excavation tools and its sidewise movement for emptying. In unstable ground excavation below the bottom of the casing to ease casing installation can lead to uncontrolled soil extraction and de-compaction in the vicinity of the pile.

11.2.2.2 Excavations supported by fluids

(1) The properties of the support fluid must correspond to the requirements of DIN EN 1536:2010-12, 6.2 in all phases of pile construction and must be monitored.

(2) The stability of the fluid-supported borehole must be analysed compliant with the provisions of DIN 4126. The fluid level in the borehole must remain

high enough in all construction phases, including when withdrawing the drilling tools, and shall be maintained always within the guide tube or the guide wall, and 1,5 m above the groundwater level.

11.2.2.3 Soil-supported boring with continuous flight augers

(1) Soil-supported boring with continuous flight augers (CFA-boring) requires ground conditions that present no risk of uncontrolled ground extraction under boring stresses. If unstable strata are present soil-supported boring is normally only allowable if the strata are no thicker than one pile diameter.

(2) Support of the borehole walls is by the loosened ground accumulated on the auger flights and of the base of the borehole by the pressure exerted through the down thrust on the drilling tool.

(3) When CFA-boring, control and assessment of the loosened ground is only possible after concreting and complete withdrawal of the auger, and also then to a limited degree only.

11.2.3 Excavation

11.2.3.1 Introduction

(1) The requirement of DIN EN 1536:2010-12, 8.2.1.1, "When constructing bored piles measures shall be taken to prevent uncontrolled inflow of water and/or soil into the bore" has highest priority for the boring procedure. This requires sufficient support of the borehole walls (DIN EN 1536:2010-12, 8.2.3 to 8.2.6), and appropriate selection and use of suitable drilling tools.

(2) For the selection of the method, in addition to groundwater levels, the permeability of the ground, the stratification, and thicknesses of ground layers and the groundwater flow conditions shall be taken into consideration.

(3) In case of discontinuous boring (e.g. grab and Kelly boring), a piston effect can occur when lifting the drilling tools from the bottom of the borehole, which can lead to loosening and heaving at the base. The risk is particularly high in fluid-supported boreholes. This effect shall be countered by the use of suitable drilling tools with sufficiently large orifices to allow percolation and appropriate operation of the tools (e.g. reduced velocity when lifting the tools from the borehole bottom).

(4) Where discontinuous boring with fluid support is employed, the risk of a piston effect extends over the entire length of the borehole.

(5) See 11.2.6 for details of continuous cased boring using continuous flight auger (double rotary boring method, cased continuous flight auger boring).

11.2.3.2 Boring below the groundwater table

(1) Cased boring in permeable ground requires a hydrostatic surcharge to generate a positive pressure in relation to the governing groundwater level

(DIN EN 1536:2010-12, 8.2.3.6). However, the positive pressure may be reduced if "a sufficient casing advancement is provided" (DIN EN 1536:2010-12, 8.2.3.7).

Note: In highly permeable ground the minimum 1,0 m difference of water heads required in DIN EN 1536:2010-12, 8.2.3.6 can mostly not be maintained.

(2) In low-permeability soils, including those with embedded permeable layers, boring without a hydrostatic head is only allowable if the ground situation is well known and under the condition that water and soil ingress is not possible due to sufficient casing advance.

(3) If dry boring is initially possible, but it can occur that water and/or soil could ingress into the bottom of the borehole before concreting begins, fluid support should be utilised during the phases of installation of reinforcement and for concreting, and concrete placement be carried out in line with the provisions for concreting in submerged conditions (DIN EN 1536:2010-12, 8.4.3). If groundwater is reliably cut off and it has been demonstrated that no water is accumulated at the bottom of the borehole, the procedures for "concreting in dry conditions" may be applied (DIN EN 1536:2010-12, 8.4.2).

(4) With regard to confined aquifers, extreme caution and early application of a hydrostatic head is required, unless the concerning layer can be safely cut off by advancing the casing.

(5) When boring under a support fluid a sufficiently high fluid level, commensurate with the acting earth and water pressures, surcharges, e.g. imposed by drilling rigs, and fluid fluctuations in the borehole caused by the tools, must be maintained (DIN EN 1536:2010-12, 8.2.4). Any necessary stability analysis of the open, fluid-supported borehole shall be calculated in line with the principles for analysis of the "open trench" and in analogy to diaphragm walls to DIN 4126.

(6) Guidance on soil-supported boring and additional information on cased boring using continuous flight augers (double rotary drilling method, cased continuous flight auger boring) is provided in 11.2.6.

11.2.3.3 Drilling tool diameter and speed of operation

(1) If the area for percolation between the drilling tool and the borehole wall or within the drilling tool is too small, and/or the drilling tool is withdrawn too quickly, the is a risk of hydraulic heave at the bottom of the borehole and, in uncased boreholes, collapse of the borehole wall, despite a hydrostatic head or a fluid support.

(2) The occurrence of hydraulic heave or collapse leads to the ingress of soil and water into the borehole. It can be recognised by the rise of the base and, in particularly serious cases, also of a sudden increase of the water level in the

borehole. Borehole wall collapse is mostly not immediately recognised during excavation, but only later by the increased concrete consumption.

(3) The drilling tool must have a sufficiently large, free inner percolation area ("ventilation channel") or an outer annulus, to allow the support fluid or water to return to below the tool when the tool is withdrawn. When withdrawal begins the pulling velocity should be increased slowly until flow through and around the drilling tool has established. It can be helpful to also rotate the drilling tool at first, until the water or support fluid has reached the underside of the drilling tool.

(4) The depth of the borehole shall be regularly checked with the aid of the rig's depth meter or by manual plumbing. In addition, the depth and any necessary casing advance shall be checked by comparing to the borehole depth. Where necessary, the drilling tool withdrawal velocity must be reduced further.

(5) The water level inside the casing should be monitored, in particular during withdrawal of the tool. If the water level suddenly rises the withdrawal speed must be immediately reduced. If necessary, a drilling tool with a greater percolation area should be used.

(6) If hydraulic heave is suspected the geotechnical expert should be consulted to assess the extent of ground de-compaction and the possible impact on the pile capacity.

(7) It is not possible to produce boreholes in soils without a certain degree of systemic ground de-compaction, see 11.2.1 (1), which can have an influence on neighbouring structures. Major ground de-compaction in the vicinity of the pile caused by soil extraction and/or hydraulic heave can be checked and extent determined using dynamic probing or cone penetration tests.

11.2.3.4 Cleaning the base of the borehole

(1) The requirement of DIN EN 1536:2010-12, 8.2.1.13 refers predominantly to the pile capacity. When boring under water or support fluid in fine-grained soils, e.g. in coarse silt or fine to medium sands, there is a further risk, that, while boring, considerable quantities of this material is held in suspension in the water or the support fluid. After the end of the excavation process, this material can "precipitate" from the water or fluid and settle at the base of the borehole. Given insufficient cleaning of the base of the borehole, desanding of the support fluid or cleaning of the water, the settled material can be whirled up at the start of the concreting procedure, be displaced to the sides or be deposited within the pile shaft.

(2) The cleaning process in the soils mentioned above should be carried out using a closely fitting reaming tool, operated especially slowly.

Note: Excavated material remaining in the tracks formed by the boring tool at the base of the borehole, and small quantities not removed during the cleaning process, do not normally affect the pile quality. This material is normally displaced to the sides of the borehole base and compacted by the concrete surcharge.

(3) Soil material settled on at the base of the borehole can have the same effect as contaminations or soft/loose soil after insufficient cleaning.

(4) In boreholes produced with fluid support the required properties of the support fluid must also be checked at the base of the borehole, also see DIN EN 1536:2010-12, 8.4.1.4. Contaminated support fluid or drilling mud might not be completely displaced by the rising concrete, such that pockets of support fluid remain which can negatively affect:

- the integrity of the shaft;
- the embedment and;
- the bond of the reinforcement.

11.2.3.5 Enlarged bases

(1) The additional provisions to be considered for enlarged bases are included in DIN EN 1536:2010-12, 1.5, 7.3.5 to 7.3.7, and 8.2.7.

Note: Grouting and deep soil mixing around the base of the pile are not classed as enlarged bases.

(2) Piles with enlarged bases are especially sensitive in terms of imperfections in the base area and any pile settlement resulting. They are often constructed using higher strength concrete in order to optimise material usage and pile capacity. Additional quality assurance measures can be necessary (e.g. control of shape and size, base cleaning, filling with concrete).

(3) The shape of enlargements should be such that sufficient cleaning of the base of the borehole is possible, concrete flow is not impaired when concreting begins and the enlargement can be completely filled with concrete. Shapes displaying a depression at the centre of the pile base should be preferred.

(4) The geometrical boundary conditions stipulated in DIN EN 1536:2010-12, 1.5 represent upper bounds. The base enlargement should not exceed a maximum size of 15 m^2. The respective dimensions shall be planned with respect to the prevalent local conditions.

(5) If base enlargements are installed in non-cohesive soils, the overhanging walls must be supported by a support fluid (possibly cement slurry). In these cases the requirements for fluid-supported boreholes must be also be met.

11.2.4 Installation of reinforcement

(1) The provisions in DIN EN 1536:2010-12, 7.5 and 8.3 primarily serve the purpose of ensuring correct installation, central positioning, unhindered concrete flow and a complete concrete embedment of the reinforcement.

(2) The minimum concrete cover required by DIN EN 1536:2010-12, 7.7 is understood as the nominal concrete cover $c = c_{min} + \Delta c$ to DIN EN 1992-1-1 (EC 2). The required concrete cover comprises the allowance value Δc, to account for the particular installation conditions (rough surface in contact with the ground, no internal concrete compaction, possible cement paste loss at the pile surface, etc., see Figure 11.1). The minimum cover c_{min} to DIN EN 1992-1-1 (EC 2) is the governing and sufficient value to guarantee the corrosion protection of the reinforcement of the completed pile.

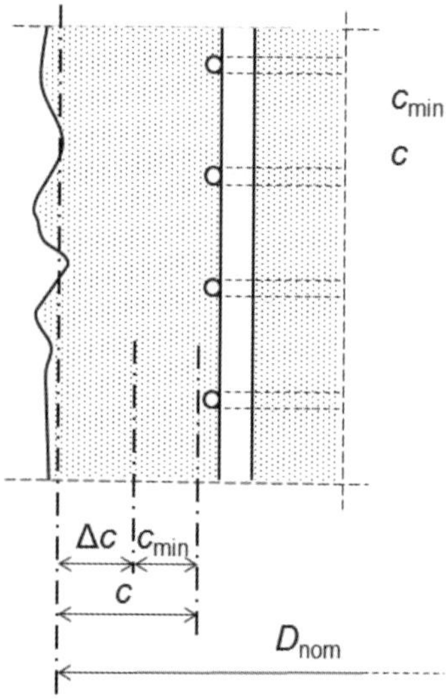

c_{min} to DIN EN 1992-1/DIN 1045-1

c as the greater value to
- DIN EN 1536 or
- DIN EN 1992-1/DIN 1045-1

Figure 11.1 Definition of the concrete cover for bored piles

(3) When using very long reinforcement cages and/or in raking pile bores there is a risk of the reinforcement sinking as a result of cage distortions or penetration into the base or wall of the borehole. This can be prevented by stiffening of the reinforcement cage, installing large spacers and an enlarged base of the cage (e.g. a steel plate or crossed flat steel bars). The free flow of concrete must however be ensured.

(4) Because depth measurement can be uncertain and to account for the installation conditions, a height tolerance of 0,15 m was specified for the upper end of the reinforcement cage of bored piles, see DIN EN 1536:2010-12, 8.1.2. It can be expedient to extend the length of the connecting reinforcement by this tolerance.

(5) If piles are reinforced only their upper section, there is a risk that reinforcement cages are lifted by the rising concrete column ("floating" of cages).

Cages can be protected against uplift, e.g. by anchoring bottom extensions. In addition, there is the risk that insufficiently supported or suspended sectional reinforcement cages can sink in the concreted pile.

(6) If the reinforcement is installed subsequent to concrete placement, the provisions of DIN EN 1536:2010-12, 8.3.6.4 to 8.3.6.7, apply.

(7) Additionally, the following should be considered for subsequent reinforcement installation

- Suitable spacers should be used to ensure that the central position of reinforcement cage.
- Complete installation of the reinforcement cage to the scheduled depth must be ensured by a suitable concrete composition, a regular construction sequence and suitable installation technique.

11.2.5 Concreting

11.2.5.1 Concrete mix

(1) The consistency ranges given in DIN EN 1536:2010-12, Table 4, refer to the time of placement of the concrete.

(2) Stable concrete mixtures meeting the requirements of DIN EN 1536:2010-12, 6.3.6.1, and the placement conditions should be preferred. All these mixtures contain excess water, which can be transferred to the ground around the pile under the placement and pressure conditions in the fresh concrete column. However, it can also travel vertically down through the fresh concrete column to the pile base and/or up to the pile head and accumulate there.

Note: This water loss is normal to a certain degree and does not lead to lower quality. We then refer to "stable concrete mixes".

(3) Insufficiently stable mixtures are those allowing rapid and excessive loss of excess water. This effect is referred to as concrete "bleeding". If the flow conditions are highly unfavourable during the loss of excess water segregation of the concrete can occur, which can also be associated with leaching of fines (e.g. cement paste) from the concrete.

Note: With regard to long piles (> 25 m), it can be expedient to examine the setting and stiffening behaviour before concreting begins, e.g. using a simple workability test ("bucket test") on site.

(4) In highly permeable soils there is a risk of premature concrete stiffening in the lower region of the fresh concrete column as result of excess water loss from unstable concrete mixes. As a consequence, a plug of stiffened concrete can form. The risk is smaller in fluid-supported boreholes, because a filter cake forms at the pile shaft-soil interface, thereby reducing the possibility of radial water loss.

(5) In cased piles a plug formation can lead to jamming of the reinforcement within the casing, such that the reinforcement cage can be lifted together with the casing. Concrete flow can also be severely hampered. In unreinforced piles a "crust" of stiffened concrete can form on the inner face of the casing.

(6) In only slightly permeable soils it is possible that excess water is lost at the shaft surface or vertically in the pile if insufficiently stable and/or concrete mixes with a high w/c-ratio are used. This can lead to the formation of flow channels, which also allow fines to be flushed out of the concrete. If the fines flushed out are predominantly cement, zones of lesser strength can occur at the shaft surface. In addition, the reinforcement's corrosion protection can be impaired.

(7) These phenomena can become even more severe in artesian conditions.

(8) To assess the risks and plan countermeasures prior to pile construction, it is necessary to also investigate possible influencing factors during the site investigation. Among others, these include:

- groundwater pressure in a confined aquifer;
- possibility of strong, local groundwater flow to a freshly concreted pile;
- percentage and distribution of permeable strata at the pile shaft, allowing radial water losses from the concrete.

(9) Necessary conditions for achieving stable concrete mixes are:

- use of binders with low water secretion. Cements with large surface areas have proven useful (Blaine value > 3800 cm^2/g);
- low total water content (e.g. ≤ 180 l/m^3);
- use of water-saving admixtures (plasticisers, superplasticisers, if possible PCE-based);
- possible use of stabilisers;
- rounded mineral aggregates (no crushed material);
- grain size distribution in the "favourable" range;
- sufficient percentage of fines below 0,125 mm (cement, admixtures such as fly ash or pulverised limestone, fines content of aggregates).

(10) If a considerable quantity of fines has been extracted from the concrete mixture, a concrete technologist should be consulted.

11.2.5.2 Concreting procedure

(1) If there is no water in the borehole, placement may in dry conditions and the concrete may be allowed to fall freely if piles are vertical. It must be ensured that the fresh concrete can reach the base of the borehole without hitting the borehole wall. A concreting hopper with an attached at least 2 m long concreting pipe must be used. If a concrete pump is used the concreting pipe must be fixed and held stable in the centre of the casing.

(2) When concreting under either water or a support fluid, the tremie pipe shall initially be placed on the pile base or on the reinforcement's bottom stiffening plate/cross and a plug (e.g. a ball, plastic plug) be placed in the pipe to separate the concrete from the water or the fluid. The tremie pipe is then completely filled with concrete and lifted only so much that concrete flow can be initiated. The bottom of the pipe shall be enclosed by the concrete and submerged within it (DIN EN 1536:2010-12, 8.4.3). Otherwise there is an increased risk of segregation in the lower shaft region, which can built up to considerable height.

(3) The immersion of the tremie pipe into the fresh concrete is only defined in terms of minimum values or qualitatively in DIN EN 1536:2010-12, 8.4.3.16, 8.4.3.17 and 8.4.4.3. Excessively large submerged tremie pipe depths can lead to blockages within the tremie pipe and/or can impede concrete flow.

(4) During withdrawal the immersion depth of the tremie pipe shall be controlled and adapted to the level of the concrete column and the lengths of the pipe sections remaining in the concrete .

(5) For piles constructed with casings, the level of the concrete column shall always be sufficiently high above the bottom of the casing and exercise sufficient pressure for stabilisation of the borehole walls. The positive concrete head must be maintained also when the casing is withdrawn and the concrete slumps for filling the annulus formed by the extracted casing.

Note: Control of the rise and slump of the concrete column during concrete placement and during withdrawal of casings, and comparison to the theoretical values serve to monitor concrete placement and can provide information on possible irregularities.

(6) If the concrete column is too high before withdrawing the first casing section, there is an enhanced risk of premature concrete set, formation of a plug or crust in the lower pile section and jamming of the reinforcement cage (cf. 11.2.5.1 (5)). In such cases a base plate or crossed flat steel bars might not be sufficient to keep the reinforcement cage in position.

(7) However, stiffening of the concrete in the lower pile section also acts favourably when withdrawing further casing sections. The reinforcement cage is held in position if the first casing section is withdrawn as early as possible.

(8) If the reinforcement cage is lifted up together with the withdrawal of the casing and cannot be stabilised in time, it should be withdrawn completely and the concrete be bored out to allow the pile to be re-constructed later in the same place. If this is not possible, remedial measures must be undertaken (e.g. construction of a replacement pile).

(9) The rising concrete column normally carries at its top a zone of segregated concrete and some mud from the base of the borehole. Concrete should therefore be cast sufficiently higher than the design pile cut-off level, see

DIN EN 1536:2010-12, 8.4.3.20 and 8.4.11.1. The slump of the concrete column caused by casing withdrawal shall be taken into consideration when defining the final concrete casting level. In case of long empty bores, the height of the casting level above the cut-off level should be increased due to the uncertainties of depth measurement, see DIN EN 1536:2010-12, 8.4.1.19.

(10) A segregated concrete zone can occur in the centre of the pile as a result of the withdrawal of the tremie pipe. This "track" has normally only a short length and usually does not affect pile integrity.

(11) If empty bores need to be filled, excessively coarse-grained material should not be used, as this might penetrate into the fresh pile concrete, see DIN EN 1536:2010-12, 8.4.1.21. Suitable are granular soils, which are slowly backfilled and compacted if necessary. The backfilling of the borehole is required to the working level and before the casing is fully withdrawn.

(12) The planned final height of the pile is usually specified for embedment in the pile head slab of around 5 cm. DIN EN 1536:2010-12, 8.4 includes rules for cutting-off the excess concrete and DIN EN 1536:2010-12, 8.1.3 contains the applicable height tolerances. After cutting-off, the upper surface of the pile needs not be planar and the pile edges may be broken. The cut-off level does not require post-treatment and edges need not be prepared by sawing. However, this assumes that the construction joint is formed according to the provisions of DIN EN 1992-1-1 (EC 2) and the necessary concrete cover can be reliably provided at all points by the concrete of the pile head slab.

11.2.6 Bored piles constructed with continuous flight augers

11.2.6.1 Introduction

(1) Bored continuous flight auger piles can be constructed using soil-supported borehole walls (continuous flight auger piles – CFA-piles) or with casing (cased continuous flight auger piles – CCFA-piles and front-of-wall piles (VDW-piles), see Table 2.1.

(2) DIN EN 1536:2010-12, 8.2.3 (cased drilling of VDW and CCFA bored piles), 8.2.5 (soil-supported boring of CFA-piles), 8.3.6.4 to 8.3.6.7 (reinforcement installation) and 8.4.6 (concreting) apply when installing bored piles using continuous flight augers. Also see 2.2.1.4 for application limits with regard to soil-supported bored piles construction.

11.2.6.2 Soil-supported auger boring

(1) Unavoidable wobbling of the continuous flight auger necessarily causes an oversize borehole and, correspondingly, a higher concrete consumption.

(2) The ground volume transported while rotating the auger downwards should generally correspond to the volume of the auger. In non-cohesive soils

a considerably larger volume is an indication for de-compaction around the boring string. The excavated material accumulated on the auger flights is to be added to the total volume transported when the auger is withdrawn.

(3) In stratified ground with major strength differentials exists the risk of excessive soil transport during boring, if intermediate, weaker layers are temporarily unstable, e.g. poorly graded, non-cohesive soils below the groundwater table.

(4) Soil transport also takes place when using partial flight augers. Partial flight augers are characterised by only possessing a helix over a few metres at the bottom of the hollow stem. To some part, the soil transport derives from auger boring into the load-bearing stratum and transporting soil upwards from there into the weaker, overlying strata. The remainder consists of the soil material accumulating on the auger flights and being extracted when the auger is withdrawn.

11.2.6.3 Cased flight auger boring

(1) The following applies in addition to 11.2.3.1 and 11.2.3.2.

(2) The augering tool and casing are simultaneously driven, with rotation in the same or in opposite directions, using a double-rotary head driving the auger and the casing. The rotary drives allow some relative movement so that the auger pilot can progress or follow slightly the casing. The augering tool serves to continuously loosen and transport the excavation material. Air or water flushing can be employed to ease transport.

(3) The borehole wall is supported by the casing. The borehole base is supported by the pressure exerted by the boring tool and any pressure imposed by the flushing air or fluid.

11.2.6.4 Concreting and installation of reinforcement

(1) The concreting process corresponds to the method described in 11.2.6.2 and 11.2.6.3. Continuous flight auger and cased continuous flight auger boring are therefore not differentiated below.

(2) During the concreting process and when withdrawing the auger, concrete must continuously and in sufficient quantity issue from the bottom of the hollow stem to fill the formed void completely with concrete. The auger withdrawal rate must be adjusted to suit the pumping rate of the concrete pump.

(3) DIN EN 1536:2010-12, 10.3 and Table 10, require that sufficient concrete pressure is documented in the piling record. Because of the dynamic effects and the difficulty attaching any transducers at the concrete issue point at the lower end of the auger, it is normally only possible to qualitatively determine

the issue pressure. The concrete pressure is normally measured at the upper pumping hose/auger connection (rotary flushing head). Because the pressure at the base of the auger always consists of the pressure measured at the top in addition to the positive pressure of the concrete column in the hollow stem, there is always a concrete pressure at the base if the hollow stem is filled. The concrete pressure shall be reduced at the end of the concreting procedure and, before the continuous auger leaves the ground, to prevent uncontrollable circulation. As long as concrete flow is reliable and controlled, sufficient pressure remains at the bottom of the auger, even if there is zero pressure at the flushing head.

(4) If an auger with a small diameter hollow stem is used, the reinforcement cage can only be installed in the concreted pile after completion. It must be installed immediately subsequent to concrete placement. The central position of reinforcement cage must be ensured.

Note: Vibratory aids are permitted for subsequent installation of reinforcement cages. A pulling-in aid can also be employed.

(5) In particular in dry, coarse-grained soils, the fresh concrete can loose water, leading to premature concrete stiffening. Subsequent installation of the reinforcement cage can be impeded or even made impossible.

Note: This situation can be mitigated by saturating the ground (adding cement slurry when boring) and using adapted, specially stabilised concretes.

11.2.7 Shaft and base grouting

(1) If shaft and/or base grouting using packers and tubes-a-manchette (multiple systems) or grouting hoses (single systems) is carried out, the tubes-a-manchette and grouting hoses shall be fixed to the outside of the reinforcement cage, in the areas to be grouted.

(2) Before grouting, the concrete cover around the valves shall be fractured with water after the concrete has set and achieved some initial hardening.

Note: Fracturing should be with relatively high pressure but little quantities of water, in order to avoid introducing excessive amounts of water into the pile/soil interface, which could lead to softening.

(3) Grouting should be carried out shortly after fracturing to allow the cracks in the concrete cover zone to be sealed again.

(4) Base grouting using flexible cells at the pile base should only be carried out once the pile concrete has achieved an initial strength and any softening of the ground in the pile/soil interface caused during pile installation has dissipated.

11.3 Displacement Piles

11.3.1 Prefabricated concrete piles – Guidance for transport, storage and installation

(1) The governing effects on prefabricated concrete piles are generally prevalent during transport, storage and installation on site.

(2) Prefabricated concrete piles are stored on site on squared timber bars or on a flat surface.

(3) Bending stresses are caused when lifting the piles for transport or under the driving helmet. The necessary longitudinal reinforcement is dependent from the pile length and is usually defined in a type-structural design.

(4) The suspension points for unloading or pulling under the driving helmet must be exactly defined (e.g. by concreted-in hooks) as the piles might otherwise be damaged during unloading due to incorrectly placed or slipping steel cables or chains .

(5) The use of lifting gear that might break edges of piles (e.g. sharp-edged chains) shall be avoided. Because of its brittleness the concrete of the prefabricated piles can break off during lifting.

(6) To protect the pile head, the condition of the pile cushion inside the pile helmet must be continuously monitored. Insufficient (too hard) or worn cushions can cause damage to the pile head and must be replaced in time.

(7) The driving rig and driving energy (hammer weight, drop height) shall be selected to suit the piles (weight, slenderness) and the local ground conditions. The use of super heavy hammers and small drop heights is normally more efficient and less damaging than driving with relatively light hammers and greater drop heights, even if the nominal driving energy is equal.

Exceptions:

- very slender piles;
- ground conditions with highly variable driving resistances;
- soft soils.

(8) When driving through soft or very soft soil strata with low penetration resistances, driving should be with small energy impact. High-energy driving leads to corresponding high penetration. The low resistances at the pile tip then generate tensile stresses in the pile, which in turn can lead to transverse cracks in the pile if the reinforcement is overstressed.

(9) If the ground contains large, impenetrable or immovable obstructions, a pile can be deflected laterally and become inclined. The rig operator must follow this inclination with the hammer and align the leader correspondingly. Otherwise this can lead to bending stresses in the pile and even to pile failure.

(10) Driving prefabricated concrete piles into hard rock or rock-like, solidified ground is only possible with limitations. The driving process shall be suspended at the latest, if no, or very little, penetration is measured at maximum driving energy. At this phase normally also the material strength limits have been reached. Further driving can lead to transgression of the material strength and to damage like spalling or fractures.

(11) The effects of the strains given above on pile integrity can be checked using integrity testing, see Section 12.

(12) See DIN EN 12699 for details of additional installation related strains.

(13) See 11.3.3 for details of displacement effects in cohesive soils.

11.3.2 Cast in place concrete displacement piles

11.3.2.1 Water/soil ingress into the drive tube

(1) When installing piles in groundwater it must be ensured that no water or water/soil mixtures have entered the drive tube prior to concreting. In practice, e.g. a small stone or similar object is dropped into the drive tube and the impact on the pile shoe or base plate (boring tip for full displacement bored piles) is checked acoustically. If in doubt, test for dryness using an appropriate plumb or sounder.

(2) If water or soil has entered into the drive tube, pile installation must be suspended. In such cases the drive tube must be withdrawn, the borehole backfilled and the pile installed anew at the same location, relocated slightly if necessary.

11.3.2.2 Concreting

(1) In principle, concreting is always in dry conditions, and well-graded, cement paste-rich concrete mixtures are used, in the F4 to F5 consistency range.

Note: No concreting or tremie pipe is required for concrete placement.

11.3.3 Displacement effect in cohesive soils

(1) While the installation of displacement piles in non cohesive soils generally leads to soil compaction (volume reduction), cohesive soils are primarily displaced (with only small change in volume) and thus moved. In addition to horizontal soil movements, ground heave can occur.

(2) In particular from large pile foundations with small pile distances can initiate major horizontal soil movements that can lead to unplanned strains on neighbouring structures (e.g. shallow foundations, pile foundations, pipelines, key walls, excavation pits).

(3) The possible effects on neighbouring structures must be considered for the survey of adjacent structures. If necessary piling works must be accompanied by monitoring of adjacent structures. It has to be considered that soil movements and heave can occur with some delay.

(4) Possible measures for preventing the above effects include pre-boring at pile locations and/or de-compaction boring.

(5) In cases of major heave tensile stresses can be caused by skin friction in the installed piles during the construction stage (i.e. before loading).

11.4 Grouted Micropiles (Composite Piles)

11.4.1 Introduction

(1) The following notes refer to common bored composite piles with a steel bearing member and a hardened cement grout cover, in particular to monobar piles, also see 2.2.6 (4) b), and tubular piles, also see 2.2.7 (2).

(2) For definition of quality assurance measures for grouted micropile installation the following objectives shall be differentiated:

- prevention of failures and;
- avoidance of reduction in capacity.

(3) Also quality characteristics like:

- corrosion protection;
- durability;
- resistance against chemical attack;

can be objectives of quality assurance measures.

11.4.2 Grouted monobar piles

(1) If piles are installed beneath structures sensitive to settlements, bored from a level below the groundwater level, or if requirements on directional accuracy are particularly high, the boring should be using double rotary drilling or the overburden drilling method.

(2) The overburden drilling method using a double rotary head with counter-clockwise operation of drilling rods and drilling pipes has proven effective, in particular in cohesive soil.

(3) Particular care should be taken when coupling sectional bars. Couplings shall be made in accordance to any technical approval regarding (direction of rotation, locking, securing against unscrewing). Incorrectly screwed connections can lead to total pile failure, in particular of tension piles.

(4) The grouting pressure should be measured as closely to the borehole mouth as possible as frictional pressure losses inside the grouting hoses can be

substantial. If this is not possible, the pressure loss should be determined and subtracted from measured pressures.

(5) For grouting a suitable rig should be used and be carried out, either continuously while withdrawing the drilling pipes, or in sections by sealing the mouth of the drilling pipes with a grouting cap.

(6) The number of grouting hoses, and the number and position of the grouting valves shall be planned to suit the length of the effective structural grout body. With regard to possible valve failure, a configuration using two valves per hose and section to be grouted should be preferred. Experience shows, that more than two valves per hose offer no advantage, because normally only the valve with the lowest resistance opens. The length of the sections to be grouted within the whole grout body should not exceed 3 m.

(7) Contaminations of and damage to the grouting hoses and valves can lead to the grouting pressure not being effective at the grouting valve and the valve does not open. Grouting hoses and valves should therefore be examined for contaminations and damage prior to installation.

(8) The diameter of the drilling pipe should be sufficiently large, to prevent damage to hoses and valves when installing the monobars with the attached grouting hoses into or when withdrawing the drilling pipe.

11.4.3 Tubular grouted piles

(1) Sufficient cement cover of the bearing element must be ensured at the whole circumference and along the whole length. This is of particular importance for working piles for long term uses. Special care should therefore be taken to install sufficient and appropriate spacers, which must be in accordance with the technical approval.

(2) The boring rate must be controlled when drilling in the bearing member. This is necessary to prevent stalling of the flushing medium and contamination of the cement cover with soil. The 'Rules for execution' of the technical approvals include detailed installation requirements; adherence to which regulations must be supervised.

(3) If the structurally required pile length cannot be achieved due to obstructions or insurmountable drilling resistance, it should be examined whether the pile performance can be demonstrated numerically or by pile load tests. Otherwise, the pile must be abandoned.

(4) The as-built pile length can be verified by integrity testing (also see Section 12).

(5) Obstacles that cannot be directly penetrated by drilling, e.g. sheet pile walls, must be perforated in an advance process using a special drilling method (coring, double rotary head or overburden drilling methods).

(6) In order to ensure that the couplings are completely screwed together, the section of the bearing member already inserted into the borehole must be held by the rig's breaking device and the next section be screwed on using the rotary head of the rig and applying sufficient torque.

11.4.4 Testing grouted micropiles

(1) The high skin friction of grouted micropiles is primarily influenced by the grouting process. This requires a high execution quality. Grouting success can be insufficient in certain ground conditions like fine-grained soils with low plasticity, in soils sensitive to liquefaction (quick sands), in banded soil strata or in mixed-grained soils.

(2) Tensile tests are recommended for quality control of working micropiles. The number of micropiles to be tested should be specified on a case-by-case basis, depending on the ground conditions and the design.

(3) Tests are also recommended for structures and foundation systems of low ductility, which can fail suddenly and without visible signs of the approach of an ultimate limit state.

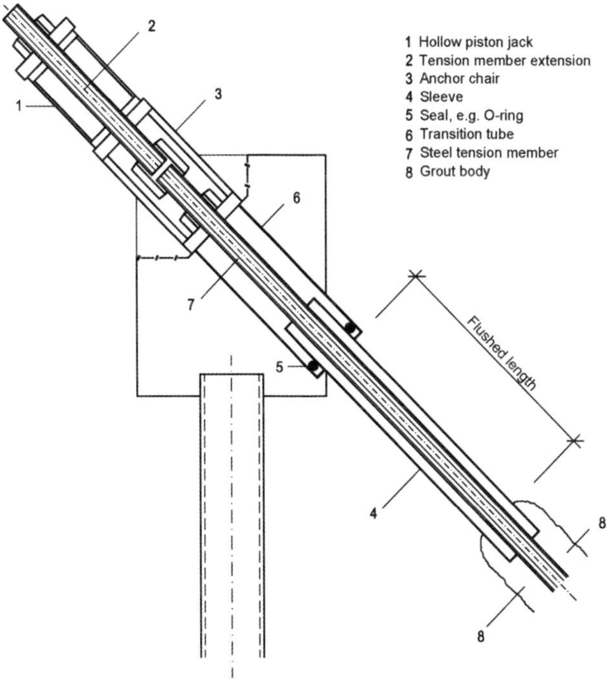

Figure 11.2 Example of a set-up for testing a raking, grouted micropile

(4) The method and the extent of tests on working piles should be specified in connection with the geotechnical design.

(5) A direct frictional connection between the reaction system and the grout body must be avoided when testing. An example configuration is given in Figure 11.2. After completion of the test any missing corrosion protection shall be complemented and the pile head structure be completed.

(6) Alternatively, if no connecting pile head structure is present at the time of testing, neighbouring piles may be employed as reaction piles via cross beams.

(7) The test load is recommended as 1,25-times the characteristic action F_k, however not more than at maximum 90% of the internal pile capacity (95% of the steel yield strength).

(8) If the testing force given in (7) cannot be unambiguously verified, a pile load test after Section 9 should be performed on the respective pile.

(9) Through testing of working piles as described in (1) to (8) evidence is gained whether comparable quality to that of load tested piles has been achieved. However, the test is not sufficient to verify the ultimate bearing capacity.

12 Pile Integrity Testing

12.1 Purpose and Procedures

(1) Quality assurance of piles is regulated in DIN EN 1536 for bored piles, in DIN EN 12699 for displacement piles and in DIN EN 14199 for micropiles. The provisions contained are predominantly relevant to the installation process. In contrast, integrity tests serve to control pile quality and geometry after pile installation.

(2) Common integrity testing methods are:

- Non-destructive "low strain" test (also known as: TNO method, impulse response method, impulse echo method): a pulse is induced to the pile head by a blow with a hand hammer and the movement of the pile head is recorded and analysed, see 12.2. The term "low strain" refers to the displacement of the pile head caused by the hammer blow, which is in the micrometre range.
- Non-destructive ultrasonic method for testing the concrete in the pile shaft: "cross-hole" method – measurement of transmitted waves between emitter and receiver probes moved in parallel, access tubes in cast-in-place concrete piles, or "single hole" ultrasonic logging with emitter and receiver probes in only one borehole or in a single access tube, see 12.3.
- Core drilling in the pile with core recovery and core testing and/or video borehole surveys or "single hole" tests, see 12.4.

(3) In the non-destructive "high strain" test the pulse is applied by a driving hammer blow to the pile head. The strains and accelerations at the pile head are measured and analysed. High strain refers to the high strain generated in the pile by the blow. This test can be used on all pile types. The test is particularly useful on coupled, driven precast concrete piles. The test is normally only performed in conjunction with a dynamic pile load test, see Section 10. Because this test does not represent an independent integrity testing method it will not be dealt with further in this section. Example graphs are shown in Annex C5.

(4) Excavating and cleaning a section of the pile, is possible or expedient as an integrity test in exceptional cases, as visual inspection and testing, e.g. of the concrete cover and quality can generally only be performed in the upper shaft region.

(5) Other specific testing methods are available for special questions, see 12.5.

(6) EC 7.9 and DIN EN 1536 use the term integrity test, without precise definitions or allocations. It can be assumed that low strain integrity testing is always referred to.

(7) The vibration tests mentioned in EC 7 (Cl. 7.9 (8)) are not defined as a method and are therefore not permitted in the scope of the EC 7-1 Handbook.

Recommendations on Piling – EA Pfähle.
Editors: German Geotechnical Society (*Deutsche Gesellschaft für Geotechnik e.V.*)

(8) Some integrity tests only cover limited areas of the pile (e.g. ultrasonic testing and coring). Because they are indirect methods, low strain and high strain integrity tests can only give an indication of quality deficiencies. Substantial quality deficiencies and, in particular, differences compared to neighbouring piles can, however, mostly be recognised.

(9) Performing and evaluating integrity tests require experience. They should therefore only be carried out by appropriately qualified persons.

12.2 Low Strain Integrity Tests

12.2.1 Low strain integrity test principles

(1) The impact wave required for low strain testing is generated by a blow, normally using a hand hammer. By monitoring the changes of the pile head movements with time, and identifying changes of movements as reflections, conclusions can be drawn about the non-visible, embedded section of the pile.

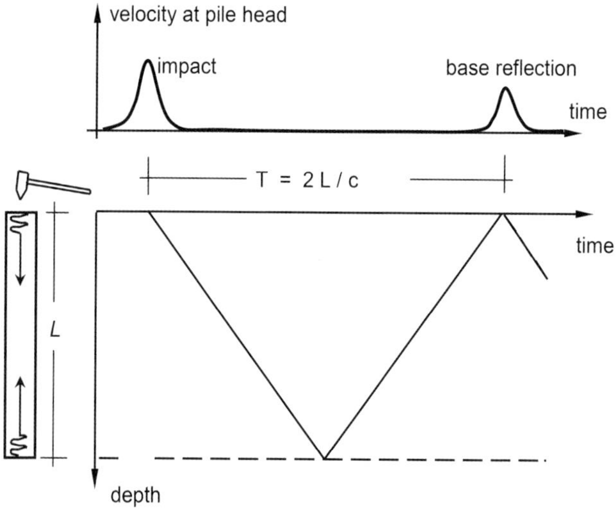

Figure 12.1 Monitoring of the pile head to identify reflections from depth

(2) The method is based on one-dimensional wave propagation [138], which describes the propagation of waves in bodies definable by a single coordinate, i.e. in which the length to thickness ratio is greater than 5. Indications of pile quality are given by the reflections of the impact wave introduced at the pile head from the pile base or from discontinuities and changes of the cross-section.

(3) If the reflections from the pile base are clear enough it is possible to state the wave velocity or the pile length, see 12.2.6.

12.2.2 Scope, number of tested piles and limitations

(1) The preferred application of this method comprises isolated, free-standing, cast-in-place concrete piles in loose soils and lengths between 5 m and 25 m. Then the compression wave generated by the pulse and the tensile wave generated by reflection at the pile base can travel the entire length of the pile and a clear signal from the base reflection can be registered at the pile head. Successful testing of longer piles depends on the individual circumstances (in particular the installation method and ground conditions).

(2) In accordance with the EC 7-1 Handbook and DIN EN 1536 the method is used to examine pile quality in cases of doubt, e.g. irregularities during pile installation, non-conformities with requirements of quality assurance, suspected deficiencies, uncertainties of existing pile foundations, etc.

(3) Integrity testing using a hand hammer allows economical and quick testing of a large number of piles per day and comparison of the results of a great number of piles tested in the same ground conditions. Any deviations from the typical signals for a given site can indicate possible defects of individual piles.

(4) The method is increasingly used as an element of quality management. The number of piles tested or the size of a random sample depends on the total number of installed piles.

(5) Of small pile foundations with up to 20 piles, all piles should be tested, where possible.

(6) For larger pile foundations containing 20 or more piles, at least 20 and thereafter 10 % of the remainder, or at least 3 piles of a given pile type, should be tested.

(7) The method should be used especially in cases where the failure of a single pile plays a decisive role in the stability of the structure.

(8) Pile testing below an existing structure requires the pile head to be exposed. It is therefore only possible on edge piles and requires a major expense.

(9) The method is only rarely successful for secant pile walls and diaphragm walls. At best, statistical conclusions can be drawn from a large number of measurements. Purposeful examination of an individual element is normally not possible. For length measurement, one of the methods outlined in 12.5 must be adopted. Defect detection is barely possible.

(10) The method cannot be used on micropiles and ground anchors.

(11) For driven steel piles and sheet piles, the use of a hand held hammer is normally not sufficient to generate a strong enough impact wave to receive a reflection from the pile base. The test can, however, be carried out as a high

strain test. Special methods as outlined in 12.5 can be employed to determine the length of sheet piles and soldier piles.

12.2.3 Pile preparation

(1) For integrity testing, the pile head must be freely accessible and cut-off to sound concrete. A smooth surface is advantageous. After cut-off, the pile head should not display any hairline cracks and shall be cleaned of any loose material. The surface of the pile head should be dry.

(2) At the time of the test the pile concrete must be sufficiently strong to accept the hammer blow without damage and allow the wave to propagate. A set-up time of seven days is generally sufficient.

12.2.4 Testing procedure

(1) The transducer shall be pressed down or fixed (glued) vertically and immovably to the pile head and parallel to the pile axis.

(2) The hammer shall be struck loosely from the wrist. No special technique is required. The signal on the screen must be checked after each blow in order to determine whether an impact wave has been initiated in the pile, or only a surface wave, or vibrations of the reinforcement have taken place.

(3) Normally, several signals per pile shall be recorded at different points on the flat surface of the pile head and be compared with each other. The number of recording and impact points depends on the pile diameter and the clarity of the measurement results.

(4) By changing the transducer points, damages in the immediate pile head area can be detected.

(5) If the piles being tested are embedded in the blinding concrete or similar, the measurement can be disturbed. The measurements can also be disturbed by protruding, vibrating, reinforcement.

(6) Measurement of piles below existing structures requires safe and dry excavation pits of at mans height. For reliable fixing of sensors and the hammer stroke, recesses (approx. 5 cm deep) shall be cut into the side of the pile shaft.

(7) In piles with a very large length to diameter ratio, it is possible that the signal from the pile base cannot be made sufficiently visible. Grinding the contact faces of the pile, increasing the weight of the hammer, or using softer hammers can improve the situation.

(8) When testing piles under an existing structure, reflections can come from the superstructure, making interpretation more difficult. Multi-channel recording can help in this situation (see 12.5.3).

(9) By additionally recording of the hammer impact force time history, extended evaluations are possible, e.g. deviations in the impedance near the pile head. Moreover, by means of a frequency transformation mobility and admittance curves can be derived, and the dynamic stiffness can be determined.

12.2.5 Measurement and instrumentation

(1) Different hammer weights with varying hammer head stiffnesses (varying pulse frequencies) can be used for the impact, to allow selecting the best suited hammer for the given situation.

(2) When the impact is made, the acceleration (or, more rarely, the velocity) is recorded as a function of time.

(3) The signals are digitised on a portable computer, processed and displayed on the screen, or stored and printed for documentation (see Figure 12.2).

(4) The measuring instruments should be able to register the signals with sufficient precision in a frequency range up to 10 000 Hz and acceleration amplitudes up to 1 000 g.

(5) Recording several signals and determining an average is recommended to eliminate high-frequency interferences.

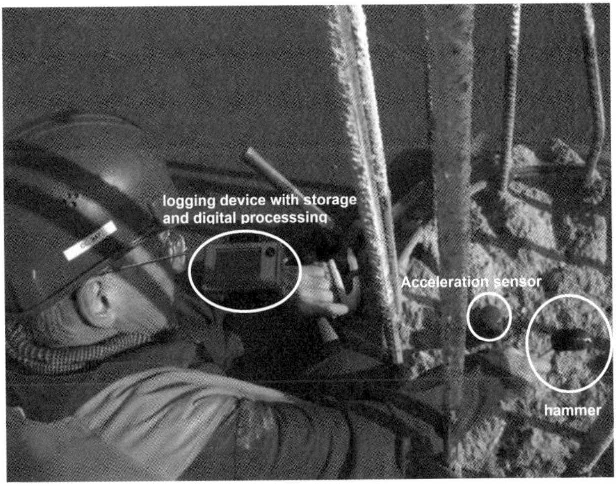

Figure 12.2 Low-strain testing

12.2.6 Evaluation of measurements

(1) The measurements and the evaluations of the impedance profile provide evidence on deviations from the planned pile shape, potential necking, cracks or changes in material properties, e.g. Young's modulus or strength.

(2) Because the method is based on one-dimensional pulse wave propagation, any conclusions on the location of a defect relate to the pile axis, i.e. the distance from the pile head. Statements on the extent of any defects perpendicular to the pile axis are possible to a very limited degree for the upper zones of large diameter piles.

(3) The recognition of the axial extent of a defect depends on the pulse duration. In order to recognise defects with a longitudinal extent of less than 50 cm especially small, hard hammers or other pulse inducing devices must be used.

(4) Contrary to the dynamic pile load testing methods (Section 10), in which forces and movements must be measured, the standard low strain integrity test only requires information on pile head movements (velocities).

(5) Normally, the velocity v(t) of the pile head, calculated from a measured acceleration, is evaluated, see Figure 12.1. It is also possible to directly evaluate the acceleration signal a(t).

(6) Conclusions on pile resistances as determined in a (dynamic) pile load test are not possible. In contrast to this, interfaces of ground layers in the ground can be determined to a certain degree, if these influence the cross-section of the pile.

(7) In the plot velocity vs. time, the time axis is directly linked to the pile axis via the wave velocity. The velocity is either drawn upwards, because it is regarded as a positive variable representing a movement in the direction of the hammer blow, or downwards to illustrate that the movement is downwards (see Figure 12.3).

(8) The transit time of the impact wave can be determined from the velocity profile, if the impact wave applied at the pile head reaches the pile base, is reflected and results in a measurable change in velocity at the pile head.

(9) For an assumed wave velocity, the pile length or the depth of a discontinuity can be directly read on the time axis at the points of reflections of the wave.

(10) The type and the extent of a discontinuity are recognisable on the ordinate by the signal deflection:

- Deviations from the planned pile cross-section can be recognised from the velocity-time history at the pile head.
- Deviations in the same direction as the input signal represent the reflection of a tensile wave, e.g. impedance reduction, Figure 12.3 left. Deviations in the opposite direction are reflections of a compressive wave, e.g. impedance increase, Figure 12.3 right.

(11) The method assumes that the Young's modulus and density of the concrete in a pile are subjected to only minor fluctuations. The changes in impedance are therefore in general assigned to changes of the cross-section. Any conclusions should be subjected to a plausibility check based on concrete quality data, pile construction documents and the results of ground investigations.

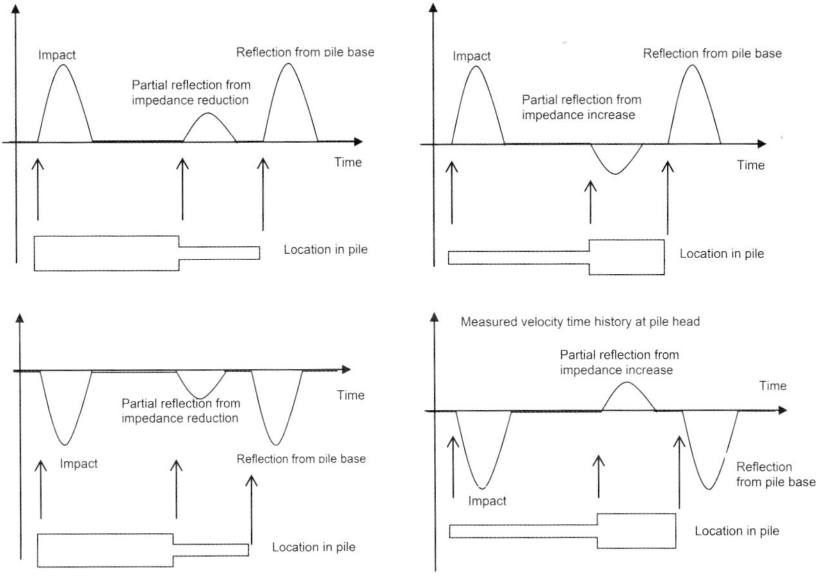

Figure 12.3 Impedance decrease (left); Impedance increase (right);
Top: Positive velocity drawn upwards;
Bottom: Positive velocity drawn downwards

(12) A signal seen to be pronounced in the majority of piles on a site can be regarded as characteristic for the local conditions and the prevalent pile-ground system. Special attention should be paid to deviations from that signal.

(13) The size of any identified discontinuities can only be determined to a limited degree. Quantitative determination of the deviation in the pile cross-section via the difference in the amplitudes of the induced and the reflected waves requires a realistic estimate of the damping in the ground and in the pile.

(14) Signal evaluation can be difficult or ambiguous if the reflection waves initiated by the pile shaft are superimposed on each other and on the reflections from the pile base.

Note: In these cases the test report shall point out that evaluation is not possible and no conclusions about quality can be drawn.

(15) The pile's impedance profile can also be determined using a signal matching procedure or on the basis of the wave theory by integrating the velocity time history. However, direct interpretation of the velocity profile should be given priority.

12.2.7 Impedance and wave velocity

(1) Any conclusions about pile quality refer to changes in the dynamic stiffness or impedance $E \cdot A/c$,

where

E Young's modulus of the pile material;
A cross-sectional area of the pile;
c the wave velocity of the pile material.

With $E = c^2 \cdot \rho$ the impedance implicitly includes also the density of the pile material.

(2) An impedance reduction is normally assigned to a reduction in cross-section or a considerable change in the structure of the concrete (honeycombed concrete, crack formation). An explanation is often found in the specific conditions of the site and by comparison to other piles (e.g. groundwater flow, changes of ground layers, interruptions of concrete placement, structures embedded in the ground). If such an explanation cannot be found, the identification of the reason for the reduction in impedance and whether it is caused by any defect can take time and be expensive, and requires destructive testing methods (coring or excavation).

(3) Given a clear base reflection, the wave velocity c is determined from the planned length L_{Plan} as an evaluation criterion:

$$c = \frac{2 \cdot L_{Plan}}{T_{gemessen}} \tag{12.1}$$

where:

L_{Plan} planned length of the pile;
$T_{measured}$ time for the reflected wave to reach the pile head.

(4) For simple, one-dimensional compression waves (also known as "waves in a rod"):

$$c = \sqrt{\frac{E_{dyn}}{\rho}} \quad [m/s] \tag{12.2}$$

where:

E_{dyn} Young's modulus of the pile material in kN/m^2;
ρ density of the pile material in t/m^3.

For detailed discussion of the influences on wave velocities in cast-in-place and prefabricated concrete piles reference is made to [113] and [127].

(5) Wave velocity is therefore dependent on the curing time and concrete quality. In addition to these material parameters it can also be influenced by

other effects (e.g. wavelength/diameter ratio, restraint in the ground, cracks, changes in wave characteristics caused by reflections, planned soil inclusions, e.g. in Atlas piles).

(6) In piles using common concrete types and strengths and tested after 7 days, the wave velocity c is normally between 3 500 m/s and 4 000 m/s. If the tested piles differ considerably in age, an increase in the wave velocity with time should be noticeable.

(7) If the wave velocity is less 3 500 m/s, it should be examined whether the piles are longer than planned, the pile heads have not yet been cut-off to the planned level or there are signs of reduced concrete quality.

(8) If the wave velocity is greater than 4 000 m/s, it should be examined whether the pile head has been cut-off below the planned level, if there are signs of enhanced concrete quality, or the piles are shorter than planned. Such deviations should be described in the report.

(9) When testing large diameter piles where the distance between the hammer blow and the sensor is greater than 20 cm, the time between the hammer contact and triggering the sensor must be taken into consideration. The time gap can be as long as several tenths of a millisecond. As a result, the wave velocity calculated as outlined in 12.1 can become too high, because the determined transit time was too short.

(10) The wave velocity limits given above should, where possible, always be examined on the basis of site-specific statistics (mean, maximum and minimum values), see Figure 12.4.

(11) Evaluation based on the criteria given above assumes that the true length of the pile corresponds to the planned length. However, this cannot always be

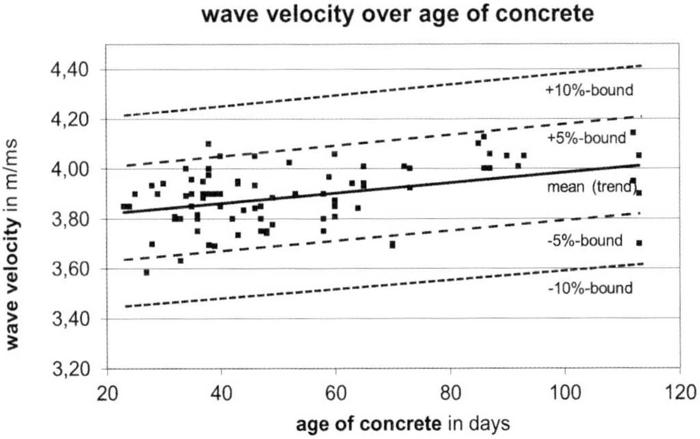

Figure 12.4 Variance in wave velocities

taken for granted, e.g. if there are doubts about execution, or in existing buildings without reliable documentation. Any deviation in the transit time can then be explained using Eq. (12.1) by either deviations in length or variations in the wave velocity. In these cases an estimated wave velocity shall be adopted (length error is then around 10%) or a calibration shall be performed based on additional information, or using complementary monitoring methods, see 12.5.

12.2.8 Assessment classes

(1) Because the velocity time history is influenced by a large number of factors and one-dimensional wave propagation according to the theory can only be recognised in very few cases, it is recommended to derive supposed pile characteristics only from clear and substantial signal characteristics.

(2) Differentiation between four result classes is recommended for the assessment:

Class A1: Measurement indicates no change in impedance.

The signal displays a clear base echo and there are no unexpected reductions in impedance in the signal along the pile axis. The impedance is as anticipated or reductions in impedance are within the tolerance bandwidth.

The wave velocity lies within the empirically anticipated range.

There are no quality-relevant anomalies.

Class A2: Measurement indicates no reduction in impedance.

Class A2 is given if the signal indicates impedance increases. The impedance is not as anticipated, but there are no reductions. The same applies if the signal displays pronounced deviation from the zero line due to the damping influence of longer piles.

The wave velocity lies within the empirically anticipated bandwidth.

There are no quality-reducing anomalies.

Note: Where the base reflection is entirely absorbed due to embedment in a solid ground layer or if it cannot reach the pile head due to high damping and great pile lengths, Class A2 can be given. It must be noted in a remark that any conclusions are therefore limited in validity or only apply to a limited depth range. A wave velocity cannot be given.

Class A3: The signal displays incomplete wave propagation due to an unexpectedly low reduction in impedance.

The magnitude of the reduction can be determined using a wave equation, if a clear base reflection is recognised, or can be estimated on an empirical basis, or compared to other piles at the same site. The impedance reduction may only amount to approximately a quarter of the anticipated impedance.

The anomalies are possibly relevant to quality.

Note: A quarter in this case is a rough figure and should not be equated with 25 %, but instead lies between 1/5 and 1/3.

Class A3 can also be allocated on the basis of a clear deviation of 5 %, up to a maximum of 10 %, in the wave velocity from the site mean. This shall be noted in a remark.

Class B: **Measurement indicates substantial defects in pile quality.**

The signal displays incomplete wave propagation due to an irregular and pronounced drop in impedance or an interrupted concrete column.

The magnitude of the reduction can be determined using a wave equation, if a clear base echo is recognised, or shall be estimated on the basis of experience. The reduction in the impedance is approximately two thirds of the planned impedance.

Note: Two thirds are not equated with 67 %, but instead lies between 1/2 and 5/6.

Class B can also be allocated on the basis of a clear deviation in wave velocity of 10 % or more from the site norm. This shall be noted in a remark.

Class 0: **The signal can not be evaluated.**

If it is necessary to summarise signals that cannot be evaluated in a single class, Class 0 should be allocated. However, the reason why evaluation is not possible should generally be given in a remark. Frequently the cause, e.g. poor concrete at the pile head or a crack in the first 50 cm immediately below the pile head, can be eradicated and a new test be performed.

(3) The examples summarised in Annex C4 for each of the classes are intended for clarification only and should not be adopted as "standard signals" for assessments.

12.2.9 Documentation and reporting

(1) During the testing, the data are stored electronically for subsequent evaluation and storage.

(2) The wave velocity adopted for the graphical representation of the velocity time history must be given in the test report, together with the reason for its adoption, if necessary.

(3) All data required to evaluate the measurements shall be contained in the test report:

- Site designation;
- Pile location;
- Pile number;
- Testing date;
- Pile installation date;
- Pile length;
- Installation method;
- Planned, installation-related changes in cross-section;
- Planned pile diameter;
- Ground profile;
- Concrete quality;
- Actual concrete volume for bored piles as given in piling log;
- Actual pile head diameter (if measured).

(4) If, for the purpose of evaluating the signals, a mean was formed from several suitable signals (accumulation), smoothing or low-pass filtering was performed, the time domain was extended or any other form of signal processing was implemented, this must be clearly documented.

(5) The test report should also include the assessment classes described in 12.2.8, if necessary with reasons or comments.

12.3 Ultrasonic Integrity Testing

12.3.1 Objective and scope

(1) Ultrasonic testing is a method of determining the homogeneity of the material of solid bodies. Any deviations in the homogeneity of the material are determined on the basis of deviations in the propagation velocity and/or the damping of ultrasonic waves. If a deviation in the propagation velocity is greater than 20% of the average value, it is characterised as an anomaly.

(2) The technical requirements for performing ultrasonic integrity tests, as well as the properties of the instrumentation, are internationally standardised, e.g. in NF P94-160-1 (France) or ASTM D6760-02 (USA). However, these regulations offer no guidance for evaluating results and the influence of any identified deviations. These Recommendations include descriptions and notes on assessing pile quality.

12.3.2 Ultrasonic integrity testing principles

(1) During an ultrasonic integrity test the transit time and signal strength of a mechanical wave in the ultrasonic range (generally 50 000 Hz) are measured using an emitter and a receiver. This wave generally runs horizontally through the pile concrete between two parallel access tubes (cross hole sonic testing) or

vertically in a single tube (single hole sonic testing). In special and rare cases inclined logging is carried out using emitters and receivers at different heights (fan shape testing).

(2) For cross hole sonic testing the emitter and receiver are placed in separate, parallel measuring tubes. The emitter and receiver are only placed above each other in the same measuring tube in the special case of single hole testing.

(3) The number of measuring tubes depends on the pile diameter. Normally, one measuring tube is installed per 25 cm pile diameter, e.g. four tubes for a 90 cm pile.

(4) For the special single hole ultrasonic testing case a single measuring tube is arranged centrally within the pile or the pile is cored for testing. With the aid of core recovery the latter option allows the results to be directly allocated to the concrete strength and quality. Supplementary borehole tests, e.g. video surveys or cylinder compression tests, can also be carried out, see 12.4.

(5) The internal diameter of the measuring tubes shall be between 38 mm and 50 mm, depending on probe diameter. A larger diameter means less precision for the determination of the measured length. If the diameter is too small there is a danger that the probe will jam. If in doubt, use of a blind probe prior to actual testing is recommended.

(6) In terms of their durability and robustness, and because of improved bonding with the surrounding concrete, steel measuring tubes are preferred to plastic ones. If plastic tubes are used they should be filled with water during concreting to prevent them losing strength as a result of the heat of hydration of the concrete. If they are filled with water after concreting they can lose their bond with the concrete.

(7) All connections must be watertight and without any reductions in the internal diameter

(8) Only that part of the pile cross-section located between the measuring tubes can be tested for integrity by the ultrasonic integrity testing. This represents between 50% and 70% of the cross-section, depending on the pile diameter and number of measuring tubes, their cover and their diameter. The ultrasonic integrity test is therefore ideally suited to large bored piles with diameters starting from around 90 cm.

(9) The section of the pile lying outside of the area covered by the measuring tubes is not included in the analysis. For instance, defects that can be recognised by low strain integrity testing, e.g. necking or imperfections at the shaft surface, can not necessarily be recognised by ultrasonic integrity testing.

(10) Certain types of defect, e.g. vertical cracks and defects near the pile head, are recognised better using the ultrasonic integrity test than the low strain method.

(11) Recognition of the longitudinal extent of a defect depends on the sampling rate (5 cm are usual distances of measurement steps) and the cross hole sonic logging is therefore more precise than the low strain method.

(12) In contrast to the low strain integrity test, for ultrasonic integrity tests no restricting limits exist regarding the testable pile length.

12.3.3 Measurement

(1) The measuring tubes must be filled with water to transmit the ultrasonic signal from the emitter into the concrete. Only longitudinal waves are transmitted through the water and the walls of the measuring tubes into the surrounding concrete.

(2) Emitter and receiver are lowered in parallel and constant rate to the end of the measuring tubes. The depth is continuously recorded via depth encoders. The ultrasonic signals from the emitter travel through the pile concrete, are amplified by the receiver, visualised and stored. The testing configuration and procedure are shown in Figure 12.5.

(3) Figure 12.6 shows testing results – on the left for an intact pile with a length of 7 m and on the right for a 32 m pile with a defect between 27 m and

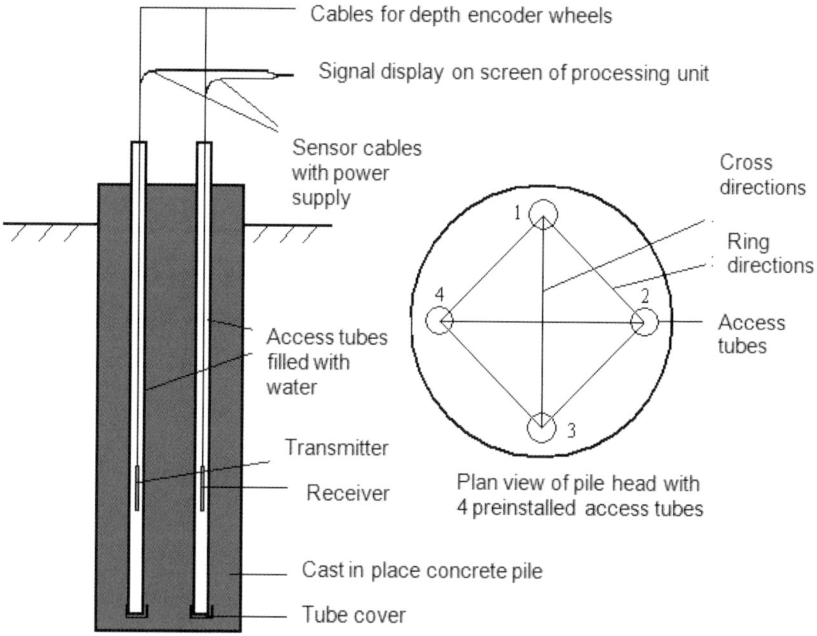

Figure 12.5 Ultrasonic integrity testing principles

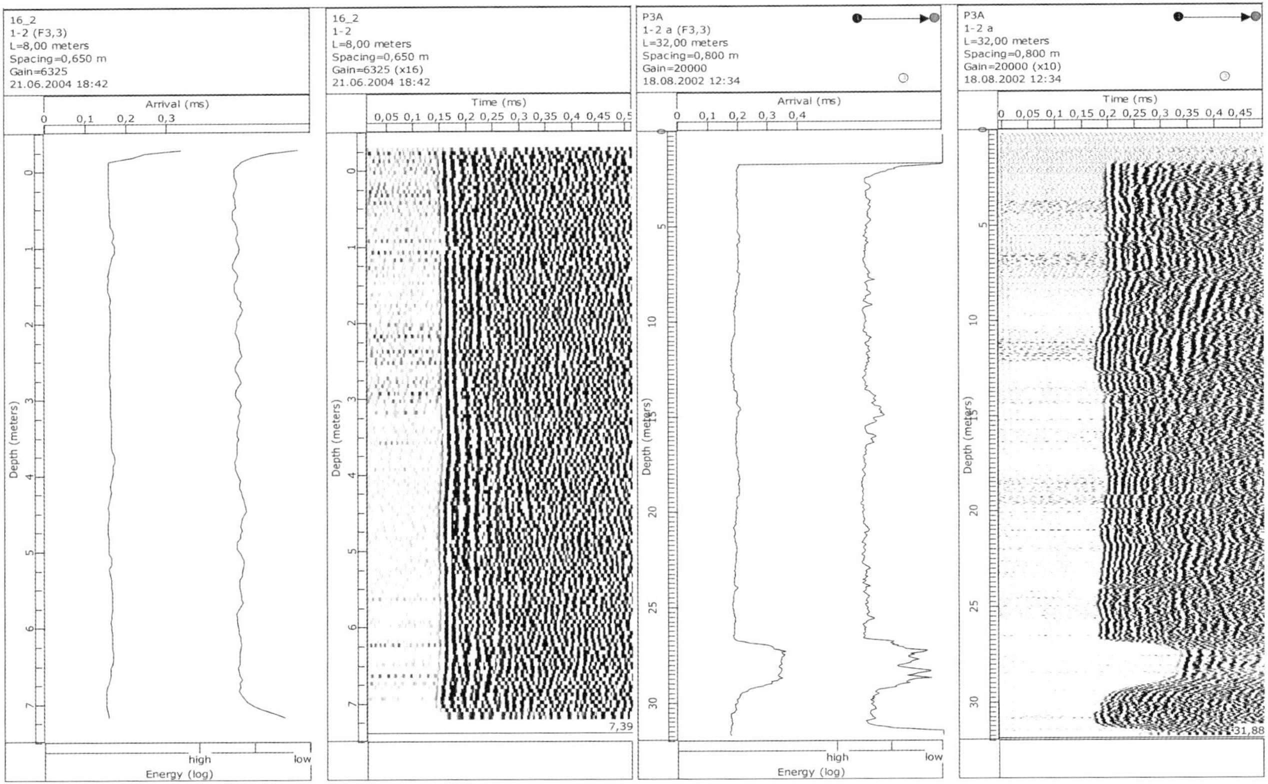

Figure 12.6 Examples of visualisation of results for an ultrasonic integrity test

29 m. The arrival times of the ultrasonic waves are calculated by the testing device, together with the energy of the received signal, and drawn as lines over the depth for each measuring direction, see Figure 12.6, left sides respectively. In addition, the respective profile of the received signal is shown on the right, also for each measured direction over depth. The profiles are shown as dotted lines for clarity, whereby each positive value is allocated a light dot and each negative value a black dot. If the signals are the same with depth and reproducible, parallel gray and black lines result (waterfall diagram or sonic map).

(4) In homogeneous concrete the velocity of an ultrasonic wave is constant.

(5) The primary objective of ultrasonic integrity testing is not to determine the absolute velocity of the ultrasonic wave, but instead to verify the uniformity of the wave velocity through the entire pile. Around ground inclusions, honeycombs or other heterogeneities in the concrete the propagation velocity falls off compared to areas of sound pile concrete, which is expressed by longer transit times, drop of energy and irregularities in the waterfall diagram.

12.3.4 Test preparation and testing procedure

12.3.4.1 Test piles

(1) It is not necessary to install separate test piles to perform ultrasonic integrity tests. In principle, all structural piles equipped with measuring tubes can be integrity-tested by ultrasonic methods.

(2) It is advisable to always equip a certain proportion of the piles at a given site (e.g. 25 %) with measuring tubes if ultrasonic integrity tests are planned for quality assurance. Whether or not all, or only some, and in that case which of the equipped piles are subjected to ultrasonic testing, can subsequently be specified e.g. by the client's representative, the engineer, the geotechnical expert or the geotechnical designer.

(3) In special cases, it can also be necessary to test piles not fitted with measuring tubes. At least one borehole (single hole testing) shall then be drilled through the pile to allow the test to be carried out, although two or three boreholes are better.

(4) The pile concrete should have reached a minimum strength at the time of testing. In contrast to the low strain integrity test, however, the strength can be relatively low. Piles can be tested in this way as early as two or three days after installation. However, longer times should be aimed for, because the concrete strength has then more and more evenly developed and, consequently, the signals will be clearer.

(5) If plastic measuring tubes are used and the waiting time is more than 10 days, it is possible that the measuring tubes detach from the concrete, which can severely impair the testing results.

12.3.4.2 Testing procedure

(1) All possible measuring directions shall be tested, e.g. three tubes – three measuring directions, four tubes – six measuring directions, five tubes – ten measuring directions, etc.

(2) Before commencing the proper measurement with the ultrasound emitter and receiver, it is advisable to check the accessibility of the measuring tubes and their depths with a suitable blind probe, and to record the positions of the tubes.

(3) The complete filling of the measuring tubes with water shall be ensured before testing.

(4) Emitter and receiver are moved parallel and at a sufficiently low speed of around 0.25 m/s within the measuring tubes. The direction of movement can be either from top to bottom or from bottom to top.

(5) The signal density with depth is given by the speed of movement and the frequency of the emitted ultrasonic pulses. The objective is to record a reproducible signal for each measuring run, e.g., every 5 cm depth. The pile can thus be considered divided accordingly into 5 cm thick test sections.

12.3.5 Evaluation

12.3.5.1 Qualitative signal evaluation

(1) Uniform concrete quality can be seen in the signal through uniform accelerations and identical arrival times, see Figure 12.6.

(2) Different explanations are possible for any anomalies recognised in the signal. It is clear from the example signal profiles shown in Figure 12.7 that it is not possible to draw an unequivocal conclusion on the type of defect from the measured arrival times. A longer transit time of the ultrasonic wave between the measuring tubes can be caused by concrete with a minor quality reduction along the whole measuring run or by concrete with more pronounced quality reduction along a shorter distance of the run, or even by larger measuring tube spacing. If local segregation occurs in the concrete the ultrasonic wave can travel around the defect, which also results in delayed arrival time.

(3) Additional information, as listed below, is required to precisely identify each individual anomaly.

- In order to rule out the existence of voids around the measuring tube connections, which can influence the signal, detailed drawings showing the locations of the connecting pieces must be available. The formation of such voids around an otherwise intact measuring tube is highly improbable, but is possible around connections.

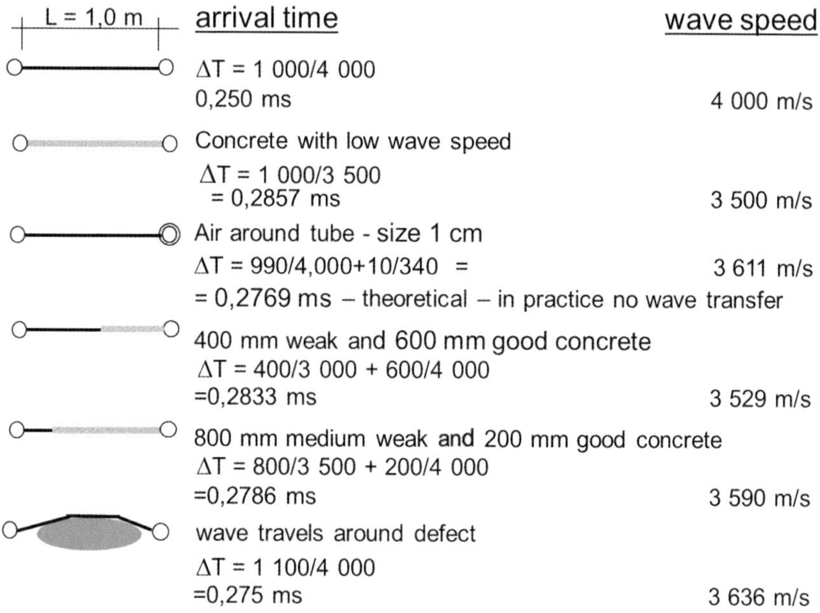

Figure 12.7 Examples of ultrasonic wave propagation, arrival time and wave velocity

- The pile installation records should be consulted to help delineate any possible defects.
- Complementary investigations can be performed using coring or low strain integrity testing to provide information on the existence and extent of a defect.

(4) Information on the shape and properties of a possible defect can be gained by comparing the received signals and their energy level. It should be possible to differentiate between wave passage through a measuring distance consisting of uniform concrete and wave passage through a section containing lower quality concrete or wave travel around a honeycomb. However, the complexity of the phenomena of spherical propagation of mechanical waves in a solid body, and in particular in a comparatively heterogeneous material like concrete, often prevents unique interpretation of the received signal. No model has yet been developed for precise characterisation of defects identified by the form of the ultrasonic signal. Only the energy level of the recorded signals can be adopted as an indicator. Low transferred energy is normally associated with low wave velocities.

(5) A defect in the pile concrete can therefore be assumed if both the energy and the wave velocity fall off in the affected cross-section.

12.3.5.2 Quantitative signal analysis

(1) The following procedure can be used to analyse ultrasonic integrity tests:

a) Determination of travel times (wave velocities) over depth and of the average transit time.
b) Deviations up to 20 % from the average wave velocities can normally be ignored, because they do not indicate significant anomalies.
c) If a deviation in the wave velocity of more than 20 % occurs in a cross-section, the energy and the signal shapes (waterfall diagram) must be examined in detail. If an anomaly is confirmed, all average values and maximum transit times are determined for all possible measuring runs at this depth.
d) Calculation of a "quality indicator" η_Q with weighted wave velocities in this cross-section with Eq. (12.2).

$$\eta_Q = \frac{\sum(\chi_{ij} \cdot a_{ij})}{\sum a_{ij}} \tag{12.2}$$

where:

$a_{i,j}$ distance between measuring tubes i and j;
$\chi_{ij} = c_{ij}/ca_{ij}$ ratio of wave velocity in the defect to the mean wave velocity for the path i–j;
c_{ij} wave velocity for the path i–j in the anomalous cross-section;
ca_{ij} mean wave velocity in the sound concrete for the path i–j.

The quality indicator η_Q can be adopted as calibration factor for the concrete quality in the cross-section under consideration and can be used to estimate a reduction of the internal pile capacity at this level, provided this procedure is covered by the analysis concept.

(2) The extent and effect of a suspected defect can be estimated by adopting a lower wave velocity bound for the defective area, e.g. the wave velocity of air. The size of the defective area can then be determined from the magnitude of the deviation in the wave velocities. For a measuring run a–b of length L_{ab} with the transit time T_{ab} the extent of the intact concrete L_1 or the defective area L_2 is given by adopting the wave velocity c_{L1} for intact concrete determined by ultrasonic testing and the lower wave velocity bound c_{L2} acquired by evaluating the linear equation system:

Geometry: $$L_1 + L_2 = L_{ab} \tag{12.3}$$

Transit time: $$L_1/c_{L1} + L_2/c_{L2} = T_{ab} \tag{12.4}$$

The possible size of the defective area also depends on the wave velocity selected for the defective area.

(3) A different evaluation approach utilises computer tomography algorithms. The analysis is performed as the inverse of the usual flow analysis using finite elements. With the aid of the known wave velocities, in turn, a quality indicator can be determined to evaluate the pile from an engineering perspective. The wave velocity distribution can be smoothed to help visualise the tomography analysis results. The result is often visualised using a false colour diagram. Further descriptions of tomography analysis are provided e.g. in [52].

(4) The determination of a quality indicator as proposed in (1) does not represent a strictly scientific derivation, but an engineering evaluation of the results of an ultrasonic integrity test.

(5) When applying the quality indicator the complexity of the actual relationships and the measuring task should be taken into consideration. The pile evaluation should be checked in conjunction with other available data relating to the pile condition, e.g. construction records.

12.3.5.3 Pile evaluation

(1) When evaluating the piles and when detecting anomalies, it shall be considered on how many measuring levels, where and to which extent anomalies are detected. In particular when small measuring steps are used the extent along the pile axis can be a governing aspect of pile evaluation.

(2) If the quality indicator is used as the reason for assuming reduced compressive concrete strength, this should be checked on cores.

12.3.6 Documentation and report

(1) All details of the ultrasonic integrity test shall be compiled in a test report. It should at least include the following data:

1. General information
   - time and place of the test; location of tested piles on the site
   - site peculiarities
   - ground conditions
2. Information on tested piles
   - pile type and number
   - installation date, concrete quality
   - dimensions (diameter, length), profile, reinforcement
   - construction method
3. Test system
   - equipment and software used
4. Test execution
   - length of measuring tubes and projection
   - test date and weather conditions
   - Sketch of measuring tube locations with reference direction (true north landmark) and spacing of tubes

- test records (transit times, energy and waterfall diagram for each measured direction)
- reasons for any deviations from standard test procedure

5. Results
   - tabled summary of results for each pile
   - where appropriate, information on the quality indicator η_Q with corresponding depth
   - where appropriate, information on the extent of defects

(2) Annex C6 includes an ultrasonic testing case study.

12.3.7 Special situations: testing secant pile walls and diaphragm walls

(1) The reinforced secondary piles of a secant pile wall and individual panels of a diaphragm wall can be tested by ultrasonic methods, analogous to the procedures described above.

(2) In individual cases, testing can also be performed between the opposing access tubes of neighbouring secondary piles. The test encompasses the intermediate primary pile and the joints between the piles.

(3) However, conclusions on the quality and size of the joints, and the bonding between the piles are only possible to a very limited degree.

12.4 Testing Piles by Core Drilling

12.4.1 Introduction

(1) Coring and tests on recovered cores, and on and within the borehole, can be carried out for the following reasons:

a) To verify the concrete compressive strength and/or the durability in the structure, e.g. in cases:
   - where the properties required by DIN EN 206/EC2 are not achieved by the samples taken for quality control, or;
   - the number of samples is insufficient.
b) To verify the quality and continuity of the pile shaft's concrete column.
c) To verify the quality of execution of a construction joint, see DIN EN 1536.
d) To verify a clean contact area and/or the material of the load-bearing stratum, e.g. for end-bearing piles on rock.
e) To provide access holes for subsequent:
   - single hole ultrasonic testing of the pile shaft using a combined probe with emitter and receiver;
   - other downhole tests, e.g. impression packer test.

12.4.2 Coring

(1) Core diameters should be as large as possible (≥ 70 mm) and cautious boring methods (e.g. double core barrels) should be used in order to keep core disturbances caused by boring small and to achieve the best possible complete core recovery.

(2) The bore should be executed using heavy equipment, stable drill rods and long core barrels in order to achieve good directional stability. Where possible the bore should be placed eccentrically within the pile to avoid the track of the tremie pipe produced during the final withdrawal of the tremie from the fresh concrete. Concrete segregation is possible within this zone.

Note: Limited segregation in the centre of the pile does not normally represent a quality deficiency.

(3) If only certain sections of the pile shaft need testing, the use of empty conduits (access tubes) sealed at the top and bottom is recommended. They are installed with the reinforcement cage and end at least 1,0 m above the zone to be investigated.

(4) The bore depth, the depth of breaking of the respective core and the length of recovered core shall be recorded for each core run.

12.4.3 Analysis

12.4.3.1 Introduction

(1) The evaluation of recognised zones of quality deficiencies and their analysis in terms of bearing capacity and durability of the pile requires specialist knowledge and experience and should be done by an expert.

(2) If quality deficiencies are recognised at the pile head or at shallow depths, deeper cut-off to sound concrete, formation of a construction joint to Eurocode EC 2 and concreting back to the design pile head level is generally the most economical solution to achieve suitable pile quality.

(3) Quality deficiencies at the pile base can normally not be rectified. Durability and settlement considerations are generally the governing assessment factors.

Note: Accumulations of coarse aggregate and reduced cement paste in the immediate base areas of bored piles are often unavoidable. They are side effects of the concrete properties and the concreting method Similar to minor inclusions of foreign matter, they are normally of little relevance to the performance of the pile.

12.4.3.2 Visual evaluation

(1) The cores recovered shall be laid out for analysis, fitted together at the fractures, where possible, and shall be measured. The cored length (run) and the core recovery shall be recorded.

Note: For depths, reference should be made to plumb measurements in the bore.

(2) The core samples shall be assessed in terms of:

- Appearance and composition of the concrete: distribution of aggregates and fines in the matrix;
- Porosity and density, where appropriate following prior wetting and drying;
- Apparent strength (e.g. using light hammer blows, sound and rebound);
- Honeycombing and segregations.

Note 1: Large pores in cast-in-place concrete piles are also normal and harmless due to the specific characteristics and concrete placement conditions. Only major areas of connected pores can indicate quality deficiencies.

Note 2: If the reinforcement is encountered or an interface penetrated (e.g. pile base) core destruction is possible due to the differing material strengths. Core assessment in these zones must be undertaken with particular care.

(3) The evaluations shall be recorded in bore logs analogous to DIN 4022-1 (DIN EN ISO 14689-1) and the cores shall be stored in robust and permanently labelled core boxes in accordance with the provisions of DIN 4021. Where samples have been taken for further investigation, the appropriate lengths shall be marked, for example by placing labelled distance pieces.

(4) Zones of lesser quality, washing out, honeycomb formation, severe segregation, inclusions of foreign matter or core loss shall de delineated and assessed in terms of their probable distribution across the pile cross-section.

Note: In such cases complementary investigations can be useful, e.g. dynamic or ultrasonic integrity tests.

12.4.4 Concrete strength and durability

(1) The specimens shall be taken from core pieces characteristic of the pile concrete and the integrity of the pile.

(2) Preparation, storage and investigation of compressive strength follow the provisions of DIN 1048-2: Testing concrete; testing of hardened concrete (specimens taken in situ).

(3) In further analysis regarding the internal pile capacity the partial factor γ_M may be reduced, if appropriate.

Note: The reduction should be performed by a concrete engineering expert.

(4) For the assessment of the durability and the corrosion protection of the reinforcement the location and extent of any suspected zone of lesser quality, the exposure class and the depth below the groundwater table must be taken into consideration.

12.4.5 Downhole tests

(1) Ultrasonic single hole testing can help to delineate the size of zones of recognised, lesser concrete quality or inclusions.

(2) Camera surveys or impression packer tests can be useful for assessment if cores do not provide reliable information on the continuity of the concrete column, zones with honeycomb formations or the quality of the concrete/ground interface at the pile base.

12.5 Other Specific Testing Methods

12.5.1 Introduction

(1) The testing methods described below do not form part of the usual scope of testing for quality control of piles, but can be used for special purposes.

12.5.2 Radiometric pile tests

(1) Radiometric pile testing utilises a γ-γ (gamma-gamma) backscatter log with an active source to determine the density around bore holes or access tubes embedded in the pile. It is used on bored piles.

(2) Foreign matter inclusions in the concrete around the probe can be detected reductions in density. The measuring range is a few centimetres only (approx. 10 cm max.).

(3) Radiometric pile testing, in contrast, e.g. to the ultrasonic method (12.3), can encompasses areas outside of the reinforcement cage.

(4) Because of the use of a radioactive source and the associated equipment and health and safety requirements and certifications, the method is only used in a few countries and even then not universally. No applications are known to date in Germany.

12.5.3 Multi-channel low strain testing

(1) The multi-channel low strain test (also referred to as "ultraseismic") is a variant of the dynamic pile testing method described in 12.2 using several sensors installed above each other near the head of the pile shaft.

(2) By analysing the pulse transit time to the various sensors, the wave velocity in the pile can be determined, allowing e.g. the length measurements to be calibrated.

(3) Where piles are parts of buildings or structures, signals from the superstructure (above the impact point) can be differentiated from pile reflections, based on the inverse arrival sequence at the sensors.

(4) The method requires access to the pile shaft along a length corresponding to at least half of the wavelength of the applied pulse (guide value: 2 m).

(5) Measurement and analysis effort is greater than for the dynamic integrity testing described in 12.2 using the low strain method.

12.5.4 Parallel seismic method

(1) The parallel seismic method is based on applying a pulse by a hammer blow to the pile head and recording the wave transit times to sensors (geophones or hydrophones) in a borehole parallel to the pile.

(2) The method allows the determination of length of all types and can also be used on pile walls, diaphragm walls and sheet pile walls, as well as foundation elements below existing structures.

(3) To facilitate analysis the measured transit times (first arrivals) are drawn in a graph against sensor depth. The location of the base of the pile is conventionally determined by the location of the break point in the transit time graph, which can lead to a considerable overestimation of the pile length, depending on the pile/borehole spacing. Accordingly, correction values or, better, model-based adaptations of the transit time graph are necessary, in which the wave velocities in the pile and the ground are incorporated. The method does not require calibration.

(4) Ground stratification, changes in pile cross-section or pile/borehole inclination all influence the result.

(5) The pile/borehole spacing should be 1 m maximum. When using model-based analyses, 2 m to 3 m are also possible.

(6) The borehole must be several metres longer than the anticipated pile length. The greater the pile/borehole spacing and the smaller the pile/ground wave velocity contrast, the greater is the required extra depth (guide value: 3 m to 5 m). The borehole casing must have a good bond to the ground. In saturated, unconsolidated materials backfilling the annulus with free-flowing material is sufficient (e.g. filter gravel). In dry, stable materials, grouting can be necessary. The casing material should be PE or PVC.

(7) Studying the entire wave information in the seismogram (visualisation of all sensor signals on top of each other) can provide additional information.

(8) The monitoring installation can be readily adopted for seismic downhole measurement for ground investigation.

12.5.5 Induction and mise-a-la-masse methods

(1) The induction and mise-a-la-masse methods are based on the introduction of an alternating or direct current into the reinforcement of a bored pile (steel pile, sheet pile) and detection of the resulting electromagnetic or electrical field by a sensor in a parallel borehole.

(2) Analysis is performed by drawing the strength of the sensor signal over the sensor depth. The signal strength falls off rapidly below the bottom end of the reinforcement.

(3) The method is only suitable for reinforced piles (with access to the reinforcement!) or steel piles/sheet piles. The length of the reinforcement in partially reinforced bored piles can be determined using this method.

(4) The pile to borehole spacing should not exceed 0,5 m. The borehole casing must be either PE or PVC. A galvanic contact to the surrounding soil is required for the mise-a-la-masse method, e.g. a slotted casing.

(5) Ground stratification, groundwater and nearby metal objects influence the signal.

12.5.6 Other borehole-based methods

(1) Other methods that can be used in boreholes sunk adjacent to the pile are available for special purposes.

(2) All of these methods require a non-metallic casing and a maximum distance from the pile of less than 0,5 m.

(3) The length of steel piles and sheet piles (and sometimes the reinforcement of concrete piles) can be determined using magnetic field probes or active electromagnetic conductivity probes, even without access to the pile.

(4) In principle, the shape and length of piles can also be determined using downhole radar. However, the method does not work in cohesive soils.

13 Bearing Capacity and Analyses of Piles under Cyclic, Dynamic and Impact Actions

13.1 Introduction

(1) Guidance on classifying pile foundations subjected to cyclic, dynamic and impact actions in accordance with the EC 7-1 Handbook [44] is provided in 4.1 and 5.9.

(2) The "substantial" actions on piles discussed in this Section 13, refer to cyclic, dynamic and impact actions in the sense of 13.2 and the EC 7-1 Handbook [44], A 2.4.2.1, A(8b).

(3) In the context of pile foundations, those actions are considered as substantial and are to be dealt with after this section, which influence the pile performance as a result of decrease of shear strength, deformation or porewater accumulation.

(4) Piles subjected to substantial cyclic, dynamic and impact actions in the sense of the EC 7-1 Handbook [44], A 2.4.2.1, A(8b) are to be classified as Geotechnical Category 3.

(5) If piles are subjected to substantial cyclic, dynamic and impact actions, depending on the boundary conditions heavily modified pile behaviour compared to static or "common" variable actions resulting from "quasi-static" loads in the sense of the EC 7-1 Handbook [44], A 2.4.2.1, A(8a) is to be expected. The EC 7-1 Handbook [44] requires that this behaviour be taken into consideration in the analysis and design of pile foundations.

(6) To date, knowledge about the load-bearing behaviour of piles subject to dynamic or impact actions is still limited, in particular concerning cyclic actions. Specific literature comprises more or less well documented results of model tests and pile load tests, with cyclic actions up to around 10^6 load cycles. An overview of the literature can be found e.g. in [98], [59], [19], [74], [141] and [142]. The results of axial, cyclic load tests are often visualised in interaction diagrams, which enable to identify failure situations.

(7) In practice, it is virtually impossible to meet the EC 7-1 Handbook [44] requirement of pile load tests for modelling realistic loading situations, in particular regarding the number of load cycles. However, even a small number of load cycles in a pile load test can provide valuable information on the different pile behaviour compared to a static load, see e.g. [53]. On the other hand, [135] reports that during axial alternating load testing of micropiles in non-cohesive soils, after completion of a few thousand cycles, the pile's displacement behaviour deteriorated disproportionately after only a few additional cycles.

Recommendations on Piling – EA Pfähle.
Editors: German Geotechnical Society (*Deutsche Gesellschaft für Geotechnik e.V.*)

13.2 Cyclic, Dynamic and Impact Actions

13.2.1 Action and loading types

(1) The actions and effects on piles are in general dealt with in 4.1.

(2) For variable actions the pile loads:

- $F_{Q,rep}$ as representative, axial actions;
- $H_{Q,rep}$ as representative actions lateral to the pile axis;
- $M_{Q,rep}$ as moments resulting from representative actions;

shall, in accordance with the EC 7-1 Handbook [44], A 2.4.2.1, A(8b), 4.1 (1) and 4.2 (4),be supplemented by the "substantial" actions from:

- cyclic loads;
- dynamic loads and;
- impact loads;

e.g. $F_{Q,rep,cyc}$, $F_{Q,rep,dyn}$ or $F_{Q,rep,imp}$, abbreviated to F_{cyc}, F_{dyn} and $F_{imp,}$. Figure 13.1 schematically demonstrates pile deflection behaviour.

(3) A clear demarcation between the terms cyclic, dynamic and impact is difficult, because of the smooth transitions. Figure 13.2 shows examples and categories of typical cyclic, dynamic and impact actions.

In the context of pile foundations the following definitions should be used:

- Cyclic actions: refers to repeated actions on the pile, where inertial forces of the pile-soil system do not need to be taken into consideration, also see 13.2.2.
- Dynamic actions: refers to repeated actions on the pile, where inertial forces of the pile-soil system need to be taken into consideration, also see 13.2.3.

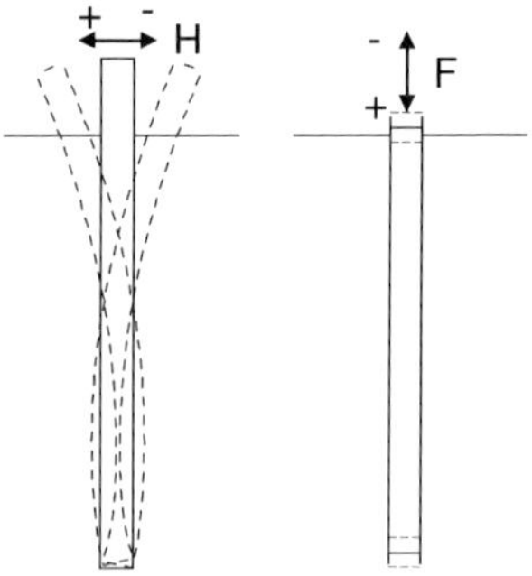

Figure 13.1 General deflection behaviour of piles subjected to cyclic, dynamic and impact actions

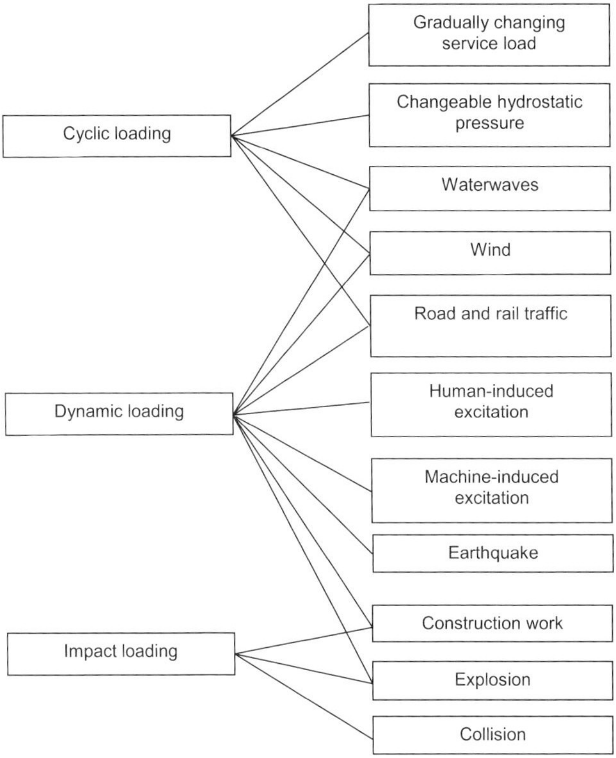

Figure 13.2 Examples of cyclic, dynamic and impact actions on pile foundations

Note: Pile foundations subjected to seismic actions are not dealt with in these Recommendations.

- Impact actions: refers to unique actions acting on the piles over a short period only. It can be necessary to take the inertial forces of the pile-soil system into consideration, also see 13.2.4.

Further definitions and relationships relating to cyclic, dynamic and impact actions on pile foundations are given in the following sections.

13.2.2 Actions from cyclic loads

(1) With regard to cyclic loads, the mean load component F_{mean} or H_{mean} and the cyclic load component F_{cyc} or H_{cyc} are differentiated, see Figures 13.3 and 13.4. The cyclic load component is also referred to as the load range. Half of the load range F'_{cyc} or H'_{cyc} defines the load amplitude.

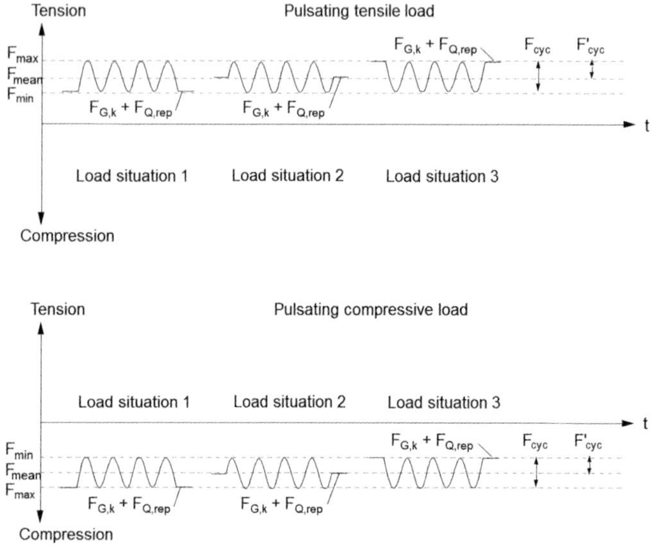

$F_{G,k}$ Permanent action

$F_{Q,rep}$ Representative changeable action

F_{mean} Mean action

F'_{cyc} Cyclic load amplitude

F_{min} Smallest absolute action

F_{cyc} Cyclic load range

F_{max} Largest absolute action

Figure 13.3 Load and action situations for cyclically loaded piles under cyclic loads; example showing harmonic load cycles, after [142]

Note: It is pointed out that F_{cyc}, H_{cyc}, F'_{cyc}, H'_{cyc}, F_{mean}, F_{mean} and F_{max} as shown in Figures 13.3 and 13.4 represent characteristic or representative actions.

(2) The load range for cyclic loading is the change between the smallest (F_{min} or H_{min}) and highest (F_{max} or H_{max}) absolute load. For an alternating load the load range is the change between the largest tensile load and the largest compressive load. For cyclically loaded piles loaded perpendicular to the pile axis, Figures 13.3 and 13.4 apply accordingly.

(3) In cyclically, axially loaded piles, the mean load level X_{mean} gives the ratio of the mean load component relative to the compressive or tensile resistance of the pile in the ultimate limit state under permanent actions imposed by cyclic loadingPiles loaded perpendicular to the pile axis are dealt with in (5) below. The cyclic load level X_{cyc} gives the ratio of the cyclic load amplitude relative

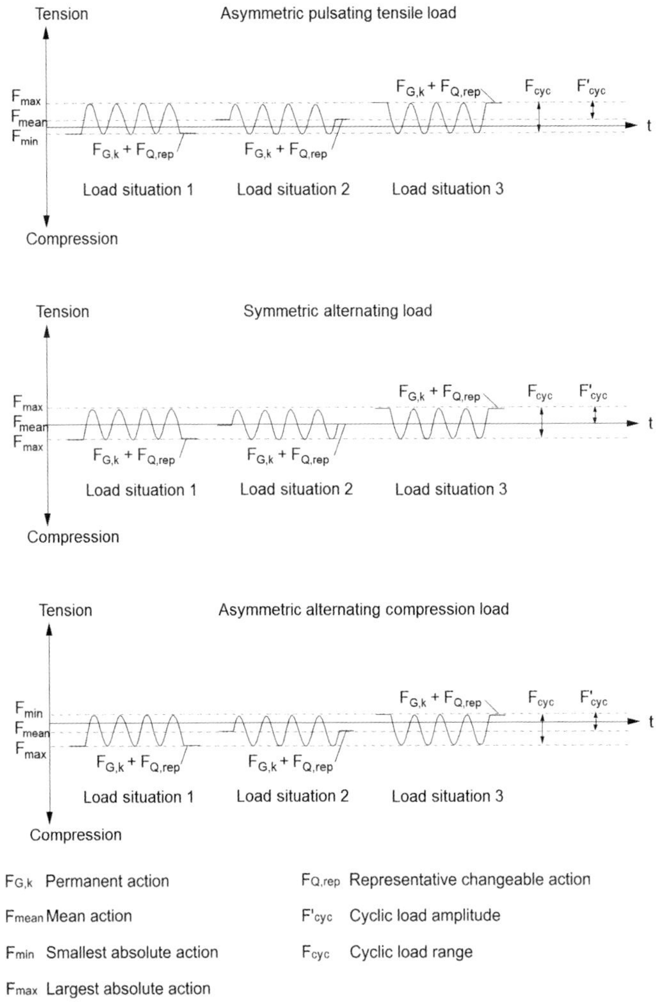

Figure 13.4 Load and action situations for cyclically loaded piles under alternating loads; example showing harmonic load cycles, after [142]

to the compressive or tensile resistance of the pile in the ultimate limit state under permanent actions.

$$X_{mean} = F_{mean} / R_{ult} \quad \text{or} \quad X_{mean} = H_{mean} / R_{ult} \tag{13.1}$$

$$X_{cyc} = F'_{cyc} / R_{ult} \quad \text{or} \quad X_{cyc} = H'_{cyc} / R_{ult} \tag{13.2}$$

(4) The reference variable $R_{ult} = R_g$ adopted in Equations (13.1) and (13.2) as the pile resistance in the ultimate limit state or at failure under static actions depends on the method adopted to determine the reference variable (static or dynamic pile load tests or empirical data). This follows the provision in the EC 7-1 handbook [44] for the use of different correlation and model factors, and partial factors for determining R_{ult}. For use of Equations (13.1) and (13.2), comparable reference variable R_{ult} are necessary. R_{ult} shall in approximation be adopted as $R_{ult} = R_{(c,m)mean}$ or $R_{ult} = R_{(t,m)mean}$ when the pile resistances are determined from pile load tests and as $R_{ult} = R_{c,k}$ or $R_{ult} = R_{t,k}$ when they are determined from empirical data, e.g. after 5.4.

(5) For flexible, long piles, loaded cyclically and perpendicular to the axis after EC 7-1 Handbook [44], 7.7.1 A(3a), it is still difficult to present data on the pile's resistances in the ultimate limit state and they would not be without ambiguity, also see 5.8. At present, equations (13.1) and (13.2) therefore only apply to piles subjected to cyclic axial loads and to short piles, cyclically loaded perpendicular to the pile axis.

(6) The pile load and action situations shown in Figures 13.3 and 13.4 aim to define the principal terms and represent the ideal case in terms of a harmonic, cyclic pile load. In practice however various cyclic load situations can act on piles, for example differing heights and sequences for wave loads. In addition, interruptions in the cyclic actions on a pile foundation are conceivable. It is therefore possible, to an extent, that cyclic pile loads can be transient actions, also see 13.2.3 (4) and (5).

(7) If different load ranges need to be taken into consideration, it may be assumed in approximation that the accumulated pile displacement is independent of the load sequence. A cyclic equivalent load, consisting of an equivalent effect $(F_{mean}, F_{cyc})_{eq}$ and a corresponding equivalent number of load cycles N_{eq}, may be derived. See 13.4 for details of possible procedures for calculating equivalent load cycle numbers.

13.2.3 Actions from dynamic loads

(1) Dynamic loads can be harmonic, periodic or transient functions of time. As such, they are deterministic, i.e. unique in physical terms.

(2) Stochastic actions on pile foundations are characterised by random fluctuations with time. They can often be related to equivalent deterministic actions.

(3) Harmonic loads as shown in Figure 13.5 can be described by Eq. (13.3).

$$F(t) = F'_{dyn} \cdot \cos\left(2 \cdot \pi \cdot \frac{t - t_0}{T}\right). \qquad (13.3)$$

where:

T duration of the period;
F'_{dyn} amplitude.

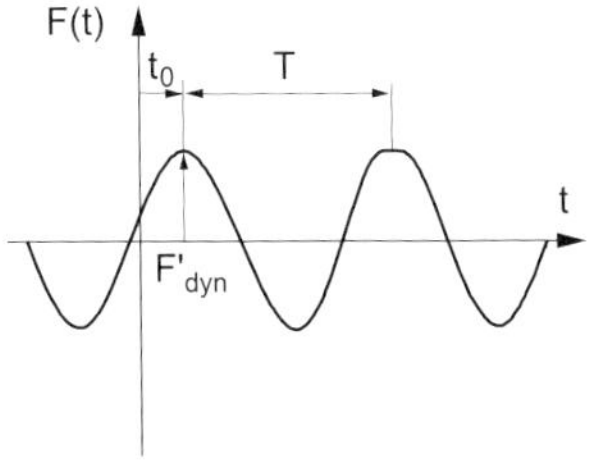

Figure 13.5 Harmonic load

(4) Periodic loading can be described as a Fourier series progression, i.e. as the sum of harmonic loads of different frequencies. In practice, it is often sufficient to adopt a few terms of the series. Examples of periodic loads include mass forces imposed by machines or water pressure imposed by waves.

(5) Transient loads are non-periodic processes of limited duration, e.g. live traffic loads. Using Fast Fourier Transformation (FFT), these can also be converted to the frequency domain.

(6) Whether the load on the piles is cyclic or dynamic and inertial forces are negligible for approximation or must be taken into consideration, can be assessed e.g. using Eq. (13.4). Inertial effects may be ignored if the inertia term in Eq. (13.4) is less than 10% of the elastic or elasto-plastic stiffness of a pile foundation found using Eq. (13.4).

$$M \cdot \omega^2 < 0{,}1 \cdot K . \tag{13.4}$$

$$\omega = 2 \cdot \pi / T . \tag{13.5}$$

where:

ω angular frequency of the dynamic load;
M effective mass of the pile foundation in t;
K linear-elastic stiffness or secant stiffness of the pile foundation in kN/m.

For single piles, the pile mass over the elastic length can be adopted as the effective mass. In pile groups the mass of the ground influenced by the piles must be taken into consideration.

13.2.4 Actions from impact loads

(1) Impact loads refer to short-duration actions on piles, see Figure 13.6. The action duration can be anywhere from milliseconds to a few seconds. Respective actions can be e.g. braking loads (brake jolting), collision and explosion loads, etc., also see DIN EN 1991-1-7.

(2) The actions on the piles can be given as an impact load function with the maximum value F_{dyn} or F'_{dyn}.

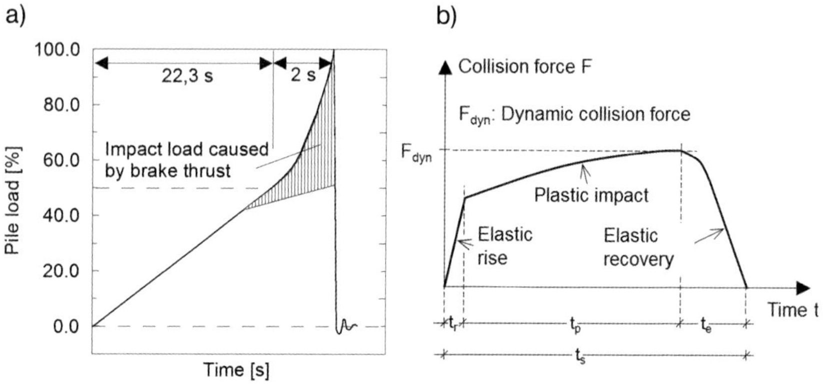

Figure 13.6 Examples of impact loads on piles; a) as result of braking loads on piled railway bridge piers, b) load-time profile for ship impact for a plastic ship response (based on DIN EN 1991-1-7)

13.3 Supplementary Geotechnical Investigations

(1) See EC 7-2 Handbook [45] and 3.2 for guidance on the fundamental requirements of geotechnical investigations on pile foundations.

(2) The objectives of the supplementary geotechnical investigations is the determination of the soil behaviour resulting from cyclic and dynamic loads and the selection of suitable investigation methods capable of suitably modelling the variable effects, as well as to derive characteristic soil parameters.

(3) The type and extent of the supplementary investigations, the required sample quality and the soil parameters to be derived shall be specified in each individual case by the geotechnical expert, taking into consideration the effects on the pile foundation and the proposed analysis model.

(4) When planning the supplementary investigations and selecting suitable investigation methods, the anticipated load amplitudes, the mean and cyclic load levels, the load direction, the number of cycles and the analysis methods after 13.7 shall be considered. If insufficient information is available at the time of the investigation, at least a sufficient number of special samples shall be retained from the relevant soil types for the purpose of subsequent supplementary laboratory testing.

(5) Common investigation methods for soils subjected to cyclic, dynamic or impact loads, separated into field and laboratory tests, are summarised in Table 13.1, together with guidance on the soil behaviour that can be determined with the respective methods. In cases where normative regulations exist for corresponding static laboratory tests, these regulations can also be transferred to the respective cyclic tests.

Table 13.1 Investigation methods for soils subjected to cyclic, dynamic and impact effects

Test suitability: ++ very good + good o limited – unsuitable	Investigated parameter					
Investigation methods	Porewater pressure development	Increase/decrease of shear strength	Volume change	Minor distortions	Material damping	Wave velocities
Laboratory testing						
Cyclic triaxial test (undrained)	++	+	–	o	o	–
Cyclic triaxial test (drained)	–	+	++	o	o	–
Cyclic simple shear test (constant volume)	++	+	–	o	o	–
Cyclic simple shear test (drained)	–	+	++	o	o	–
Cyclic box shear test (drained)	–	+	+	o	o	–
Cyclic box shear test (drained) under CNS (constant normal stiffness) conditions	–	+	+	o	–	–
Cyclic uniaxial compression test with lateral constraint (oedometer test)	–	o	+	–	o	–
Cyclic torsion test (undrained)	+	+	–	o	o	–
Cyclic torsion test (drained)	–	+	+	o	o	–
Resonant column test	–	+	o	++	++	++
Bender (piezo) element test, tension-compression (piezo) element test	–	–	–	+	–	++
Field investigations						
Shallow geophysics (reflection/refraction seismics, harmonic resonator)	–	–	–	+	–	+
Downhole geophysics (cross-hole, down/up-hole)	–	–	–	+	–	+
Seismic piezocone (SCPTU)	+	–	–	+	–	+
Cyclic dilatometer test	–	+	–	–	o	–

(6) An overview of general soil behaviour under cyclic and dynamic loads, as well as explanations of the investigation methods given in Table 13.1 can be found e.g. in [150], [140] and [126]. The DGGT's Recommendations of the

Ground Dynamics Committee [18] contain empirical data and correlations for dynamic soil parameters.

13.4 Bearing Behaviour and Resistances under Cyclic Loads

13.4.1 Introduction

(1) "Substantial" cyclic pile loads in accordance with 4.1 (1) and 13.1 (2) can lead to completely altered pile performance. It should be noted that for soils, contrary to other materials, few major actions are more relevant than numerous minor actions. The loading sequence may be ignored, in approximation, if there is sufficient distance to the ultimate limit state and porewater pressure developments are irrelevant.

(2) The different performance of cyclically loaded piles is characterised by the increase (accumulation) of plastic axial displacements $s_{cyc}(N)$ or horizontal displacements $y_{cyc}(N)$ of the pile with the number of load cycles N and by the different the post-cyclic pile resistance $R_{ult}(N)$. The influences of these two phenomena are relevant for the stability analyses after 13.7 and 13.8.

(3) The magnitude of the plastic displacements and any change of the pile behaviour after N numbers of load cycles depend on the load parameters, soil parameters and pile characteristics. Guidance on the general cyclic pile behaviour and on determining characteristic pile bearing capacity are given in 13.4.2 and 13.4.3, for the two principle load directions.

13.4.2 Axial loads

(1) A summarising evaluation of the the numerous cyclic axial pile tests documented in the literature has shown that the pile capacity can be greatly altered by cyclic loading, if the cyclic load range F_{cyc} after EC 7-1 Handbook [44], A 2.4.2.1, A(8b) is greater than 20% or the cyclic load amplitude F'_{cyc} is greater than 10% of the characteristic static pile resistance R_{ult} in the ultimate limit state or the failure state. See 13.2.2 (4) for details of adopting R_{ult}.

Note: Previously, a somewhat more conservative boundary was considered relevant, e.g. in DIN 1054:2005-01 or [59].

(2) If these boundary conditions are prevalent, after EC 7-1 Handbook [44] cyclic loading must be taken into consideration in the ultimate and serviceability limit state pile analyses, also see 13.7.

(3) The pile performance under "substantial" cyclic load components, when comparing with the pile performance under static loads or loads treated as "quasi-static" (curve 1 in Figure 13.7a), is generally characterised by increased deflections $s_{cyc}(N)$ in the serviceability limit state, and by decreased (curve 2) or, where applicable, increased (curve 3) compressive or tensile resistances of the pile in the ultimate limit state $R_{ult}(N)$.

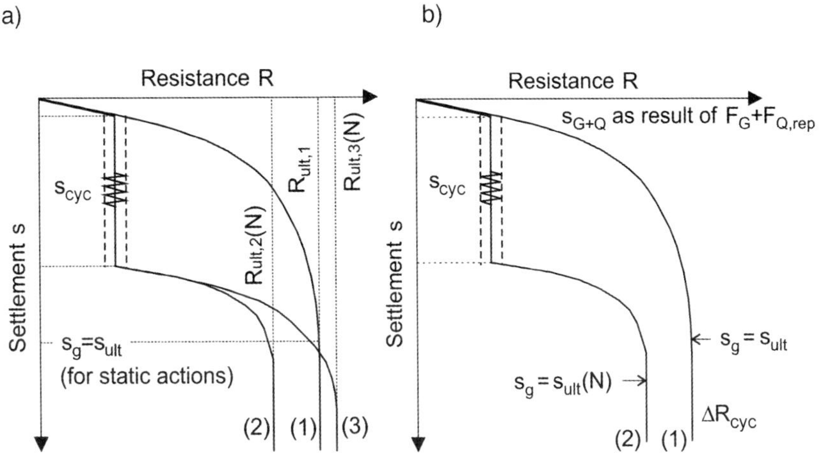

Figure 13.7 a) Examples of resistance-settlement curves without cyclic loading (1) and after cyclic loading (post-cyclic loading) (2) and (3); b) determining ΔR_{cyc}

(4) If an increase in the post-cyclic pile behaviour resulting from cyclic actions is identified or forecast (see curve 3 of Figure 13.7), pile performance corresponding curve 1 shall conservatively be adopted for analysis of the pile foundation in the ultimate limit state after 13.7.1.

(5) If post-cyclic pile behaviour is identified or forecast similar to curve 2 of Figure 13.7a and Figure 13.7b, the respective altered behaviour (ΔR_{cyc} = cyclic capacity reduction) can be considered to be predominantly the result of a reduction in shaft resistance. It can generally be assumed that the pile base resistance is not changed by cyclic loading.

(6) Determination of the reduction in capacity as a result of cyclic loading and verification of the 20% criterion for the load range or the 10% criterion for the load amplitude after (1) may be limited to the shaft resistance component, unless a reduction in the pile base resistance was determined by means of suitably instrumented cyclic pile load tests.

(7) When comparing the resistance-settlement curves of statically loaded piles to post-cyclic loaded piles, it is generally anticipated that the limit settlement s_g is different for the two load types, see Figure 13.7b. If no other criteria are relevant, the limit settlement of cyclically loaded piles can be assessed as

$$s_g = s_{ult}(N) \leq 0{,}1D + s_{cyc} \qquad (13.6)$$

(8) Piles under symmetrical or asymmetrical alternating cyclic loads with a low mean load level normally display smaller plastic displacements in the first load cycles than piles subjected to cyclic ("swelling") loads at similar cyclic load levels. In these cases extrapolating the displacements is not permissible,

because a disproportionate increase in plastic displacements at higher cycle numbers can occur, in conjunction with pile failure. Piles subjected to alternating loads normally display lower post-cyclic pile capacities. Depending on the number of load cycles and the cyclic load range, and on the drainage conditions, porewater pressure can accumulate, which, in extreme cases, can lead to a reduction in effective stresses, resulting in liquefaction of the ground.

(9) Cyclic actions can lead to changes of the density of the soil and to rearrangement of the soil's grains in the vicinity of the pile, especially in non-cohesive soils, which influence the pile performance [60]. In non-cohesive soils normally more unfavourable cyclic pile behaviour than in overconsolidated, cohesive or mixed-grained soils should be anticipated ([4] and [143]).

(10) Given the current state of knowledge, the analysis methods available for describing pile performance under cyclic actions after (1) require that a single-stage load spectrum is specified from $F'_{cyc,eq}$ or $F_{cyc,eq}$, $F_{mean,eq}$ and N_{eq} that is equivalent to the true effect, see Figure 13.8 and 13.2.2 (6). A possible method is described in Annex D4.1.1 and a corresponding example given in Annex D4.2.1.

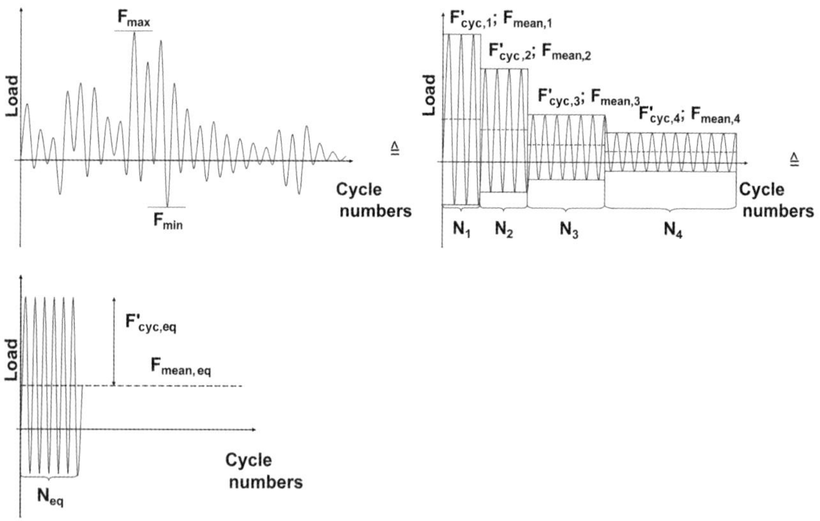

Figure 13.8 Example for deriving an equivalent single-stage load spectrum from the true effect history on the piles

(11) The characteristic displacement behaviour of the pile after N load cycles in the serviceability limit state (SLS) should be derived from cyclic pile load tests. If this is not possible, or not possible to a sufficient extent, also see 13.1

(7), the plastic displacements $s_{cyc}(N)$ may be assessed based on analysis models calibrated against tests.

(12) The change in the characteristic compressive or tensile resistance of the pile in the ultimate limit state (ULS) after N load cycles $R_{ult}(N)$ should be derived from post-cyclic, static pile load tests. This refers to load tests in which the test is brought to the ultimate limit state under a static load, immediately after the requisite number of load cycles was applied. If this is not possible, analogous to (11), analysis models calibrated against tests may also be adopted for the pile resistance $R_{ult}(N)$. The characteristic pile resistance $R_{ult}(N)$ may be determined independent of settlement, i.e. $s_{ult} \neq s_{ult}(N)$, see Figure 13.7b.

(13) A selection of analysis models used to describe pile behaviour after (11) and (12) is provided in Annex D2, along with analysis examples.

(14) The resistance of the pile under alternating loads must be taken into consideration for both directions, because $R_{c,ult}(N) \neq R_{t,ult}(N)$.

13.4.3 Lateral loads

(1) Pile bearing behaviour under "substantial" cyclic load components lateral to the pile axis after the EC 7-1 Handbook [44] is, compared to "usual" variable or "quasi-static" actions, normally characterised by a change in the soil properties or the subgrade reaction. In analogy to axially loaded piles, this leads to the accumulation of pile deflections with load cycles, see Figure 13.9. In addition, a reduction in pile capacity can occur, e.g. if excess porewater pressures are generated by cyclic effects.

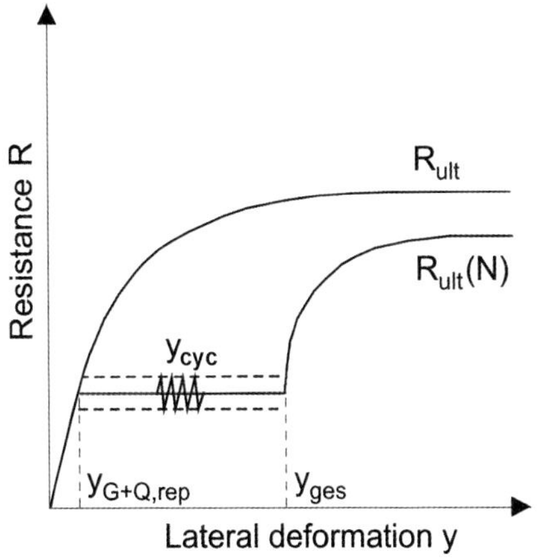

Figure 13.9 Resistance-deflection curves for static lateral loads with and without cyclic preloading

(2) In laterally loaded piles, cyclic loads generally lead to larger accumulated deformations than symmetrical alternating loads, assuming a similar load level.

Note: According to investigations documented in [74], an asymmetrical alternating load can generate considerably larger cyclic deformations than a cyclic load. After [20] considerably increased deformations can result from a minor change in the load direction.

(3) A number of the analysis methods available for describing pile behaviour under cyclic lateral loads require that a single-stage load spectrum is specified equivalent to the true effect history $H_{cyc,eq}(N_{eq})$, analogous to Figure 13.8 and 13.2.2 (7). One possible method is described in Annex D4 together with an analysis example.

(4) The characteristic deflection behaviour of the pile after N load cycles should be derived from cyclic pile load tests. If this is not possible, or not possible to a sufficient extent, also see 13.1 (7), a numerical estimate of pile deflections can be made for cyclic loads using analysis methods allowing a forecast of deflections based on cyclic laboratory tests. Empirical approaches can also be adopted for the estimate, if similar boundary conditions apply for the tests on which the approaches are based, i.e. pile properties, pile installation, soil and loading parameters.

(5) A numerical estimate of the bearing behaviour of cyclically, laterally loaded piles can be based on p-y curves, for example, see e.g. [75]. The shape of the p-y curves and the magnitude of the limit value of the acceptable subgrade stress shall be determined as function of depth and stress and in dependence of the cyclic effect and the porewater pressure accumulation in the ground. These values shall be determined by adopting the results of cyclic element tests or shall be based on well-established empirical data acquired under comparable conditions. The characteristic p-y curves adopted to analyse cyclically, laterally loaded piles must be confirmed by the geotechnical expert.

(6) Guidance and approaches for reducing the subgrade stresses and passive earth pressures resulting from cyclic loads can e.g. be found in DIN EN ISO 19902, [2] and [19].

(7) A selection of analysis models used to describe pile bearing behaviour under cyclic, lateral loads can be found in Annex D3, along with calculation examples.

(8) Numerical methods may be employed for more reliable analyses if system behaviour under cyclic loads is realistically described. These analysis models should also be calibrated against comparable pile load tests.

(9) For an initial estimate of the pile length under cyclic loads the embedment depth of a flexible pile can also be assessed by studying the pile deflection curve. Stable pile behaviour under cyclic loads may be assumed if the dis-

placement of the pile base remains small and the pile deflection curve is only slightly influenced by a calculative increase in the pile length.

(10) Not withstanding (9), more accurate investigations of the magnitude of the anticipated deformations shall be performed and the influences on the action effects under cyclic loads be considered.

(11) For short, practically rigid piles in noise abatement walls along railway lines, an indirect serviceability analysis derived from test results is described in [119]. It bases on an analysis of increased global factors or reduced utilisation factors.

13.5 Bearing Behaviour and Resistances under Dynamic Loads

(1) A dynamic pile load in accordance with 13.2.3 can lead to heavily changed pile bearing behaviour, also see 13.1 (5).

(2) Inertial and damping forces become effective in the pile resistance under dynamic loads. The resistance can either increase or reduce. This effect is particularly pronounced in pile groups. It is caused by the dynamic pile-soil-pile interaction between the piles and the enclosed soil mass, and by the influence of wave interferences in the ground.

(3) Guidance on the vibration behaviour and analysis methods of single piles and pile groups can be taken from [18] and [152].

(4) The soil parameters governing the analysis (stiffness modulus, shear modulus, material damping) depend on the magnitude of the dynamic shear strains. They are normally small compared to the static strains. The soil stiffnesses adopted for dynamic analysis are therefore generally considerably larger than the static soil stiffnesses. In contrast, a fundamental differentiation between static and dynamic parameters, often encountered in more dated literature, is not expedient.

(5) In saturated, non-cohesive soils, an increase in porewater pressure caused by the dynamic action can lead to a decrease in the ground resistance up to failure due to soil liquefaction.

(6) Basically, the same considerations apply with regard to bearing behaviour as for a cyclic load. If the dynamic load range is large enough that shear effects are produced in the ground which result in structural modifications in the soil matrix, an accumulation of plastic displacements and a decrease in the characteristic compressive or tensile pile resistance R_{ult} must be anticipated. Guidance on the limit values of shear effects (shear strains) can e.g. be taken from [151].

13.6 Bearing Behaviour and Resistances under Impact Loads

13.6.1 Introduction

(1) An impact pile load as described in 13.2.4 can lead to heavily modified pile performance, also see 13.1 (5).

(2) Inertial and damping forces become effective in the pile resistance under impact loads. In general terms, the resistance is increased compared to a static pile load.

(3) Normally a direct dynamic analysis of the entire structure consisting of foundation and superstructure is necessary. In certain cases also a quasi-static investigation of the pile foundation can be performed, in which a static equivalent load is determined by means of a dynamic load factor or a static equivalent force to DIN EN 1991-1-7 (or φ_{dyn} to DIN 1055-9 and technical reports).

(4) Under impact loads a pile foundation can behave both elasto-dynamically and plastically, whereby one of the two components often dominates depending on the intensity and direction of action of the load.

(5) At low intensity no significant plastic displacements are anticipated (e.g. machine foundations with operational impact loads). The pile foundation can be regarded as being under a high-frequency dynamic load, see 13.5. The elastic soil parameters can be adopted in the same way, as function of the shear strain.

(6) At high intensities, on the other hand, significant plastic displacements are anticipated (e.g. vehicle collisions). Here, the strength of the ground is the decisive factor for the magnitude of the plastic displacements.

(7) The elastic soil parameters and the material damping governing a dynamic analysis depend on the magnitude of the dynamic shear strains. A fundamental differentiation between static and dynamic parameters, often encountered in more dated literature, is not expedient.

(8) In the case of saturated soils the porewater causes radiation damping, which in turn causes an effective increase in stiffness for short-term loads.

13.6.2 Axial loads

(1) For axially loaded piles under impact loads, until further notice analogous procedures as for laterally loaded piles may be adopted, see 13.6.3.

13.6.3 Lateral loads

(1) Previously it was common practice and permissible for bored piles to DIN 4014:1990-03 when subject to horizontal impact actions from collision loads, not to carry out further analyses but to increase the horizontal subgrade moduli by up to a factor of 3. Recent investigations [73] have shown that under

impact loads in dry, poorly-graded sand actions resulting from mass inertia forces and effects of soil structure redistributions can occur leading to softer pile behaviour, in particular in the pile head region.

(2) Under certain boundary conditions and impact loads the ground around the pile reacts stiffer than under static effects, although this can be compensated for by the increased loads imposed by the inertial forces and by soil structure redistributions.

(3) In saturated soils the porewater causes soil damping and a numerical increase in stiffness for loads of short durations.

(4) Taking these effects into consideration and in cases where static equivalent loads after 13.6.1 (3) are adopted, the same subgrade reaction modulus may, in simplification, be adopted for impact actions in dry and naturally moist soils as for static actions. In saturated soils, in simplification three times the static subgrade reaction modulus may be adopted. This approach is allowed only, if the impact duration is short or the interval between multiple impacts is sufficient to allow excess porewater pressures to dissipate.

(5) At high load intensities and therefore significant plastic effects, the maximum deformations and pile foundation action effects should be determined in a more precise non-linear, dynamic analysis, not least for reasons of economy. E.g. Sliding bodies can be adopted and their inertial forces taken into consideration.

13.7 Stability Analyses of Cyclic, Axially Loaded Piles

13.7.1 Analysis of the bearing capacity of an isolated pile

(1) The procedure for analysing the "external" capacity (load transfer to the ground) of a cyclically, axially loaded, single pile for the governing design situations in the ultimate limit state based on 6.2.1 (1), is as follows:

a) The characteristic, axial actions F_k at the pile head are determined as foundation loads imposed by the superstructure on the proposed system. The system consists of the selected pile type and dimensions, the pile head and superstructure and, where applicable, any geotechnical actions (e.g. negative skin friction), distinguishing between permanent and variable actions.
b) Taking cyclic axial actions F'_{cyc} into consideration, the design values of the actions are given by the characteristic actions on the piles using:

$$F_d = F_{max,d} = F_{G,k} \cdot \gamma_G + F_{Q,rep} \cdot \gamma_Q + \lambda \cdot F'_{cyc} \cdot \gamma_Q \qquad (13.7)$$

where:

λ valued at 0, 1 or 2, such that the largest absolute action in accordance with Figures 13.3 and 13.4 results and using the corresponding partial factors γ_G and γ_Q in accordance with EC 7-1 Handbook [44] or Annex A2. In approximation, it is assumed that load situations 1 to 3 idealised in Figures 13.3 and 13.4 cause the same pile bearing behaviour.

c) The design values of the pile resistances after (2) are given by the pile resistances subjected to cyclic actions $R_{c,ult}$ and $R_{t,ult}$, $R_{c,ult}(N)$ and $R_{t,ult}(N)$ determined after 13.4.2 2.

(2) When determining the design values of the pile resistances, taking the cyclic actions into consideration, the proceeding is as follows, whereby it is assumed that the static, dynamic or cyclic pile load tests are performed on-site under comparable conditions, e.g. similar ground stratification, pile dimensions, cyclic effects, etc.:

a) If cyclic pile load tests are performed:

$$R_{c,d}(N) = R_{c,k}(N)/\gamma_t \quad \text{for compression pile resistance} \tag{13.8a}$$

and

$$R_{t,d}(N) = R_{t,k}(N)/\gamma_{s,t} \quad \text{for tension pile resistance} \tag{13.8b}$$

where:

$$R_{c,k}(N) = R_{ult}(N)/\xi_i = R_{c,m}(N)/\xi_i \tag{13.8c}$$

and

$$R_{t,k}(N) = R_{ult}(N)/\xi_i = R_{t,m}(N)/\xi_i \tag{13.8d}$$

If comparable conditions for the cyclically loaded test piles and the structural piles as stated above are not prevalent or the cyclic pile load tests and the working piles are on different sites, e.g. cyclic pile load tests onshore and working piles offshore but in similar ground conditions, the cyclic pile load tests may be used for site-related adjustment of the analysis method to determine ΔR_{cyc}. Supplementary dynamic or static pile load tests are recommended, in addition to the cyclic tests on the test field.

b) If $R_{c,ult}$ or $R_{t,ult}$ are determined from static or dynamic pile load tests:

$$R_{c,d}(N) = R_{c,k}/\gamma_t - \eta_{cyc} \cdot \Delta R_{cyc} \tag{13.9a}$$

and

$$R_{t,d}(N) = R_{t,k}/\gamma_{s,t} - \eta_{cyc} \cdot \Delta R_{cyc} \tag{13.9b}$$

where:

$$R_{c,k} = R_{c,ult}/\xi_i = R_{c,m}/\xi_i \tag{13.9c}$$

and

$$R_{t,k} = R_{t,ult}/\xi_i = R_{t,m}/\xi_i \tag{13.9d}$$

where ΔR_{cyc} is estimated with the aid of analysis methods, see e.g. Annex D.

c) If $R_{c,k}$ or $R_{t,k}$ are derived from empirical data, e.g. after 5.4:

$$R_{c,d}(N) = R_{c,k}/\gamma_t - \eta_{cyc} \cdot \Delta R_{cyc} \quad (13.10a)$$

and

$$R_{t,d}(N) = R_{t,k}/\gamma_{s,t} - \eta_{cyc} \cdot \Delta R_{cyc} \quad (13.9b)$$

where ΔR_{cyc} is estimated with the aid of analysis methods, see e.g. Annex D.

(3) The correlation factors ξ_i are determined in accordance with the EC 7-1 Handbook [44] and Annex A4. They shall be determined separately and depending on the number of respective tests, i.e. for cyclic, static and dynamic pile load tests. The same correlation factors as for static pile load tests may be employed for cyclic pile load tests, see Annex A4.1. The provisions of 5.3 (9) shall be observed when executing both static and dynamic pile load tests on site. The partial factors γ_t or $\gamma_{s,t}$ are given by the EC 7-1 Handbook [44], Table A2.3 or Annex A3.2. The partial factor for compression piles describes both the base and the shaft resistance.

(4) If the analysis method to determine ΔR_{cyc} after Annex D is selected in accordance with (2b)) and (2c)) above was subjected to site-related adjustment based on cyclic pile load tests in accordance with (2a)), a model factor $\eta_{cyc} = 1$ may be adopted. This also applies if comparable cyclic pile load tests were performed, where the comparability is confirmed by the geotechnical expert. A model factor $\eta_{cyc} > 1$ shall be specified if approximation methods are used which are not adjusted by cyclic pile loading tests, see e.g. Annex D2. When adopting any method other than those given in Annex D2 the model factor must be confirmed by the geotechnical designer or the geotechnical expert.

(5) Using the determined axial design actions and resistances, the limit state condition Equation (13.11a) or (13.11b) must be fulfilled for each governing action combination in accordance with the EC 7-1 Handbook [44] or 6.2 (2).

$$F_{c,d} \leq R_{c,d}(N) \quad \text{for compression pile resistances} \quad (13.11a)$$

$$F_{t,d} \leq R_{t,d}(N) \quad \text{for tension pile resistances} \quad (13.11b)$$

For analysis, the maximum compressive force $F_{c,d}$ and the maximum tensile force $F_{t,d}$ given by the load and action combinations in Figures 13.3 and 13.4 are the governing factors. If there is no harmonic load, the extreme values of the effect profile F_{max} and F_{min} govern $F_{c,d}$ and $F_{t,d}$, see Figure 13.8. Hereby the pile resistances $R_{c,d}(N)$ and $R_{t,d}(N)$ are derived regardless of this, taking an equivalent single-stage load spectrum after 13.4.2 into consideration.

(6) If the limit state conditions are not fulfilled the pile dimensions must be increased accordingly.

(7) Safety against structural failure (“internal” pile capacity) is analysed in accordance with 6.3.3, taking 5.10 into consideration.

(8) Annex D2.2 contains examples of analysis of the "external" capacity of cyclically, axially loaded isolated piles.

13.7.2 Analysis of the serviceability of a single pile

(1) If an appropriate examination reveals that the deflections of the pile foundation are relevant to the structure as a whole, an analysis of the serviceability limit state (SLS) must be performed for the governing effect situations.

(2) Analysis may be performed using the allowable settlement allow. s_k or heave allow. $s_{h,k}$ as provided by the structural designer, assuming characteristic effects on the pile foundation in the serviceability limit state, as follows:

$$\text{vorh. } s_k \leq \text{zul. } s_k \quad \text{or} \quad \text{vorh. } s_{h,k} \leq \text{zul. } s_{h,k} \qquad (13.12)$$

and

$$\text{vorh. } s_k = s_{G,k+Q,k} + s_{cyc,k} \quad \text{or} \quad \text{vorh. } s_{h,k} = s_{G,k+Q,k} + s_{cyc,k} \qquad (13.13)$$

where:

$s_{G+Q,k}$ displacement from permanent and variable action;
$s_{cyc,k}$ displacement from cyclic action.

(3) Annex D2.2 contains examples of the serviceability analysis of cyclically, axially loaded single piles.

13.8 Stability Analyses of Cyclical, Laterally Loaded Piles

13.8.1 Analysis of the bearing capacity of a single pile

(1) Limit values of the mobilisable passive earth pressure and the subgrade reaction $e_{ph,k,cyc}$ must first be defined to analyse the external capacity of a cyclically, laterally loaded pile. To what degree a reduction should be anticipated compared to the limit values for static loads depends on the pile system and geometry, soil type, load frequency and number of load cycles. See 13.4.3 (5) and (6) for additional guidance.

(2) Otherwise, the bearing capacity of flexible piles can be analysed using the cyclic passive earth pressure and adopting subgrade stiffnesses for cyclic loads, as described in 6.3.2 (1) for static loads:

- Calculation of the characteristic normal stresses $\sigma_{h,k}$ between the pile and the soil, and the characteristic action effects. It shall be proved that the characteristic normal stress $\sigma_{h,k}$ does not exceed the limit value of the mobilisable passive earth pressure $e_{ph,k,cyc}$ at any point:

$$\sigma_{h,k} \leq e_{ph,k,cyc} \qquad (13.14)$$

Note: If a modulus of subgrade reaction method with non-linear p-y curves (p-y method) is adopted, this condition is normally met automatically, if the p-y curves include a strength limitation.

- Determine the characteristic support force $B_{h,k}$ to the depth of the pile pivot point (displacement zero point) by integrating the characteristic normal stresses over the pile skin area down to the pivot point.
- Analyse the external capacity (ground resistance):

$$B_{h,d} \leq R_{ult,d,cyc} \tag{13.15}$$

Where $R_{ult,d,cyc}$ is the design value of the mobilisable passive earth pressure down to the pile pivot point, given by integrating the mobilisable passive earth pressure stresses $e_{ph,k,cyc}$ divided by the partial factor $\gamma_{R,e}$ after EC 7-1 Handbook [44], Table A2.3, and Annex A3.2.

- Analyse the internal capacity by comparing the design values of the effects (action effects) to the design values of the material resistance:

$$E_d \leq R_{M,d} \tag{13.16}$$

(3) The bearing capacity of short, practically rigid piles can be analysed as described in 6.3.2 (3) for static loads, adopting the cyclic limit values of the passive earth pressure $e_{ph,k,cyc}$:

- The characteristic support forces and action effects are determined using an estimated pile length and adopting the equilibrium conditions.
- The characteristic support forces $B_{h,k}$ are converted to design forces by multiplication with the partial factors and then compared to the design values of the mobilisable passive earth pressures $R_{ult,d,cyc}$ as shown in Eq. (13.15).

Analysis of internal capacity is performed by comparing the design values of the effects (action effects) to the design values of the material resistance as shown in Eq. (13.16), taking modified subgrade behaviour for cyclic actions into consideration.

(4) Annex D3.2 contains an example for determining the subgrade degradation of cyclically, laterally loaded piles.

13.8.2 Analysis of the serviceability of a single pile

(1) If the deformations in the pile foundation under lateral loads are relevant to the structure as a whole, it must be demonstrated that the horizontal deflections y_k and rotations θ_k of the pile after Equations (13.17) and (13.18), when accumulating under cyclic loads during the design lifetime, remain smaller than the project-specific established limit values.

$$\text{exist. } y_k \leq \text{allow. } y_k \tag{13.17}$$

and

$$\text{exist. } \theta_k \leq \text{allow. } \theta_k \tag{13.18}$$

where exist. y_k and exist. θ_k represent the total horizontal deflections and rotations resulting from the permanent, variable and cyclic actions.

(2) Analysis after (1) can be performed using either criteria based on experience or by numerical investigations, where necessary taking the results of cyclic, cyclic laboratory tests into consideration. Also see 13.4.3 (4) to (8).

(3) For short, practically rigid piles, given the boundary conditions are taken into consideration in principle investigations, an indirect serviceability limit state analysis can be made. This analysis should prove that a project-specific, reduced utilisation factor μ is adhered to, also see 13.4.3 (11):

$$\text{exist. } \mu = B_{h,d}/R_{ult,d,cyc} \leq \text{req. } \mu \qquad (13.19)$$

where:

$B_{h,d}$ design value of the characteristic bearing force determined using the equilibrium condition, see 13.8.1. (3), and

$R_{ult,d,cyc}$ design value of the mobilisable passive earth pressure.

(4) Annex D3.2 contains examples for determining the accumulated deflection of cyclically, laterally loaded piles.

13.9 Stability Analyses of Dynamic or Impact-loaded Piles

(1) No supplementary regulations are currently available for analysing the stability of piles subjected to dynamic or impact loads. Piles loaded in this manner are analysed in the context of the Eurocodes, taking national regulations and literature references into consideration.

Annex A
Terms, Partial Safety Factors and Principles for Analysis

A1 Definitions and notations

Notation	Description	Unit
Latin letters		
A	Pile cross-section	m^2
A_b	Nominal pile base area	m^2
a	Pile centre distances in a pile group	m
a_L	Length of the longest side of the pile base area	m
a_s	Pile width for square cross-section or length of the shortest side of the pile base area	m
$A_{s,i}$	Nominal pile skin area in layer i	m^2
b_F	Flange width of H-pile section	mm – cm – m
B_h	Horizontal ground reaction force	kN – MN
c	Wave propagation velocity	m/s
C	Attenuation coefficient	$(MN/m^2)/(m/s)$
c'	Effective cohesion	kN/m^2
C_b	Pile base attenuation constant	$(MN/m^2)/(m/s)$
$c_{f,max}$	Maximum resistance in vane shear test	kN/m^2
$c_{p,i}$	(spring) stiffness of pile i	MN/m
C_s	Pile skin attenuation constant	$(MN/m^2)/(m/s)$
c_u	Shear strength of the undrained soil	kN/m^2
d	Pile embedment depth in load-bearing ground	m
D	Pile diameter, general	m
D	Density of the non-cohesive soil	–
D_a	External diameter of the auger for partial displacement bored piles	m
D_b	Pile base diameter	m
D_{eq}	Equivalent pile diameter	m
D_i	Diameter of the hollow stem of partial displacement bored piles	m
D_s	Pile shaft diameter	m

Editors: German Geotechnical Society (*Deutsche Gesellschaft für Geotechnik e.V.*)

Notation	Description	Unit
E	Young's modulus	MN/m^2
e_a	Active earth pressure	kN/m^2
E	Effect	kN – MN
E_E	Young's modulus of a single pile	MN/m^2
$E_{G,k}$	Characteristic value of the effect of permanent actions	kN – MN
$E_{Q,k}$	Characteristic value of the effect of variable actions	kN – MN
$E_{Q,rep}$	Representative value of the effect of variable actions	kN – MN
e_p	Passive earth pressure	kN/m^2
E^r_{ph}	Three-dimensional, horizontal passive earth pressure	kN/m^2
E_s	Stiffness modulus	MN/m^2
E·I	Flexural stiffness	MNm^2
F(t)	Impact force at pile head as a function of time	kN – MN
F_1	Impact force at time t_1	kN – MN
F_2	Impact force at time t_2	kN – MN
$F_a(t)$	Inertial force	kN – MN
F	Axial action on a pile	kN – MN
H	Action normal to the pile axis	kN – MN
M	Moment action	kNm – MNm
F_G	Vertical action on the whole pile group	kN – MN
F_n	Action from negative skin friction	kN – MN
g	Gravitational acceleration	m/s^2
G_E	Weight of the attached soil block	kN – MN
G_R	Group factor in terms of pile resistance	–
$G_{R,i}$	Resistance-related group factor (pile i in the group)	–
G_s	Group factor in terms of pile settlements	–
h	Height of H-pile section	M
H	Horizontal load	kN – MN
H_G	Horizontal action on a group pile	kN – MN
H_i	Horizontal action on a pile group	kN – MN
I	Moment of inertia	m^4
I_c	Consistency index	–

Notation	Description	Unit
I_D	Relative density	–
I_I	Moment of inertia in condition I (cross-section uncracked)	m^4
I_p	Plasticity index	–
I_W	Effective moment of inertia in condition II (cross-section cracked)	m^4
J_c	Attenuation coefficient	–
K_0	Coefficient of at-rest earth pressure	–
K_a	Coefficient of active earth pressure	–
K_p	Coefficient of passive earth pressure	–
k_s	Creep	
$k_{s,k}$	Characteristic modulus of subgrade reaction	MN/m^3
L	Pile length	m
l_a	The greater grid dimension	m
l_b	The smaller grid dimension	m
L_E	Elastic length	
L_i	Length of a diaphragm wall panel	m
N	Number of load cycles	–
N_{10}	Blow count per 10 cm penetration of the heavy dynamic penetration probe (DPH)	–
N_{30}	Blow count per 30 cm penetration of the standard penetration probe (SPT)	–
n_G	Number of piles in a group	–
p_f	Flow pressure	kN/m
q_b	Pile tip/base resistance	kN/m^2 – MN/m^2
q_c	CPT cone resistance	MN/m^2
q_s	Pile skin friction	kN/m^2 – MN/m^2
q_u	Unconfined compression strength	MN/m^2
R	Bearing resistance of a single pile, general	kN – MN
R_b	Base resistance of a single pile	kN – MN
R_c, R_c(ULS), R_{ult}	Bearing resistance of a compression pile in the ultimate limit state	kN – MN
R_c(SLS)	Bearing resistance of a compression pile in the serviceability limit state	kN – MN

Notation	Description	Unit
$R_{c,m,mean}$	Compression pile resistance relative to the mean value acquired from a number of pile tests as a measured or tested value in the ultimate limit state	kN – MN
$R_{c,m,min}$	Compression pile resistance relative to the smallest value acquired from a number of pile tests as a measured or tested value in the ultimate limit state	kN – MN
R_E	Resistance of a single pile in contrast to a group pile	kN – MN
$R_{E,s=0.1D}$	Pile resistance of a single pile for a settlement of $s = 0{,}1 \cdot D$	kN – MN
R_G	Overall resistance of a pile group	kN – MN
$R_{G,n}$	Group pile resistance (n^{th} pile)	kN – MN
$R_{m,n}$	Measured value or value acquired from testing of the resistance of the nth pile	kN – MN
R_s	Skin resistance of a single pile	kN – MN
R_t, $R_t(ULS)$, R_{ult}	Bearing resistance of a tension pile in the ultimate limit state	kN – MN
$R_t(SLS)$	Bearing resistance of a tension pile in the serviceability limit state	kN – MN
$R_{t,m,mean}$	Tension pile resistance relative to the mean value acquired from a number of pile tests as a measured or tested value in the ultimate limit state	kN – MN
$R_{t,m,min}$	Tension pile resistance relative to the smallest value acquired from a number of pile tests as a measured or tested value in the ultimate limit state	kN – MN
s_g	Limit settlement or settlement at failure	mm – cm
s_{ult}	Settlement of pile in the ultimate limit state	mm – cm
s(SLS)	Settlement of pile in the serviceability limit state	mm – cm
s_E	Settlement of a single pile	mm – cm
s_G	Mean settlement of a pile group	mm – cm
s_{PR}	Settlement of a (combined) piled raft foundation	mm – cm
s_N	Standard deviation	–

Notation	Description	Unit
s_{sg}	Limit settlement for the settlement-related, characteristic pile skin friction	mm – cm
$s_{sg,z}$	Limit heave	mm – cm

Greek letters

α_{PR}	Piled raft coefficient	–
α_0	Pile head rotation	
α, β	Factors for specifying the value of the characteristic negative skin friction for non-cohesive and cohesive soils	–
γ_G	Partial safety factor for permanent actions	–
γ_Q	Partial safety factor for unfavourable, variable actions	–
γ_t	Partial safety factor for compression pile resistance	–
$\gamma_{s,t}$	Partial safety factor for tension pile resistance	–
γ	Unit weight of wet soil	kN/m^3
γ_r	Unit weight of saturated soil	kN/m^3
γ'	Buoyancy unit weight of soil	kN/m^3
δ	Inclination angle of earth pressure	°
ε	Strain	–
η	Model factor	–
μ	Utilisation factor	–
ρ	Density	t/m^3
σ_h	Horizontal stress in front of pile	kN/m^2 – MN/m^2
σ'_v	Effective vertical stress	kN/m^2
$\tau_{n,k}$	Characteristic negative skin friction	kN/m^2
φ'	Effective soil friction angle	°
φ_u	Friction angle of the undrained soil	°
ξ	Pile test correlation factor	

Indexes

k	Characteristic values
d	Design values

Further definitions and notations are provided in the EC 7-1 Handbook [44].

A2 Partial safety factors γ_F[1)] and γ_E[2)] for actions and effects from EC 7-1 Handbook [44], Table A 2.1

Action or effect	Notation	Design situation		
		DS-P	DS-T	DS-A
HYD and UPL: Limit state of failure by hydraulic heave and buoyancy				
Destabilising permanent actions[a)]	$\gamma_{G,dst}$	1,05	1,05	1,00
Stabilising permanent actions	$\gamma_{G,stb}$	0,95	0,95	0,95
Destabilising variable actions	$\gamma_{Q,dst}$	1,50	1,30	1,00
Stabilising variable actions				
Seepage force in favourable subsoil	$\gamma_{Q,stb}$	0	0	0
Seepage force in unfavourable subsoil	γ_H	1,35	1,30	1,20
	γ_H	1,80	1,60	1,35
EQU: Limit state of loss of static equilibrium				
Unfavourable permanent actions	$\gamma_{G,dst}$	1,10	1,05	1,00
Favourable permanent actions	$\gamma_{G,stb}$	0,90	0,90	0,95
Unfavourable variable actions	γ_Q	1,50	1,25	1,00
STR and GEO-2: Limit state of failure of the structure, structural elements and the ground				
Effects of permanent actions in general[a)]	γ_G	1,35	1,20	1,10
Effects of favourable permanent actions[b)]	$\gamma_{G,inf}$	1,00	1,00	1,00
Effects of permanent actions from at-rest earth pressure	$\gamma_{G,E0}$	1,20	1,10	1,00
Effects of unfavourable variable actions	γ_Q	1,50	1,30	1,10
Effects of favourable variable actions	γ_Q	0	0	0
GEO-3: Limit state of failure by loss of overall stability				
Permanent actions[a)]	γ_G	1,00	1,00	1,00
Unfavourable variable actions	γ_Q	1,30	1,20	1,00
SLS: Serviceability limit state				
γ_G = 1,00 for permanent actions or effects γ_Q = 1,.00 for variable actions or effects				

a) Including permanent and variable water pressure

b) Only in the special case dealt with in EC 7-1 Handbook [44], 7.6.3.1 A (2).

1) The coefficient γ_F is a generic for the respective, individual cases of the partial safety factors relative to the actions F.

2) The coefficient γ_E is a generic for the respective, individual cases of the partial safety factors relative to the effects E.

Note 1: In contrast to DIN EN 1990 the partial safety factors γ_G and γ_Q for the effects of permanent and unfavourable, variable actions for the DS-A design situation have been increased from $\gamma_G = \gamma_Q = 1{,}00$ to $\gamma_G = \gamma_Q = 1{,}10$ in order to retain the proven previous level of safety.

Note 2: The partial safety factors $\gamma_{G,E0}$ are reduced compared to the factors γ_G, because the at-rest earth pressure already decreases to a lower value for minor relaxing movements, and to the considerably smaller active earth pressure in the limit case.

Note 3: DIN EN 1990 prescribes no partial safety factors for the DS-E design situation.

A3 Partial Safety Factors for Geotechnical Parameters and Resistances from EC 7-1 Handbook [44], Tables A 2.2 and A 2.3

A3.1 Partial safety factors γ_M[1)] for geotechnical parameters

Soil parameter	Notation	Design situation		
		DS-P	DS-T	DS-A
HYD and UPL: Limit state of failure by hydraulic heave and buoyancy				
Friction coefficient tan φ′ of the drained soil and friction coefficient tan φ_u of the undrained soil	γ_{φ}', $\gamma_{\varphi u}$	1,00	1,00	1,00
Cohesion c′ of the drained soil and shear strength c_u of the undrained soil	γ_c', γ_{cu}	1,00	1,00	1,00
GEO-2: Limit state of failure of the structure, structural elements and the ground				
Friction coefficient tan φ′ of the drained soil and friction coefficient tan φ_u of the undrained soil	γ_{φ}', $\gamma_{\varphi u}$	1,00	1,00	1,00
Cohesion c′ of the drained soil and shear strength c_u of the undrained soil	γ_c', γ_{cu}	1,00	1,00	1,00
GEO-3: Limit state of failure by loss of overall stability				
Friction coefficient tan φ′ of the drained soil and friction coefficient tan φ_u of the undrained soil	γ_{φ}', $\gamma_{\varphi u}$	1,25	1,15	1,10
Cohesion c′ of the drained soil and shear strength c_u of the undrained soil	γ_c', γ_{cu}	1,25	1,15	1,10

1) The coefficient γ_M is a generic for the partial safety factors relative to the respective, individual cases.

Note: DIN EN 1990 prescribes no partial safety factors for the DS-E design situation.

A3.2 Partial safety factors γ_R[1)] for resistances

Resistance	Notation	Design situation		
		DS-P	DS-T	DS-A
STR and GEO-2: Limit state of failure of the structure, structural elements and the ground				
Ground resistances				
– Passive earth pressure and bearing resistance	$\gamma_{R,e}$, $\gamma_{R,v}$	1,40	1,30	1,20
– Sliding resistance	$\gamma_{R,h}$	1,10	1,10	1,10
Pile resistances from static and dynamic pile testing				
– Base resistance	γ_b	1,10	1,10	1,10
– Skin resistance (compression)	γ_s	1,10	1,10	1,10
– Overall resistance (compression)	γ_t	1,10	1,10	1,10
– Skin resistance (tension)	$\gamma_{s,t}$	1,15	1,15	1,15
Pile resistances based on empirical data				
– Compression piles	γ_b, γ_s, γ_t	1,40	1,40	1,40
– Tension piles (in exceptional cases only)	$\gamma_{s,t}$	1,50	1,50	1,50
Pull-out resistances				
– Soil and rock nails	γ_a	1,40	1,30	1,20
– Grouted body of ground anchors	γ_a	1,10	1,10	1,10
– Flexible reinforcement elements	γ_a	1,40	1,30	1,20
GEO-3: Limit state of failure by loss of overall stability				
Shear Strength				
– See Table A 3.1				
Pull-out resistances				
– See STR and GEO-2				

[1)] The coefficient γ_R is a generic for the partial safety factors relative to the respective, individual resistance cases.

Note 1: The partial safety factors for the material resistance of the steel tendon of prestressing steel and reinforcing steel is given in DIN EN 1992-1-1 for the limit states GEO-2 and GEO-3 as

$\gamma_M = 1,15$.

Note 2: The partial safety factors for the material resistance of flexible reinforcement elements are given in EBGEO [28] for the limit states GEO-2 and GEO-3.

Note 3: DIN EN 1990 prescribes no partial safety factors for the DS-E design situation.

A4 Correlation Factors ξ_i for Determining the Characteristic Pile Resistances for the Ultimate Limit State Acquired from Tested or Measured Data of Static and Dynamic Pile Tests acc. to the EC 7-1 Handbook

A4.1 Correlation factors from static pile tests

(1) The characteristic pile resistances $R_{c,k}$ and $R_{t,k}$ are determined from the data tested or measured in static pile tests by dividing by the correlation factors ξ_i given in Table A4.1 (EC 7-1 Handbook [44], Table A7.1). The following application rules shall be observed.

(2) Eq. (A4.1) for the ultimate compressive resistance must be fulfilled for structures incapable of redistributing loads from "flexible" to "stiff" compression piles.

$$R_{c,k} = \mathrm{MIN}\left\{\frac{(R_{c,m})_{mean}}{\xi_1}, \frac{(R_{c,m})_{min}}{\xi_2}\right\} \tag{A4.1}$$

ξ_1 and ξ_2 are correlation factors which depend on the number of pile tests performed, and are applied to the mean $(R_{c,m})_{mean}$ or the smallest value $(R_{c,m})_{min}$ of $R_{c,m}$.

(3) If structures do possess sufficient stiffness to redistribute loads from "flexible" to "stiff" compression piles, the numerical values of ξ_1 and ξ_2 may be divided by 1,1, assuming that ξ_1 never becomes smaller than 1. "Flexible" compression piles are anticipated e.g. in cases where single piles are acting independently. If the structural loads can e.g. be distributed by a rigid capping beam on several piles , the piles are considered as "rigid" compression piles.

Table A4.1 Correlation factors ξ_i for deriving characteristic pile resistances from static pile testing on compression and tension piles

n	1	2	3	4	≥ 5
ξ_1	1,35	1,25	1,15	1,05	1,00
ξ_2	1,35	1,15	1,00	1,00	1,00
n is the number of tested piles.					

(4) Eq. (A4.2) applies in analogy to Eq. (A4.1) for the limit value of the pull-out resistance of a tension pile.

$$R_{t,k} = \mathrm{MIN}\left\{\frac{(R_{t,m})_{mean}}{\xi_1}, \frac{(R_{t,m})_{min}}{\xi_2}\right\} \tag{A4.2}$$

There is no such differentiation between "flexible" and "stiff" tension piles like that for compression piles.

A4.1 Correlation factors from dynamic pile tests

(1) The characteristic pile resistance $R_{c,k}$ is determined from the data tested or measured in impact or dynamic pile tests by dividing by the ξ_I correlation factors. The base values of the $\xi_{0,i}$ correlation factors are used to calculate the ξ_i correlation factors, together with the associated surcharge factors $\Delta\xi$ and model factors η_D from Table A4.2 (EC 7-1 Handbook [44], Table A7.2) and Figure A4.1 (EC 7-1 Handbook [44], Figure A7.1).

(2) Also see 5.3 for details of the plausibility of the test results and data acquired from dynamic pile testing.

(3) Eq. (A4.3) for the ultimate compressive resistance must be fulfilled for structures incapable of redistributing loads from "flexible" to "stiff" compression piles.

$$R_{c,k} = \mathrm{MIN}\left\{\frac{(R_{c,m})_{mean}}{\xi_5}, \frac{(R_{c,m})_{min}}{\xi_6}\right\} \qquad (A4.3)$$

ξ_5 and ξ_6 are correlation factors which depend on the number of pile tests performed, and are applied to the mean $(R_{c,m})_{mean}$ or the smallest value $(R_{c,m})_{min}$ of $R_{c,m}$.

(4) Table A4.2 d) and Figure 4.1 apply when considering "flexible" and "stiff" compression piles in analogy to A4.1 (3).

Table A4.2 Base values $\xi_{0,i}$ with corresponding increase factors and model factors for correlation factors ξ_5 and ξ_6 used to derive characteristic values from impact or dynamic pile tests

$\xi_{0,i}$ for n =	2	5	10	15	≥ 20
$\xi_{0,5}$	1,60	1,50	1,45	1,42	1,40
$\xi_{0,6}$	1,50	1,35	1,30	1,25	1,25

– n is the number of tested piles;
– Intermediate values of $\xi_{0,5}$ and $\xi_{0,6}$ for n = 2 to 20 may be linearly interpolated;

a) To calculate the correlations factors ξ_i : $\xi_i = (\xi_{0,i} + \Delta\xi) \cdot \eta_D$, also see Figure A 4.1.

b) For the surcharge value $\Delta\xi$:

→ $\Delta\xi = 0$: for calibrating dynamic evaluation methods with static pile test results on the same site;
→ $\Delta\xi = 0{,}10$: for calibrating dynamic evaluation methods with static pile test results on a comparable construction project;
→ $\Delta\xi = 0{,}40$: for calibrating dynamic evaluation methods based on documented or common empirical data for pile resistances. Adoption of a direct method such as the Case or TNO method is not permitted.

c) The following apply to the model factor η_D for consideration of the evaluation method

→ $\eta_D = 1{,}00$: for direct evaluation methods;
→ $\eta_D = 0{,}85$: For extended methods with complete modelling.

d) If structures possess sufficient stiffness and strength to redistribute loads from "flexible" to "stiff" piles, the numerical values of ξ_5 and ξ_6 may be divided by 1,1.

c) The following apply to the model factor η_D for consideration of pile driving formulae:

→ $\eta_D = 1{,}05$: if the wave equation method is adopted;
→ $\eta_D = 1{,}10$: if a pile driving formula with measurement of the quasi-elastic pile head movement under driving impact is adopted;
→ $\eta_D = 1.20$: if a pile driving formula without measurement of the quasi-elastic pile head movement under driving impact is adopted.

f) If different piles are used in the foundation, groups of similar piles should be considered separately when selecting the number n of test piles. This also applies to areas of similar ground conditions within the same site.

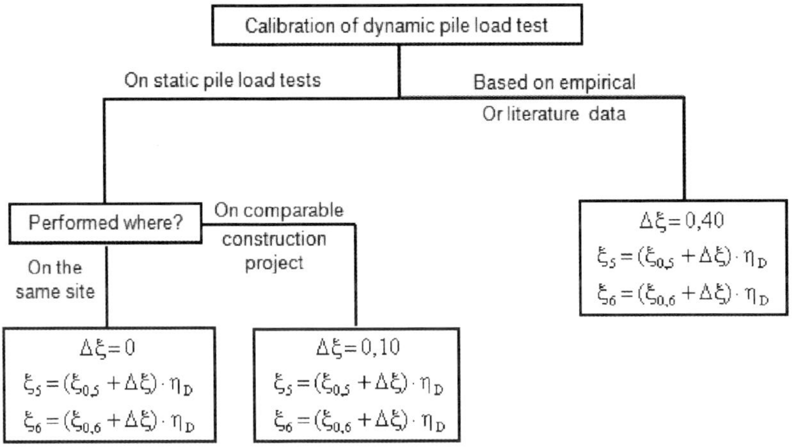

- The model factor is:
  - γ_D = 1,00 for the direct method,
  - γ_D = 0,85 for the advanced method with complete modelling,
  - γ_D = 1,05 / 1,10 / 1,20 when adopting driving formulae corresponding Table A 4.2, Footnote e

- For structures able to redistribute loads from "flexible" to "stiff" piles,
  ξ_5 and ξ_6 may be divided by 1,10.

Figure A4.1 Procedure for deriving the correlation factors ξ_5 and ξ_6 as a function of the calibration based on Table A4.2

A5 Procedure for Determining the Resistance of Piles Against Buckling Failure in Soil Strata with Low Lateral Support (informative)

A5.1 Guidance notes

(1) When adopting the following method and that outlined in [149] to determine the characteristic resistance against pile buckling in soft ground it shall be noted that the method has only been evaluated in the context of a research project and that no extensive project-related empirical data exist. It therefore does not represent generally recognised best practice, which is to be observed by the user or any respective execution project.

(2) The method presented hereafter is of informative character only, because established and proven methods are not available at present.

A5.2 Ground support

(1) Suitable models, capable of taking the non-linear deformation resistances into consideration, should be adopted for considering the lateral support of the ground. In particular, the plasticisation of the surrounding ground must be taken into account for the analysis. The adoption of a subgrade reaction with deformation-independent constant stiffness is not allowed for modelling ground support. Time-dependent effects resulting from the viscosity of the ground and/or the consolidation of the ground around the pile, and leading to an anticipated reduction in the support effect, must be taken into consideration. Suitable assumptions must be made with regard to loading velocity and the parameters governing the consolidation of the ground (compressibility, dilatancy, permeability and drainage path). Established models confirmed by testing are not available at present. Consequently, accurate specification of the lateral support of the ground transferable to all common boundary conditions is not feasible. Test-supported design using pile tests should therefore be given preference when in doubt, see 5.10.1 (3). If buckling failure occurs during testing of trial piles the boundary conditions should be precisely documented and be reported to facilitate the development of suitable, long-term models of the mobilisable in-situ support effect.

(2) Until detailed, scientific investigations or well-documented empirical data together with the respective boundary conditions are available from construction practice, the ground support acting against buckling should be taken into consideration as follows:

a) Upon achieving a critical lateral deformation w_f a maximum limiting stress p_f is mobilised in the ground, which cannot be increased with increasing deformations.

b) The deformation w_f can be estimated after [90] for fine-grained soils with soft or very soft consistency using the following relationship as a function of the outside diameter of the pile shaft D_s:

$$w_f = 0{,}1 \cdot D_s \tag{A5.1}$$

In firmer ground a smaller mobilisation path w_f may be taken into consideration in the design.

c) The limit value of the ground support p_f, which acts on a pile shaft of diameter D_s, can be calculated for piles completely enclosed in the ground and in the undrained condition by (see also [121]):

$$p_f = 11 \cdot c_{u,k} \cdot D_s \text{ [kN/m}^2 \cdot \text{m]} \tag{A5.2}$$

Equation A5.2 implies that the limit state occurs due to the ground flowing around the pile shaft. Equation A5.2 does not describe the resistance that occurs as a result of a shallow rupture of the corresponding soil stratum. This resistance is normally smaller and must be determined as described in paragraph (d) below. For use of Equation (A5.3) and undrained conditions, the ground parameters $K_q = 0$ and $c'_k = c_{u,k}$ apply.

d) If the velocity of load application is sufficiently gradual and/or the soil stratum in question is sufficiently permeable, the maximum ground support p_f mobilisable in the drained case can be adopted. The models used for calculating the three-dimensional earth pressure to DIN 4085 can also be adopted to calculate p_f. Simplified, p_f can be calculated using:

$$p_f = K_q \cdot \sigma'_z + K_c \cdot c'_k \tag{A5.3}$$

as a function of the effective friction angle φ'_k and the effective cohesion c'_k [12]. The effective vertical stress σ'_z is determined as a function of depth, corresponding to the soil's self-weight. The value of the mean vertical effective stress σ'_z may be adopted when determining the buckling resistance at the governing buckling length L_{cr} within a soft soil stratum. The variables K_q and K_c can be determined as a function of the depth z/D_s as shown in Figure A5.1.

(3) Saturated, loose soil strata, in particular consisting of fine sands and silt, tend to liquefy during earthquakes or under other dynamic actions. In the limit case the ground support p_f is lost. This case is highly unfavourable with regard to numerical resistance against buckling and must therefore be taken into consideration when designing pile foundations in critical regions.

(4) Any impacts of the viscosity of the soft ground should be incorporated by reducing the undrained shear strength $c_{u,k}$ in cases of gradual load application. Where the load is applied quickly, for example as the result of impacts, higher undrained shear strength $c_{u,k}$ design values can be adopted.

(5) If the soft soil stratum consists of soil types other than clays or silts, the flow pressure p_f must be determined by specific investigations.

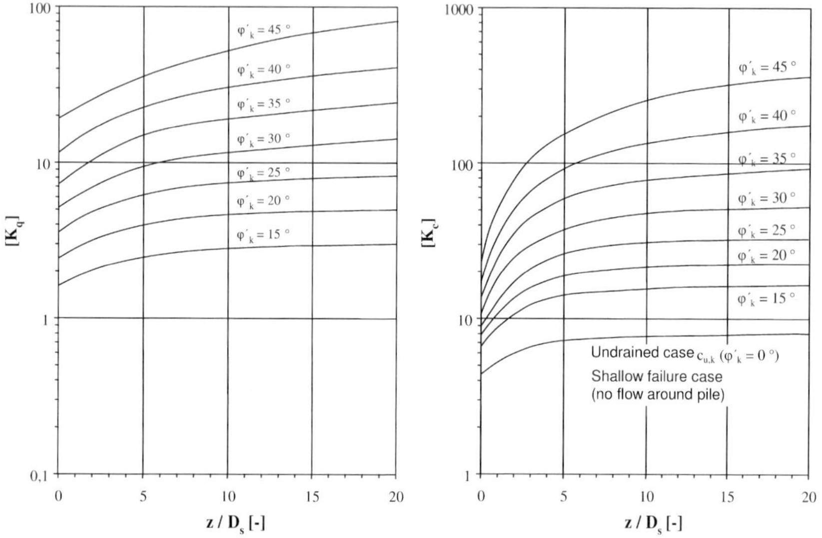

Figure A5.1 Parameters K_q and K_c after [12] for calculating the limit value of the ground support p_f, in the drained case and for the case of shallow ground failure (the ground does not flow around the pile shaft)

(6) In particular, determining a characteristic undrained shear strength $c_{u,k}$ is subject to uncertainties with regard to organic soils and other soils with structural strength. The uncertainty has a direct effect on the adopted limit stress p_f and mobilisation path w_f.

A5.3 Static system and equilibrium conditions using second-order theory (inclusion of lateral deflections)

(1) A pile with a lateral subgrade is considered, for which bilinear, elastic-plastic behaviour is modelled (see Figure A5.2). The maximum ground resistance p_f in kN/m and the mobilisation path w_f are calculated as described in A5.2. The values for the linear springs $k_l = p_f/w_f$ are generally in a range $40 \cdot c_u \leq k_l \leq 180 \cdot c_u$, which is compatible with literature data, cf. [156] and [157].

(2) The deformation figure of the buckling pile, unsupported along the length L, with hinged bearings (pin ends) at the head and base, has the shape of a simple half wave. The buckling length L_{cr} then corresponds to the length L of the unsupported length. With regard to elastically supported piles, in contrast, buckling modes (eigenmodes) with greater wavelengths are generally governing, depending on the soil stiffness, see [149]. The buckling loads for different wavelengths shall be determined. The smallest value determined is the governing buckling

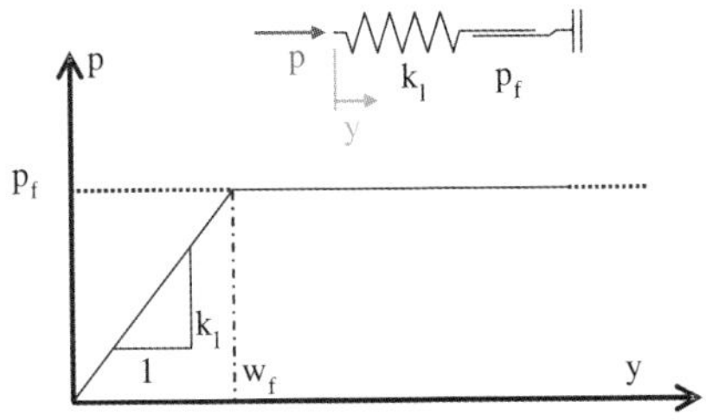

Figure A5.2 Idealised lateral soil support for pile susceptible to instability, after [149]

load. Conservatively, infinitely long piles are adopted for the method presented, in which the buckling modes can freely develop as a wave line. The effective forces and geometrical data can be visualised for a half wave cut from the wave line, taking a sinusoidal, geometrical imperfection e_0 into consideration as shown in Figure A5.3. The length of the half wave then corresponds to the buckling length L_{cr}.

(3) The possible types of buckling failure and possible equilibrium conditions given various boundary conditions are shown Figure A5.4. The compressive force in the pile N is drawn on the ordinate and the lateral deflection y necessary for equilibrium on the abscissa. The deflection y is given for the centre of the buckling length, The following cases are differentiated:

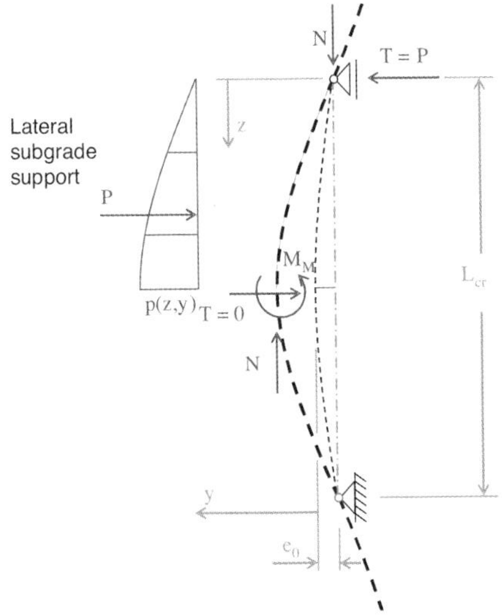

Figure A5.3 Structural system after [149] for calculating the characteristic buckling resistance of piles

a) The simple "Euler" case of a perfect strut without lateral support is characterised by a vertical line up to the buckling load. When the buckling load is reached indifferent equilibrium states are possible for all deflections (horizontal line).
b) If the free compression strut is predeformed, a moment is generated under compressive load and thus a deflection occurs under even minor compressive loads, increasing with additional loads. The equilibrium curve approaches the "Euler" load asymptotically.
c) The perfect elastically supported strut is characterised by "Engesser's" solution. The buckling load increases to a greater value than that after "Euler" as a function of the lateral support. An indifferent system exists after reaching the bifurcation load.
d) An imperfect, elastically supported strut has an equilibrium curve that asymptotically approaches the "Engesser" load.
e) A bilinear, elastic-plastic supported compression strut behaves identically to an elastically supported compression strut at small deflections, both for a perfect and an imperfect strut. However, as soon as the deformation w_f is reached, where the soil support p can no longer be increased, the curve falls off again and converges on the "Euler" solution.

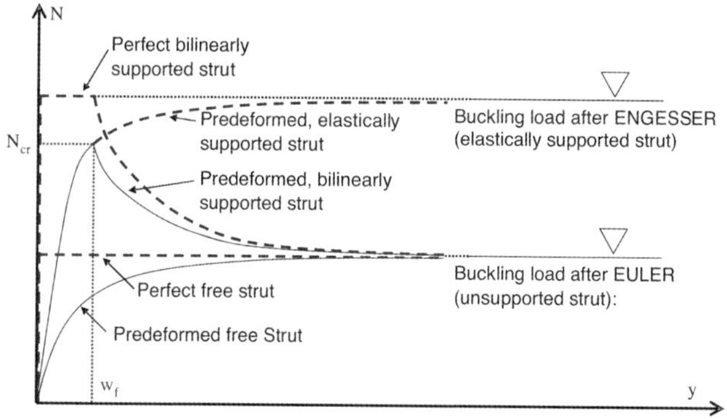

Figure A5.4 Equilibrium conditions after second order theory

A5.4 Requirements for the application of the analysis method

(1) The method for calculating the characteristic buckling resistance described below assumes that lateral soil stresses (soil resistances) initially occur when a pile bends and that they act as a stabilising factor. It is absolutely imperative that these resistances remain available for the lifetime of the structure. In contrast to this, lateral soil stresses can also be actions (lateral pressure on piles after Section 4.5) and are to be considered as such, which then impacts extremely unfavourably on the performance and the stability.

(2) In composite cross-sections of concrete, mortar or cement, the flexural stiffness changes with increasing curvature of the pile axis and accompanying crack formation. A reduction in flexural stiffness must be taken into consideration in the analysis method.

(3) Effects related to the viscosity of soft soils are not covered in the method proposed for determining the buckling resistance and must be taken into consideration by a reduction in the buckling load calculated using a elastic-plastic subgrade reaction.

(4) When taking imperfections into consideration it is assumed that the mode of the imperfection is affine to the eigenmode of the buckling strut.

(5) Determining the ultimate load of slender piles is a complex process, because of the material and geometrical non-linearities involved. It requires deeper understanding of the mechanical relationships and is usually performed electronically due to the iterative computation required. A specimen worksheet (MS Excel) can be found at http://www.gb.bv.tum.de, for example, although the author accepts no responsibility for correct programming or the correctness of the calculated loads.

A5.5 Determining the characteristic resistance against pile buckling

(1) The ideal bifurcation load N_{cr} representing the characteristic resistance against pile buckling is calculated using:

$$N_{cr} = \frac{w_f \cdot (E \cdot I)_p \cdot \left(\frac{\pi}{L_{cr}}\right)^2 + p_f \cdot \left(\frac{L_{cr}}{\pi}\right)^2}{w_f + e_0} \quad \text{(A5.4)}$$

Where:

N_{cr}	[kN]	ideal bifurcation load;
L_{cr}	[m]	buckling length; if the infinitely long strut is considered, the governing half-wave of the buckling mode for determining its ideal bifurcation load is given when N_{cr} approaches a minimum. This can be solved for L_{cr} with the aid of the first derivative of Eq. (A5.4), which is best achieved in practice by numerical zero finding for the derived function.
$(E \cdot I)_p$	[kNm^2]	flexural stiffness of the pile cross-section; see Section A5.5 (6) for more information;
w_f	[m]	lateral pile deflection in the centre of buckling length L_{cr}, at which the soil plasticises;
e_0	[m]	pre-deformation as a result of imperfection.

(2) For the perfect strut where $e_0 = 0$, the solution converges on "Engesser's", which in turn is reduced from "Euler's" for $p_f = 0$.

(3) Investigations in [149] demonstrated that the bifurcation loads of a few test results were estimated approximately correctly using the bilinear approach and adopting $w_f = 0{,}1 \cdot D_s$ and $p_f = 6 \cdot c_u \cdot D_s$, as well as an imperfection e_0 corresponding to a curvature radius relative to L_{cr} of 200 m.

(4) In addition, with regard to the capacity of an imperfect strut, it must be examined whether the pile fails before reaching the ideal bifurcation load N_{cr} as a result of its limited material strength. Use of the interaction diagram, in which the combinations of M and N at which the strut material plasticises are represented, is useful here. The following equation can be adopted to describe the interaction curve of the maximum, simultaneously acting action effects N and M necessary to achieve plasticisation of the cross-section:

$$M = M_{pl} \cdot \left(1 - \left(\frac{N}{N_{pl}}\right)^{\alpha}\right) \quad \text{(A5.5)}$$

(5) The pile's fully plastic action effects N_{pl} and M_{pl} can be calculated for steel sections or be looked up in tables. In composite cross-sections consisting of concrete and steel computer-aided methods are recommended to determine N_{pl} and M_{pl}. The variable α in Equation (A5.5) is a shape coefficient depending on the geometry of the pile cross-section. For steel sections it is calculated from the ratio of the plastic to the elastic resistance moment. The exponent α of the interaction relationship can be conservatively adopted at 1,0.

(6) Where piles have a composite cross-section, the flexural stiffness should be determined on the at maximum to the middle fractured cross-section (Condition II for reinforced concrete cross-sections). For piles with a single, central bearing member with surrounding mortar or concrete cover, the flexural stiffness is conservatively adopted for the central bearing member. However, the laterally supporting soil resistance always acts on the full width of the pile. The tests carried out to date on composite piles with cement mortar cover could not clarify whether the use of corrugated plastic sleeves, enveloping the bearing member and parts of the cement stone body, guarantee full bonding and thus an increase in flexural stiffness and the fully plastic actions effects. Nor have any tests or numerical investigations been performed to date on grouted micropiles with several steel bars as bearing member or on micropiles with external steel tubes.

(7) The relative deflection of monobar micropiles resulting from geometrical and material imperfections within the buckling length to be analysed should be calculated from a curvature radius of at least 200 m. Geometrical drilling deviations and the not precisely cylindrical shape of the grout body should be taken into consideration. If the tendons are connected by threaded sleeves, the imperfections can increase considerably.

(8) In elastically supported slender struts comparatively small buckling lengths L_{cr} often result due to the greater waviness of the buckling mode. Regardless of other boundary conditions, determining the buckling length for an infinitely long pile is recommended. Only if the buckling length is substantially larger than given by the thickness of a soft stratum and the restraint boundary conditions above and below this stratum, a buckling length corresponding to the thickness of the soft stratum could be investigated. However, in this case all shorter buckling lengths must still be investigated. The governing characteristic buckling load is always the smallest buckling load calculated from the various possible buckling lengths.

(9) If settlements or horizontal deflections are anticipated in the soft stratum, any ensuing actions must also be taken into consideration in the analysis.

A6 Bonding Stress in Grouted Displacement Piles (informative)

A6.1 Guidance notes

(1) When adopting the values below (taken from [61] and [62]) for the bonding stress of grouted displacement piles it should be noted that the values are only derived from a single research project and have not yet been verified full-scale on real construction projects. They therefore do not represent generally recognised best practice. This fact should be noted by the user in the course of appropriate execution projects. However, experience with grouted displacement piles has now been collected over decades, e.g. in the harbours of Hamburg, where no damages were identified to date which could result of insufficient bonding in the pile shaft/grout interface.

(2) The method presented hereafter is of informative character only, because established and proven methods are not available at present.

A6.2 Characteristic and design values of bonding stresses

(1) Table A6.1 contains empirical values of bonding stresses for analysing the pile shaft/grout interface of grouted displacement piles to DIN EN 12699. The following terms are introduced to define and describe surfaces, also see Figure A6.1:

- *Braced surfaces*: Braced surfaces refers to opposing section surfaces (parallel or rotated against each other) where the intervening space is completely filled with grout, e.g. the inner flanges of H-piles.
- *Unbraced surfaces*: Unbraced surfaces refers to all surfaces not defined as braced surfaces.

Table A6.1 Empirical values for the characteristic boding stress $\tau_{R,k}$ in the pile shaft/grout body interface of grouted displacement piles

Area	Characteristic values of bonding stress $\tau_{R,k}$ in the pile shaft/grout body interface [N/mm²]	
	Grout with swelling agent	Grout without swelling agent
Braced surfaces	0,60	0,45
Unbraced surfaces	0,30	0,30

Note: The data in Table A6.1 are the result of investigations after [61] and [62].

(2) The prerequisite for adopting the data given in Table A6.1 is a cube compressive strength $f_{c,cube} \geq 25$ N/mm² for the grout employed and a minimum grout thickness of $d_{VM} = 25$ mm.

(3) The developed circumference of the steel tendon is adopted as the circumference of the pile shaft/grout body interface.

(4) For pressure-grouted piles the bonding stress differs between the section's flange and web faces. The pile skin surfaces are therefore separated into braced and unbraced surfaces as shown in Figure A6.1

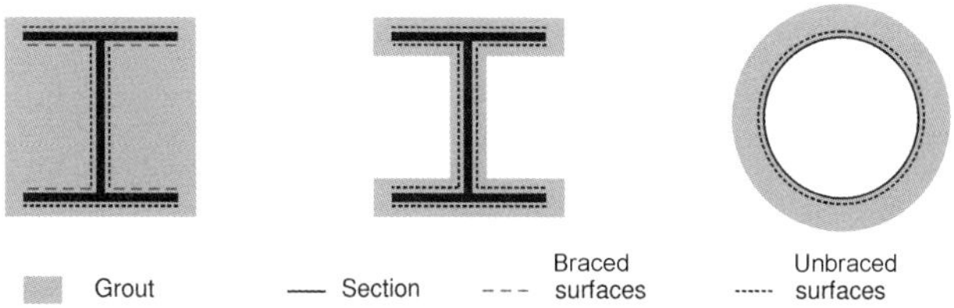

Figure A6.1 Definition of terms braced and unbraced surfaces

Annex B
Example Calculations for Pile Resistance Analysis and Verifications

Note: Some of the following examples were taken from [58]. Further examples can be found in [64].

B1 Determining the Axial Pile Resistances from Static Pile Load Tests, and Ultimate and Serviceability Limit State Analyses

B1.1 Objectives

Figure B1.1a shows a foundation situation with a pile of diameter D = 1,2 m and a permanent load $F_{G,k} = 1,5$ MN, in addition to a variable load $F_{Q,rep,k} = 1,0$ MN. Two static pile load tests were executed, the results of which are included in Figure B1.1b and Table B1.1 as R_{m1} and R_{m2}. The ultimate settlement is defined as $s_g = s_{ult} = 0,1 \cdot 120$ cm = 12 cm using Eq. (5.1). Because the static pile load tests were only executed up to a settlement s = 10 cm, the ultimate settlement was extrapolated.

The analysis comprises the characteristic pile resistances in the ultimate limit state (ULS) for "flexible" and "stiff" piles after 5.2.2 and the characteristic boundary lines in the serviceability limit state (SLS) after 5.2.3, and the external capacity and serviceability for the specified pile load. An allowable

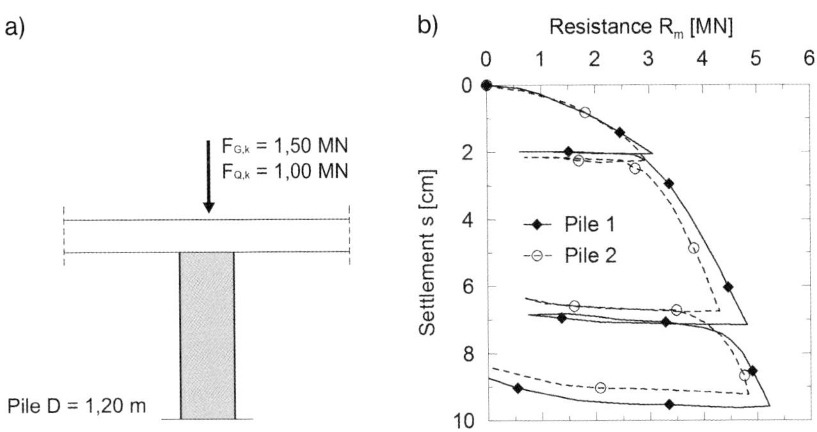

Figure B1.1 a) System and effect; b) Logged R_m values for both static pile load tests

Recommendations on Piling – EA Pfähle.
Editors: German Geotechnical Society (*Deutsche Gesellschaft für Geotechnik e.V.*)

pile settlement allow. $s_k = 2{,}0$ cm is specified by the structural design for the serviceability limit state (SLS).

The static pile load tests can be taken from [154].

B1.2 Deriving the characteristic pile resistances in the ultimate and serviceability limit states

The characteristic, ultimate pile resistance $R_{c,k}$ is given by the lesser of either the mean value $(R_{c,m})_{mean}$ or the minimum value $(R_{c,m})_{min}$ of the pile load test results using Equation (A4.1) as follows:

$$R_{c,k} = \text{MIN}\left\{\frac{(R_{c,m})_{mean}}{\xi_1}; \frac{(R_{c,m})_{min}}{\xi_2}\right\}$$

The correlation factors ξ_1 and ξ_2 depend on the number of static pile load tests performed and are selected after Annex A4. The correlation factors given there apply to “flexible” compression piles. If “stiff” compression piles are used, the correlation factors may be divided by 1,1, assuming that ξ_1 does not become smaller than 1,0.

For the range of small pile settlements, after 6.4, Eq. (6.13), characteristic boundary lines were derived for the serviceability limit state analysis using the κ values (based on [59]). For the present case κ = 0,15 was adopted to relate to the average of the measured values.

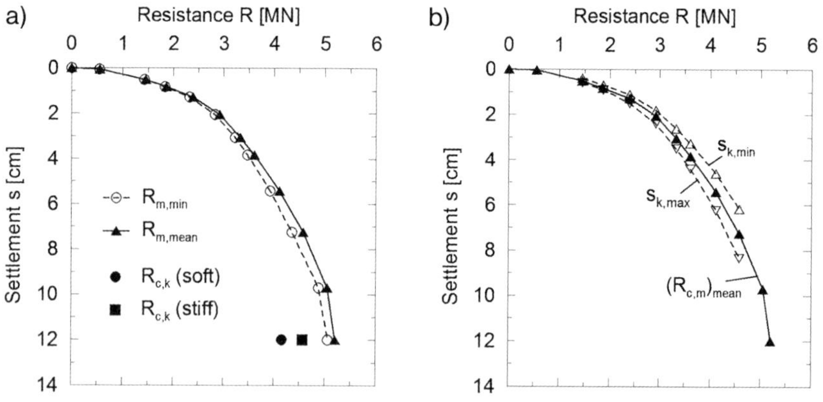

Figure B1.2 a) Characteristic pile resistances in the ultimate limit state (ULS) for flexible and stiff piles derived from the $R_{m,i}$ values recorded in static pile load tests; b) Section of the resistance-settlement curves in the serviceability limit state (SLS) and derivation of characteristic boundary curves

Note: The method adopted here for the characteristic boundary lines in the service load range represents only one possible option. Other reasoned procedures are also possible.

Figure B1.2 and Table B1.1 show the results for the determination of $R_{c,k}$ (SLS) and $R_{c,k} = R_{c,k}$ (ULS).

Table B1.1 Results of two static pile load tests and derivation of the characteristic resistance-settlement curves for "flexible" and "stiff" piles

Settlement s [cm]	R_{m1} [MN]	R_{m2} [MN]	$(R_{c,m})_{mean}$ [MN]	$(R_{c,m})_{min}$ [MN]	$\Delta s_k = \kappa \cdot s_k$			
					$s_{k,min}$ [cm]		$s_{k,max}$ [cm]	
0	0	0	0	0	0		0	
0,51	1,483	1,424	1,454	1,424	0,4		0,6	
0,83	1,891	1,831	1,861	1,831	0,7		1,0	
1,28	2,458	2,321	2,390	2,321	1,1		1,5	
2,06	3,015	2,830	2,922	2,830	1,8		2,4	
3,05	3,427	3,230	3,329	3,230	2,6		3,5	
3,83	3,750	3,469	3,610	3,469				
5,42	4,301	3,924	4,112	3,924				
7,26	4,803	4,354	4,579	4,354	Flexible		Stiff	
9,71	5,222	4,881	5,051	4,881	ξ_1	ξ_2	ξ_1	ξ_2
12,0	5,347	5,060	5,204	5,060	1,25	1,15	1,14	1,05

Using the results from Table B1.1, the characteristic pile resistance $R_{c,k}$ in the serviceability limit state (ULS) can be determined for "flexible" and "stiff" piles in Table B1.2.

Table B1.2 Deriving the characteristic pile resistance $R_{c,k}$ in the ultimate limit state for "flexible" and "stiff" piles

	$\frac{(R_{c,m})_{mean}}{\xi_1}$ [MN]	$\frac{(R_{c,m})_{min}}{\xi_2}$ [MN]	$R_{c,k}$ [MN]
"flexible" piles	4,163	4,400	4,163
"stiff" piles	4,565	4,819	4,565

B1.3 Bearing capacity analysis

The limit state condition after 6.2 and 6.3:

$$F_{c,d} \leq R_{c,d}$$

must be adhered to for ultimate limit state (ULS) analysis.

a) For single piles acting independently ("flexible" piles):

$$F_{c,d} = F_{G,k} \cdot \gamma_G + F_{Q,rep,k} \cdot \gamma_Q = 1{,}500 \cdot 1{,}35 + 1{,}000 \cdot 1{,}50 = 3{,}525 \text{ MN}$$

$$R_{c,d} = R_{c,k}/\gamma_t = 4{,}163/1{,}10 = 3{,}785 \text{ MN}$$

$$F_{c,d} = 3{,}525 \text{ MN} < R_{c,d} = 3{,}785 \text{ MN}$$

b) For load distribution by means of a rigid capping slab ("stiff" pile):

$$F_{c,d} = 3{,}525 \text{ MN}$$

$$R_{c,d} = R_{c,k}/\gamma_t = 4{,}565/1{,}10 = 4{,}150 \text{ MN}$$

$$F_{c,d} = 3{,}525 \text{ MN} < R_{c,d} = 4{,}150 \text{ MN}$$

B1.4 Serviceability analysis

When determining the pile resistances R(SLS) in the serviceability limit state, differentiation after 6.4 is required whether minor or major (adopted here) differential pile settlements are to be expected. To this end, the characteristic boundary curves in the service load range were derived from the recorded pile load test data in Table B1.1 and Figure B1.2b.

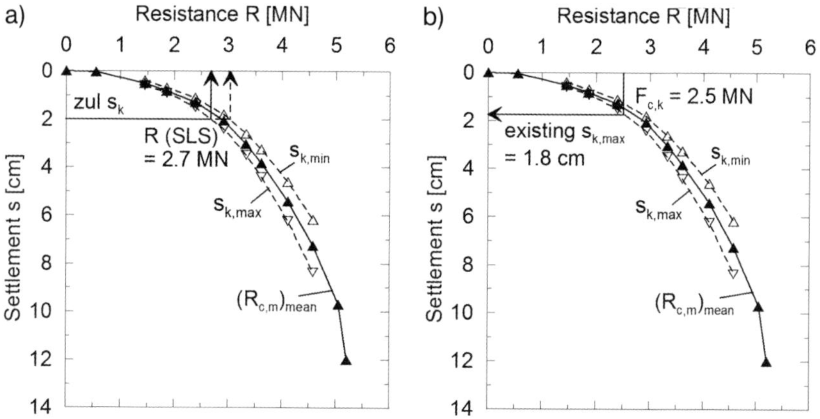

Figure B1.3 a) Analysing serviceability via the characteristic pile forces; b) Analysing serviceability via existing and allowable settlements

The specified allowable settlement (e.g. specified in the structural design) in the example is allow. $s_k = 2$ cm.

Using Figure B1.3a, serviceability is demonstrated via pile forces after 6.3, Eq. (6.11) as:

$$F_d\,(SLS) = F_k = 2{,}500\ MN < R_d\,(SLS) = R_k\,(SLS) = 2{,}7\ MN.$$

Analysis by comparing settlements in accordance with Figure B1.3b after 6.3, Eq. (6.12) results to:

$$\text{exist. } s_{k,max} = 1{,}8\ cm < \text{allow. } s_k = 2{,}0\ cm$$

B2 Characteristic Axial Pile Resistances from Dynamic Load Tests

B2.1 Objective

Dynamic pile load tests were performed to determine the ultimate capacity of reinforced concrete displacement piles and then calibrated against a static load test from a different, but comparable site. Evaluation using the direct method (CASE equation) resulted in the following measured data for the compressive resistance of the ground against the pile in the ultimate limit state:

$R_{c,m,1} = 0,875$ MN
$R_{c,m,2} = 0,950$ MN
$R_{c,m,3} = 1,050$ MN
$R_{c,m,4} = 1,100$ MN
$R_{c,m,5} = 1,225$ MN.

The respective characteristic ultimate limit state resistances for "flexible" and "stiff" compression piles shall be derived from the measured data of the dynamic pile load tests.

B2.2 Characteristic pile resistances

The characteristic pile resistance in the ultimate limit state (ULS) $R_{c,k}$ is given by the lesser of the mean value $(R_{c,m})_{mean}$ or the minimum value $(R_{c,m})_{min}$ of the pile load test results using Equation (A4.3) as follows:

$$R_{c,k} = \mathrm{MIN}\left\{\frac{(R_{c,m})_{mean}}{\xi_5};\frac{(R_{c,m})_{min}}{\xi_6}\right\}$$

The correlation factors ξ_5 and ξ_6 to be adopted are given in Annex A4.2 and are derived from the base values of the correlation factors $\xi_{0.5}$ and $\xi_{0.6}$, the increased value $\Delta\xi$ and the model factor η_D. The correlation factors ξ_5 and ξ_6 may be divided by 1.1 for "stiff"compression piles.

The correlation factors ξ_5 and ξ_6 are provided in Table B2.1 and the determination of the characteristic compression pile resistances $R_{c,k}$ in Table B2.2.

Table B2.1 Correlation factors ξ_5 and ξ_6 derived from the five dynamic pile load tests for "flexible" and "stiff" compression piles

$(R_{c,m})_{min}$	$(R_{c,m})_{mean}$	$\xi_{0,5}$	$\xi_{0,6}$	$\Delta\xi$	η_D	flexible		stiff	
						ξ_5	ξ_6	ξ_5	ξ_6
[MN]	[MN]	[–]	[–]	[–]	[–]	[–]	[–]	[–]	[–]
0,875	1,040	1,50	1,35	0,10	1,00	1,60	1,45	1,45	1,32

Table B2.2 Derivation of the characteristic pile resistances $R_{c,k}$ in the ultimate limit state for "flexible" and "stiff" compression piles

flexible			stiff		
$(R_{c,m})_{mean}/\xi_5$	$(R_{c,m})_{min}/\xi_6$	$R_{c,k}$	$(R_{c,m})_{mean}/\xi_5$	$(R_{c,m})_{min}/\xi_6$	$R_{c,k}$
[MN]	[MN]	[MN]	[MN]	[MN]	[MN]
0,650	0,603	0,603	0,717	0,663	0,663

B3 Determining the Characteristic Axial Pile Resistances from Empirical Data for a Bored Pile

B3.1 Objective

Figure B3.1 (example taken from DIN 4014:1990-03) summarises the information on soil type, ground strength and pile geometry necessary for the determination of the axial pile resistance $R_{c,k}(s)$ based on empirical data.

The characteristic resistance-settlement curve shall be determined using the table data after 5.4.6 (Tables 5.12 to 5.15).

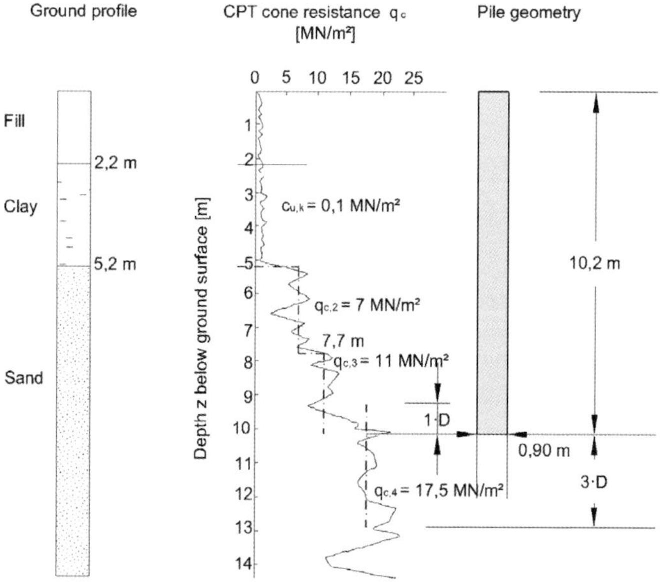

Figure B3.1 Ground profile, penetration test diagram and dimensions for an example calculation of the resistance-settlement curve; D = 0,9 m, U = 2,83 m, A = 0,64 m^2

B3.2 Analysis for lower and upper table values

Note: Reference is made to the application principles and limitations in 5.4.3, in particular with regard to the upper table values. In the example presented here both the lower and the upper table values are used as examples (not as a rule).

B3.2.1 Determining the pile shaft resistance $R_{s,k}$

The ultimate limit state skin friction values for the sand and the clay are given in Tables 5.13 and 5.15 in 5.4.6.2. By adopting the associated pile skin areas, the ultimate limit state pile shaft resistances $R_{s,k,i}$ are provided in Table B3.1.

The settlement s_{sg}, in cm, is calculated as follows, adopting the ultimate limit state pile shaft resistance $R_{s,k}$ in MN:

$$s_{sg} = 0{,}50 \cdot R_{s,k} + 0{,}50.$$

Using the figures from the example the pile head settlement is:

$s_{sg} = 0{,}50 \cdot 1{,}243 + 0{,}50 = 1{,}1$ cm for the lower table values and

$s_{sg} = 0{,}50 \cdot 1{,}726 + 0{,}50 = 1{,}4$ cm for the upper table values.

Table B3.1 Ultimate pile shaft resistance for the lower and upper table values

Stratum i [m]	$A_{s,i}$ [m²]	$c_{u,k,i}$ or $q_{c,i}$ [MN/m²]	$q_{s,k,i}$ [MN/m²]	$R_{s,k,i}$ [MN]
2,20 to 5,20	8,48	0,10	0,039–0,051	0,331–0,432
5,20 to 7,70	7,07	7,00	0,051–0,075[a)]	0,361–0,530
7,70 to 10,20	7,07	11,00	0,078–0,108	0,551–0.764
[a)] Extrapolated data				$R_{s,k}$ = 1,243–1,726 MN

B3.2.2 Determining the pile base resistance $R_{b,k}$

A mean soil strength is adopted in a region from $1 \cdot D$ (0,9 m) above and $3 \cdot D$ ($3 \cdot D = 2{,}70$ m) below the pile base to determine $R_{b,k}$. For this zone a mean cone resistance $q_{c,m} = 17{,}5$ MN/m² is shown in the penetration test diagram in Figure B3.1.

The pile base capacity can be calculated by adopting the figures from Table 5.12 in 5.4.6.2 and taking the previously determined value of $q_{c,m}$ into consideration. Table B3.2 reproduces the calculated figures.

Table B3.2 Pile base resistance for the lower and upper table values

Relative settlement s/D	$q_{b,k}$ [MN/m²]	$R_{b,k}(s)$ [MN]
0,02	1,225–1,625	0,784–1,040
0,03	1,575–2,088	1,008–1,336
0,10	3,250–4,325	2,080–2,768

B3.2.3 Characteristic resistance-settlement curve

The pile resistances calculated from the pile base and pile shaft resistances are listed in Tables B3.3 and B3.4 as a function of the pile head settlement and are given for the lower and upper values. The settlement of the pile head for each value of the pile resistance $R_{c,k}$ is given by the characteristic resistance-settlement curve in Figure B3.2.

Table B3.3 Pile resistance as a function of pile head settlement (lower values)

Relative settlement s/D	Pile head settlement [cm]	$R_{s,k}(s)$ [MN]	$R_{b,k}(s)$ [MN]	$R_{c,k}(s)$ [MN]
s_{sg}	1,1	1,243	0,479	1,722
0,02	1,8	1,243	0,784	2,027
0,03	2,7	1,243	1,008	2,251
0,10	9,0	1,243	2,080	3,323

Table B3.4 Pile resistance as a function of pile head settlement (upper values)

Relative settlement s/D	Pile head settlement [cm]	$R_{s,k}(s)$ [MN]	$R_{b,k}(s)$ [MN]	$R_{c,k}(s)$ [MN]
s_{sg}	1,4	1,726	0,809	2,535
0,02	1,8	1,726	1,040	2,766
0,03	2,7	1,726	1,336	3,062
0,10	9,0	1,726	2,768	4,494

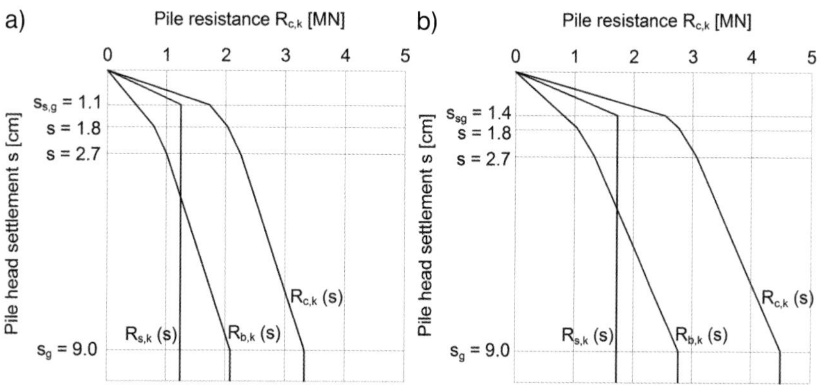

Figure B3.2 Resistance-settlement curve; a) Lower values, b) Upper values

B4 Determining the Characteristic Axial Pile Resistances from Empirical Data for a Prefabricated Driven Pile

B4.1 Objective

Figure B4.1 summarises the information on soil type, ground strength and pile geometry required to determine the axial pile resistance $R_{c,k}(s)$ based on empirical data.

The characteristic resistance-settlement curve shall be determined using the table data after 5.4.4 (Tables 5.1 to 5.4).

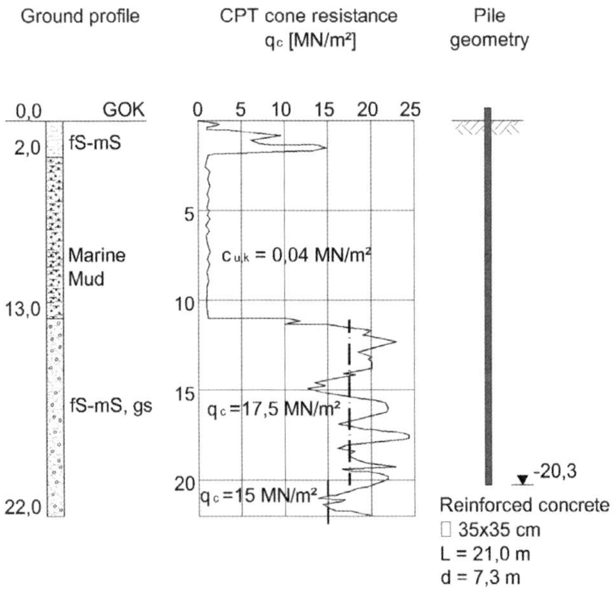

Figure B4.1 Ground profile, penetration test diagram and dimensions for an example calculation for the resistance-settlement curve

B4.2 Characteristic axial pile resistance from empirical data for lower and upper table values

Note: Reference is made to the application principles and limitations in 5.4.3, in particular with regard to the upper table values. In the example presented here both the lower and the upper table values are used as examples (not as a rule).

B4.2.1 Determining the pile shaft resistance $R_{s,k}$

The empirical skin friction values in the zones of the load-bearing non-cohesive soil and the weak cohesive soil are given by Tables 5.2 and 5.4 in Section 5.4.4. Together with the corresponding pile skin areas, taking the correlation factor for the skin area from Table 5.5 in Section 5.4.4 into consideration, the pile shaft resistance upon mobilisation of the ultimate limit state state $R_{s,k}$ (s_{sg^*}) is given in Table B4.1 and the pile shaft resistance $R_{s,k}$ (s_g) at failure in Table B4.2.

Table B4.1 Pile shaft resistance upon mobilisation of the ultimate limit state for the lower and upper table values

Stratum i [m]	$A_{s,i}$ [m²]	$q_{c,i}$ [MN/m²]	$q_{s,k,i}$ (s_{sg^*}) [MN/m²]	η_s [–]	$R_{s,k,i}$ (s_{sg^*}) [MN]
13,0 to 20,3	10,22	17,50	0,070–0,098	1,0	0,715–1,002
					$R_{s,k}$ (s_{sg^*}) = 0,715–1,002 MN

Table B4.2 Shaft resistance at failure for the lower and upper table values

Stratum i [m]	$A_{s,i}$ [m²]	$q_{c,i}$ [MN/m²]	$q_{s,k,i}$ (s_{sg^*}) [MN/m²]	η_s [–]	$R_{s,k,i}$ (s_{sg^*}) [MN]
13,0 to 20,3	10,22	17,50	0,103–0,134	1,0	1,053–1,370
					$R_{s,k}$ (s_g) = 1,053–1,370 MN

Upon mobilisation of the failure state the settlement in cm for the skin friction s_{sg^*}, adopting $R_{s,k}(s_{sg^*})$ in MN, is determined using the following equation for the pile shaft resistance $R_{s,k}(s_{sg^*})$:

$$s_{sg^*} = 0{,}50 \cdot R_{s,k}(s_{sg^*}).$$

Using the figures from the example the pile head settlement is:

$s_{sg^*} = 0{,}50 \cdot 0{,}715 = 0{,}4$ cm for the lower table values and

$s_{sg^*} = 0{,}50 \cdot 1{,}002 = 0{,}5$ cm for the upper table values.

B4.2.2 Determining the pile base resistance $R_{b,k}$

For determination of $R_{b,k}$ a mean soil strength is adopted from $4 \cdot D_{eq}$ below to $1 \cdot D_{eq}$ above the pile base..

The equivalent pile diameter of a square prefabricated driven pile is determined using:

$$D_{eq} = 1{,}13 \cdot a_s.$$

Using the dimensions of the example the equivalent pile diameter is:

$D_{eq} = 1{,}13 \cdot 0{,}35 = 0{,}40$ m.

The nominal value of the square pile base area in this case is:

$A_b = a_s^2 = 0{,}35^2 = 0{,}123$ m².

The penetration test diagram in Figure B4.1 displays a mean characteristic cone resistance along the respective length of:

$$q_{c,m} = \frac{1 \cdot 17{,}5 + 4 \cdot 15{,}0}{5} = 15{,}5\,\mathrm{MN/m^2}.$$

Using the figures in Table 5.1 of these Recommendations and referring to the previously determined value of $q_{c,m}$ and the correlation factor for the pile base area in Table 5.5 (5.4.4), the pile base resistance can be calculated. Table B4.3 contains the determined numerical values.

Table B4.3 Pile base resistance for the lower and upper table values

Relative settlement s/D	$q_{b,k}$ [MN/m²]	η_b [–]	$R_{b,k}(s)$ [MN]
0,035	4,025–6,550	1,0	0,495–0,806
0,100	7,658–10,265	1,0	0,942–1,263

B4.2.3 Characteristic resistance-settlement curve

Tables B4.4 and B4.5 contain the pile resistances calculated for the lower and upper values from the pile base and shaft resistances as a function of the pile head settlement. The settlement of the pile head for each value of the pile resistance $R_{c,k}$ is given by the characteristic resistance-settlement curve in Figure B4.2.

Table B4.4 Pile resistance as a function of pile head settlement (lower values)

Relative settlement s/D	Pile head settlement [cm]	$R_{s,k}(s)$ [MN]	$R_{b,k}(s)$ [MN]	$R_{c,k}(s)$ [MN]
s_{sg^*}	0,4	0,715	0,141	0,856
0,035	1,4	0,809	0,495	1,304
0,100	4,0	1,053	0,942	1,995

Table B4.5 Pile resistance as a function of pile head settlement (upper values)

Relative settlement s/D	Pile head settlement [cm]	$R_{s,k}(s)$ [MN]	$R_{b,k}(s)$ [MN]	$R_{c,k}(s)$ [MN]
s_{sg^*}	0,5	1,002	0,288	1,290
0,035	1,4	1,097	0,806	1,903
0,100	4,0	1,370	1,263	2,633

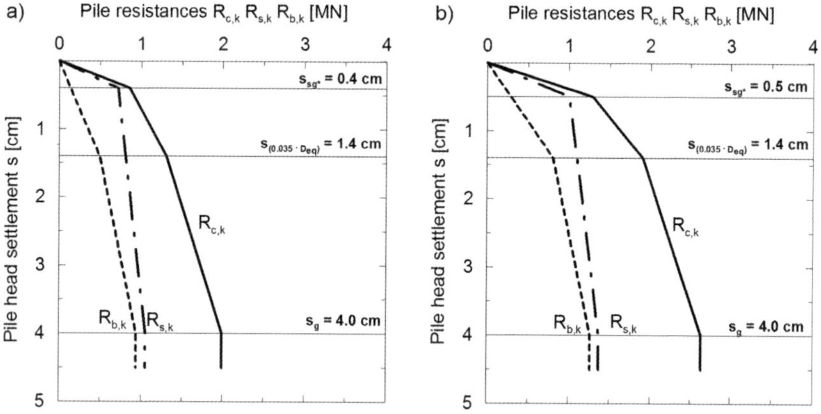

Figure B4.2 Resistance-settlement curve; a) Lower values, b) Upper values

B5 Determining the Characteristic Axial Pile Resistances from Empirical Data for a Fundex Pile

B5.1 Objective

Figure B5.1 summarises the information on soil type, ground strength and pile geometry required for determination of the axial pile resistance $R_{c,k}(s)$ based on empirical data.

The characteristic resistance-settlement curve using the table data after 5.4.8.2 (Tables 5.24 and 5.25) shall be determined.

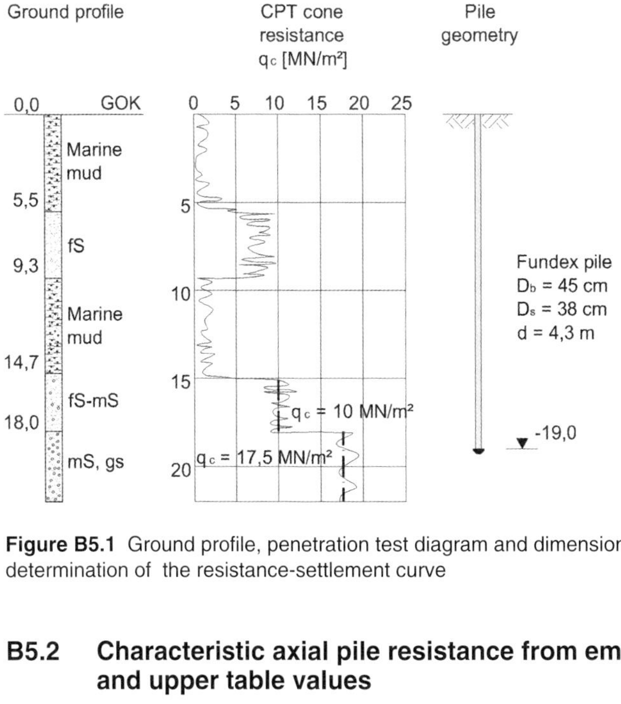

Figure B5.1 Ground profile, penetration test diagram and dimensions for an example for determination of the resistance-settlement curve

B5.2 Characteristic axial pile resistance from empirical lower and upper table values

Note: Reference is made to the application principles and limitations in 5.4.3, in particular with regard to the upper table values. In the example presented here both the lower and the upper table values are used as examples (not as a rule).

B5.2.1 Determining the pile shaft resistance $R_{s,k}$

The ultimate limit state skin friction values in the load-bearing, non-cohesive soil are contained in Table 5.25 in 5.4.8.2. Adopting the associated pile skin areas, the ultimate pile shaft resistance $R_{s,k}$ is provided in Table B5.1.

The settlement in cm for the skin friction s_{sg} with $R_{s,k}$ in MN is determined using the following equation for the pile shaft resistance $R_{s,k}$:

$$s_{sg} = 0{,}5 \cdot R_{s,k} + 0{,}5.$$

Using the figures from the example the pile head settlement is:

$s_{sg} = 0{,}50 \cdot 0{,}316 + 0{,}5 = 0{,}7$ cm for the lower table values and

$s_{sg} = 0{,}50 \cdot 0{,}430 + 0{,}5 = 0{,}7$ cm for the upper table values.

Table B3.1 Ultimate pile shaft resistance for the lower and upper table values

Stratum i [m]	$A_{s,i}$ [m²]	$q_{c,i}$ [MN/m²]	$q_{s,k,i}$ [MN/m²]	$R_{s,k,i}$ [MN]
14,7 to 18,0	3,94	10,00	0,052–0,072	0,205–0,284
18,0 to 19,0	1,19	17,50	0,093–0,123	0,111–0,146
				$R_{s,k}$ = 0,316–0,430 MN

Note: Load-bearing strata with $q_c \geq 7{,}5$ MN/m² are not considered if they are underlain by weak strata.

B5.2.2 Determining the pile base resistance $R_{b,k}$

To determine $R_{b,k}$ a mean soil strength $q_{c,m}$ = 17,5 MN/m² is adopted from $4 \cdot D_{eq}$ below to $1 \cdot D_{eq}$ above the pile base.

The nominal value of the pile base area in this case is:

$$A_b = D_b^2 \cdot \pi/4 = 0{,}45^2 \cdot \pi/4 = 0{,}159 \text{ m}^2.$$

Adopting the figures from Table 5.24 in 5.4.8.2 and taking the previously determined value of $q_{c,m}$ into consideration, the pile base capacity can be calculated. Table B5.2 contains the determined figures.

Table B5.2 Pile base resistance for the lower and upper table values

Relative settlement s/D_b	$q_{b,k}$ [MN/m²]	$R_{b,k}(s)$ [MN]
0,02	2,788–3,413	0,443–0,543
0,03	3,600–4,350	0,572–0,692
0,10	7,475–9,100	1,189–1,447

B5.2.3 Characteristic resistance-settlement curve

The pile resistances calculated from the pile base and pile shaft resistances are given in Tables B5.3 and B5.4 as a function of the pile head settlement for the lower and upper values. The settlement of the pile head for each value of the pile resistance $R_{c,k}$ is derived from the characteristic resistance-settlement curve in Figure B5.2.

Table B5.3 Pile resistance as a function of pile head settlements (lower values)

Relative settlement s/D	Pile head settlement [cm]	$R_{s,k}(s)$ [MN]	$R_{b,k}(s)$ [MN]	$R_{c,k}(s)$ [MN]
s_{sg}	0,7	0,316	0,345	0,661
0,02	0,9	0,316	0,443	0,759
0,03	1,4	0,316	0,572	0,888
0,10	4,5	0,316	1,189	1,505

Table B5.4 Pile resistance as a function of pile head settlements (upper values)

Relative settlement s/D	Pile head settlement [cm]	$R_{s,k}(s)$ [MN]	$R_{b,k}(s)$ [MN]	$R_{c,k}(s)$ [MN]
s_{sg}	0,7	0,430	0,422	0,852
0,02	0,9	0,430	0,543	0,973
0,03	1,4	0,430	0,692	1,122
0,10	4,5	0,430	1,447	1,877

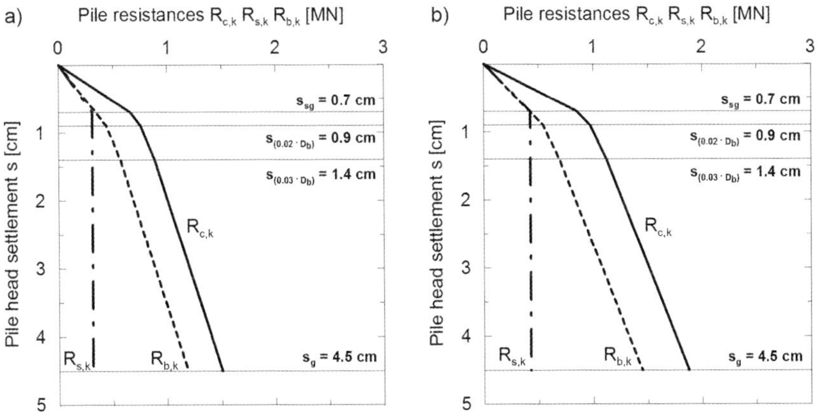

Figure B5.2 Resistance-settlement curve; a) Lower values, b) Upper values

B6 Principle of the Evaluation of a Static Pile Load Test Using a Prefabricated Driven Pile shown on an Example and Comparison with Empirical Data after 5.4.4.2

B6.1 Objective

Figure B6.1 summarises the tested or measured value $R_{c,m}$ from a static load test for the resistance-settlement curve of a prefabricated reinforced concrete driven pile. It also contains the information on soil type, ground strength and pile geometry required to determine the axial pile resistance $R_{c,k}(s)$ based on empirical data after 5.4.4.2.

For the comparison with the results of the static load test, the characteristic resistance-settlement curve shall be determined using the empirical data after 5.4.4.2 (Tables 5.1 to 5.4). In addition, the design resistance in the ultimate limit state (ULS), based on both static load tests and on empirical data, is determined.

Note: Comparison of the axial pile performance derived from a static pile load test with the empirical data after 5.4 requires that the pile skin friction is adopted over the entire length of the pile. The calculation of the resistance-settlement curve after 5.4.4.2 deviates from the previous procedure for determining characteristic axial pile resistance (see example B4) inasmuch as non-load-bearing or weak strata with $q_c \leq 7{,}5$ MN/m² or $c_{u,k} \leq 60$ kN/m² (approximate extrapolation of table values) are taken into consideration.

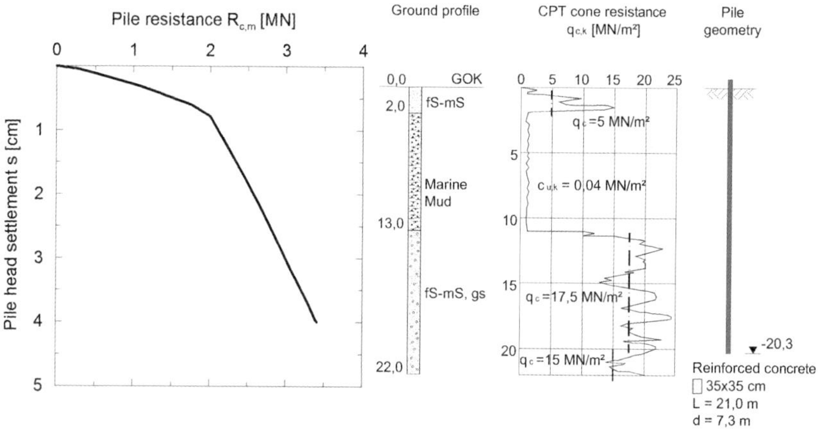

Figure B6.1 Resistance-settlement curve from static pile load tests, ground profile, penetration test diagram and dimensions

B6.2 Characteristic axial pile resistance from empirical lower and upper table values

Note: Reference is made to the application principles and limitations in 5.4.3, in particular with regard to the upper table values (not as a rule). In this example the upper table values are adopted to evaluate the static pile load tests.

B6.2.1 Determining the pile shaft resistance $R_{s,k}$

The empirical skin friction values in the zone of the load-bearing non-cohesive and the weak cohesive soil are given by Tables 5.2 and 5.4 in Section 5.4.4.2 and, together with the corresponding pile skin areas, taking the correlation factor for the skin area from Table 5.5 into consideration, the pile shaft resistance upon mobilisation of the ultimate limit state $R_{s,k}$ (s_{sg^*}) is given in Table B6.1 and the pile shaft resistance $R_{s,k}$ (s_g) at failure in Table B6.2.

Table B6.1 Pile shaft resistance upon mobilisation of the ultimate limit state for the upper table values

Stratum i [m]	$A_{s,i}$ [m²]	$c_{u,k,i}$ or $q_{c,k,i}$ [MN/m²]	$q_{s,k,i}$ (s_{sg^*}) [MN/m²]	η_s [–]	$R_{s,k,i}$ (s_{sg^*}) [MN]
0,0 to 2,0	2,80	5,00	0,027	1,0	0,076
2,0 to 13,0	15,40	0,04	0,020	1,0	0,308
13,0 to 20,3	10,22	17,50	0,098	1,0	1,002
					$R_{s,k}$ (s_{sg^*}) = 1,386 MN

Table B6.2 Pile shaft resistance at failure for the upper table values

Stratum i [m]	$A_{s,i}$ [m²]	$c_{u,k,i}$ or $q_{c,k,i}$ [MN/m²]	$q_{s,k,i}$ (s_g) [MN/m²]	η_s [–]	$R_{s,k,i}$ (s_g) [MN]
0,0 to 2,0	2,80	5,00	0,040	1,0	0,112
2,0 to 13,0	15,40	0,04	0,023	1,0	0,354
13,0 to 20,3	10,22	17,50	0,134	1,0	1,370
					$R_{s,k}$ (s_g) = 1,836 MN

Upon mobilisation of the failure state the settlement in cm for the skin friction s_{sg^*}, adopting $R_{s,k}(s_{sg^*})$ in MN, is determined using the following equation for the pile shaft resistance $R_{s,k}(s_{sg^*})$:

$$s_{sg^*} = 0{,}50 \cdot R_{s,k}(s_{sg^*}).$$

Using the figures from the example analysis the pile head settlement is:

$s_{sg*} = 0{,}50 \cdot 1{,}386 = 0{,}7$ cm.

B6.2.2 Determining the pile base resistance $R_{b,k}$

For determination of $R_{b,k}$ a mean soil strength is adopted in a region from $4 \cdot D_{eq}$ below to $1 \cdot D_{eq}$ above the pile base..

The equivalent pile diameter of a square prefabricated driven pile is determined using the following equation:

$D_{eq} = 1{,}13 \cdot a_s$.

Using the dimensions of the example the equivalent pile diameter is:

$D_{eq} = 1{,}13 \cdot 0{,}35 = 0{,}40$ m.

The nominal value of the square pile base area in this case is:

$A_b = a_s^2 = 0{,}35^2 = 0{,}123$ m².

The penetration test diagram in Figure B6.1 displays a mean characteristic cone resistance along the respective length of:

$$q_{c,m} = \frac{1 \cdot 17{,}5 + 4 \cdot 15{,}0}{5} = 15{,}5 \text{ MN/m}^2 .$$

Using the figures given in Table 5.1 of 5.4.4.2 and taking the previously determined value of $q_{c,m}$ and the correlation factor for the pile type as given in Table 5.5 into consideration, the pile base resistance can be calculated. Table B6.3 contains the determined figures.

Table B6.3 Pile base resistance for the upper table values

Relative settlement s/D	$q_{b,k}$ [MN/m²]	η_b [–]	$R_{b,k}(s)$ [MN]
0,035	6,550	1,0	0,806
0,100	10,265	1,0	1,263

B6.2.3 Characteristic resistance-settlement curve for empirical data compared to tested or measured values

The pile resistance calculated for the upper values from empirical data of the pile base and pile shaft resistance is given in Table B6.4 as a function of the pile head settlement. The settlement of the pile head for each value of the pile resistance $R_{c,k}$ is derived from the characteristic resistance-settlement curve in Figure B6.2 as an upper empirical value $R_{c,k}(s)$ and is listed for comparison with the tested or measured data from the load tests $R_{c,m}(s)$.

Table B6.4 Pile resistances as a function of the pile head settlement (upper empirical values and tested or measured data)

Relative settlement s/D	Pile head settlement [cm]	$R_{s,k}(s)$ [MN]	$R_{b,k}(s)$ [MN]	$R_{c,k}(s)$ [MN]	$R_{c,m}(s)$ [MN]
s_{sg^*}	0,7	1,386	0,403	1,789	1,889
0,035	1,4	1,481	0,806	2,287	2,298
0,100	4,0	1,836	1,263	3,099	3,408

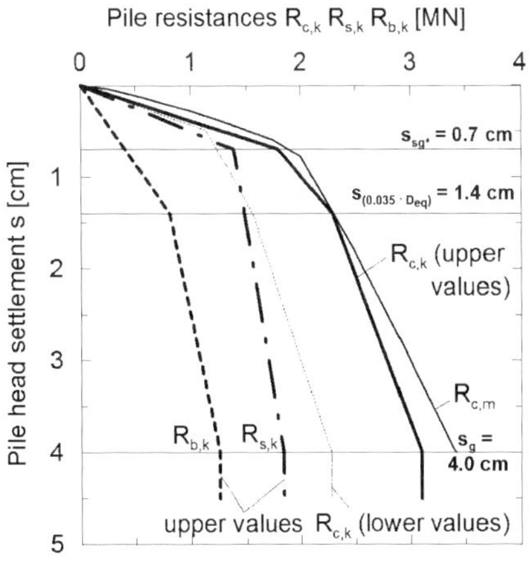

Figure B6.2 Resistance-settlement curve for the upper empirical values compared to the values measured in load tests; the RSC for the lower empirical values is included for orientation

B6.3 Characteristic axial pile resistance from static load tests

The characteristic pile resistance $R_{c,k}$ for the ultimate limit state (ULS) shall be derived from $R_{c,m}$ using Equation (A4.1) and adopting the correlation factors ξ_1 and ξ_2 from Table A4.1.

$$R_{c,k} = \text{MIN}\left\{\frac{(R_{c,m})_{mean}}{\xi_1}; \frac{(R_{c,m})_{min}}{\xi_2}\right\}$$

Where only one pile load test is used, the correlation values are $\xi_1 = \xi_2$. If the structural system is sufficiently stiff, "stiff" compression piles can be adopted by dividing the correlation factors by 1,1, whereby ξ_1 may never become smaller than 1. The determined correlation factors are:

- $\xi_1 = \xi_2 = 1{,}35$ for flexible systems;
- $\xi_1 = \xi_2 = 1{,}23$ for stiff systems.

The characteristic pile resistances $R_{c,k}$ are given by the measured pile resistance in the ultimate limit state (ULS) $R_{c,m}$ = 3,408 MN after Table B6.4 using:

- $R_{c,k}$ = 2,524 MN for flexible systems;
- $R_{c,k}$ = 2,771 MN for stiff systems.

B6.4 Design values of pile resistances in the ultimate limit state

a) From empirical data

The design resistances in the ultimate limit state (ULS) based on empirical data $R_{c,d}$ are determined by dividing the characteristic pile resistances $R_{c,k}$ by the partial factor γ_t = 1,4 after the table in Annex A3.2.

The design values of the upper and lower table values are derived and summarised in Table B6.5. The lower table values are of comparative character only.

Table B6.5 Design values in the ultimate limit state (ULS) for upper and lower table values

	$R_{c,k}$ [MN]	γ_t	$R_{c,d}$ [MN]
Upper table values	3,099	1,40	2,214
Lower table values	2,271	1,40	1,622

b) From the pile load test

To derive the design values of the pile resistances $R_{c,d}$ in the ultimate limit state from the pile load test, the characteristic pile resistances $R_{c,k}$ after B6.3

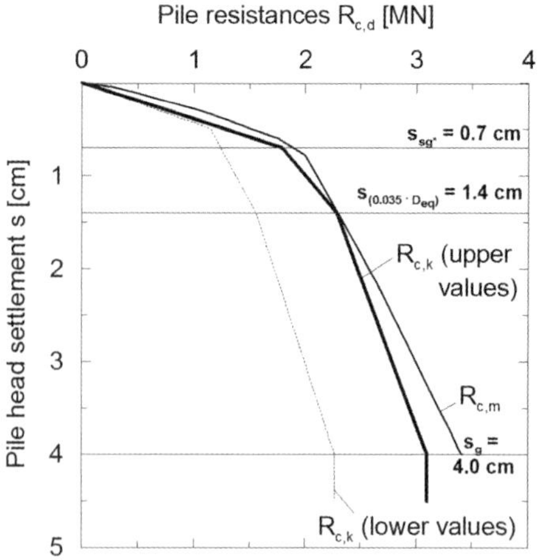

from experienced datas:
□ $R_{c,d}$ (lower values) = 2.214 MN
○ $R_{c,d}$ (upper values) = 1.622 MN
from pile load tests:
■ $R_{c,d}$ (stiff systems) = 2.519 MN
● $R_{c,d}$ (soft systems) = 2.295 MN

Figure B6.3 Comparison of the design values in the ultimate limit state (ULS) from empirical data and those from the static pile load test are divided by the partial factor $\gamma_t = 1,1$ from Table A3.2. The design values are given by:

- $R_{c,d} = 2,524$ MN/1,1 = 2,295 MN for flexible systems;
- $R_{c,d} = 2,771$ MN/1,1 = 2,519 MN for stiff systems.

The design values in the ultimate limit state (ULS) from empirical data and those from the static pile load test are compared in Figure B6.3.

B7 Preliminary Design and Analysis of the Ultimate Limit State of Franki Piles Based on Empirical Data and Comparison to a Pile Load Test Result

B7.1 Objective

Figure B7.1 summarises the information on soil type, ground strength and pile geometry required to determine the axial pile resistance $R_{c,k}(s)$ based on empirical data after 5.4.5.3.

The pile base volume necessary to accept a characteristic action F_k based on empirical data is required for the preliminary design of a Franki pile. In this example, initially the lower empirical values after 5.4.5.3 are adopted, whereby the pile base is designed for a characteristic permanent action $F_{G,k}$ = 1,20 MN and a characteristic variable action $F_{Q,k}$ = 0,40 MN.

Analysis of the ultimate limit state (ULS, GEO-2) must be performed for the necessary pile base volume determined in the preliminary design for the characteristic action, adopting the upper and lower empirical values, based on the actual driving energy expended during pile installation as shown in Figure B7.2.

To check the pile resistance based on empirical data after 5.4.5.3 the pile resistance in the ultimate limit state is compared to the static load test in Fig-

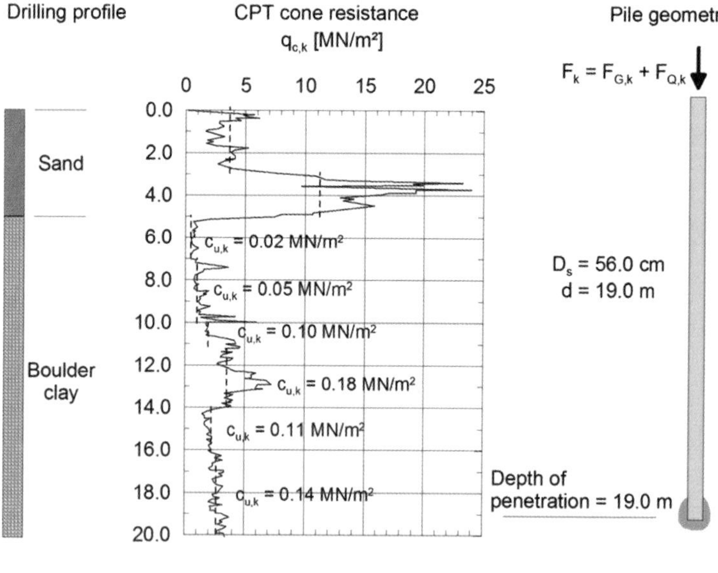

Figure B7.1 Ground profile, penetration test diagram and dimensions ure B7.2. To this end, the characteristic pile resistance must be determined using the upper table values and taking into consideration the non-load-bearing strata, and compared to the result of the static load test.

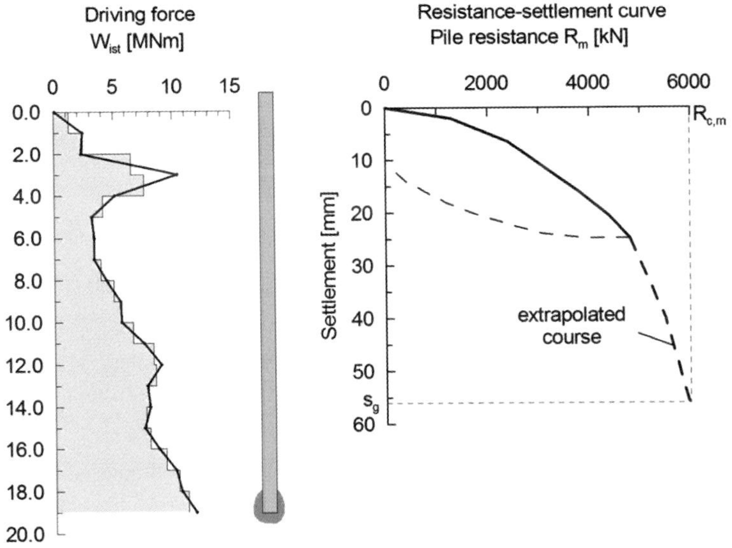

Figure B7.2 Driving energy and logged data from the static load test

B7.2 Determining the base volume from empirical data

B7.2.1 Determining the pile shaft resistance $R_{s,k}$

The empirical skin friction values in the zone of the load-bearing boulder clay from a depth of 0,80 m above driving depth and, adopting the associated pile skin areas, the ultimate pile shaft resistance $R_{s,k}$ after Table B7.1, are given in Table 5.11 in 5.4.5.3.

Table B7.1 Ultimate pile shaft resistance for the lower table values

Stratum i [m]	$A_{s,i}$ [m²]	$c_{u,k,i}$ [MN/m²]	$q_{s,k,i}$ [MN/m²]	$R_{s,k,i}$ [MN]
10,0 to 11,0	1,76	0,100	0,044	0,077
11,0 to 14,0	5,28	0,180	0,060	0,317
14,0 to 16,0	3,52	0,110	0,046	0,162
16,0 to 18,2	3,87	0,140	0,053	0,205
				$R_{s,k}$ = 0,761 MN

Note: The non-load-bearing strata and the strata overlying them were not taken into consideration for the preliminary design. The empirical ultimate skin friction for cohesive soils after Table 5.11 can be used for the boulder clay.

B7.2.2 Determining the pile base volume of a Franki pile

The Franki pile is designed for a characteristic action $F_k = F_{G,k} + F_{Q,k}$. For Loading Case 1 this means that a characteristic axial pile resistance of:

$$R_{c,k} = \left(F_{G,k} \cdot \gamma_G + F_{Q,k} \cdot \gamma_Q\right) \cdot \gamma_t = (1{,}20 \cdot 1{,}35 + 0{,}40 \cdot 1{,}50) \cdot 1{,}40$$
$$= 3{,}108 \text{ MN}$$

must be verified. The necessary pile base resistance is given in Table B7.2.

A mean soil strength is adopted in the zone from $3 \cdot D_s$ below to $2 \cdot D_s$ above driving depth to determine $R_{b,k}$. A mean undrained shear strength $c_{u,k} = 0{,}140$ MN/m² is given for this zone by the penetration test diagram in Figure B7.1.

Adopting the undrained shear strength $c_{u,k}$ and the necessary characteristic pile base resistance $R_{b,k}$ from Table B7.2 gives the pile base volume for the lower empirical values in Figure 5.10. The necessary base volume of the Franki pile for the lower empirical values is given in Table B7.2.

Table B7.2 Necessary characteristic pile base resistance and pile base volume

$R_{c,k}$ [MN]	$R_{s,k}$ [MN]	$R_{b,k}$ [MN]	V [m³] (from Figure 5.9)
3,108	0,761	2,347	1,05

B7.3 Analysis of the ultimate limit state (ULS, GEO-2) by means of the driving energy expended during pile installation

B7.3.1 Characteristic pile resistance $R_{c,k}$ after applying the lower empirical values

The ultimate pile shaft resistance after Table B7.1 is:

$$R_{s,k} = 0{,}761 \text{ MN.}$$

The driving energy for the final 2 m is given by the driving energy diagram in Figure B7.2 as:

$$W_{actual} = 10{,}410 + 11{,}183 = 21{,}593 \text{ MNm}.$$

The norm driving energy after 5.4.5.3, Table 5.9, for a Franki pile with $D_s = 56{,}0$ cm is:

$$W_{norm} = 30{,}469 \text{ MNm}$$

and the norm driving energy ratio:

$$W = \frac{W_{actual}}{W_{norm}} = 0{,}71 .$$

The pile base resistance $R_{b,k}$ is given using the standard driving work ratio and the existing pile base volume after Table B7.2 from Figure 5.10:

$$R_{b,k} = 2{,}675 \text{ MN}.$$

The axial compression pile resistance in the ultimate limit state from empirical data is given by:

$$R_{c,k} = R_{b,k} + R_{s,k} = 2{,}675 + 0{,}761 = 3{,}436 \text{ MN}.$$

B7.3.2 Characteristic pile resistance $R_{c,k}$ after applying the upper empirical values

Note: Reference is made to the application principles and limitations in 5.4.3, in particular with regard to the upper empirical values (not as a rule).

The ultimate limit state pile shaft resistance using the upper values from Table 5.11 is given in Table B7.3.

Using the norm driving energy ratio:

$$W = \frac{W_{ist}}{W_{norm}} = 0{,}71$$

and the existing pile base volume after Table B7.2, the pile base resistance $R_{b,k}$ is given by Figure 5.11.

$$R_{b,k} = 3{,}890 \text{ MN}.$$

Table B7.3 Ultimate pile shaft resistance for the upper table values

Stratum i [m]	$A_{s,i}$ [m²]	$c_{u,k,i}$ [MN/m²]	$q_{s,k,i}$ [MN/m²]	$R_{s,k,i}$ [MN]
10,0 to 11,0	1,76	0,100	0,056	0,099
11,0 to 14,0	5,28	0,180	0,076	0,401
14,0 to 16,0	3,52	0,110	0,059	0,208
16,0 to 18,2	3,87	0,140	0,067	0,259
				$R_{s,k} = 0{,}967$ MN

The axial compression pile resistance in the ultimate limit state from empirical data is:

$$R_{c,k} = R_{b,k} + R_{s,k} = 3{,}890 + 0{,}967 = 4{,}857 \text{ MN}.$$

B7.3.3 Ultimate limit state analysis

The design value of the compressive pile resistance is given by:

$$R_{c,d} = \frac{R_{c,k}}{\gamma_t}$$

and the design value of the action by:

$$F_{c,d} = F_{G,k} \cdot \gamma_G + F_{Q,k} \cdot \gamma_Q \, .$$

Where $\gamma_t = 1{,}40$, $\gamma_G = 1{,}35$ and $\gamma_Q = 1{,}50$ for the DS-P design situation.

$$F_{c,d} \leq R_{c,d}$$

must be adhered to for analysis of the ultimate limit state. The utilisation factors are given in Table B7.4.

Table B7.4 Utilisation factors for upper and lower table values

	$R_{c,d}$ [MN]	$F_{c,d}$ [MN]	μ [–]
Lower table values	2,454	2,220	0,91
Upper table values	3,476	2,220	0,64

B7.4 Comparison of the axial pile resistance based on empirical data with static load tests

B7.4.1 Characteristic axial pile resistance from empirical data

The characteristic axial pile resistance based on empirical data from a static load test is examined.

Note: Comparison of the axial pile performance derived from a static pile load test with the empirical data after 5.4.5.3 requires that the pile skin friction is adopted over the entire length of the pile. The calculation of the pile resistance deviates from the previous procedure for determining characteristic axial pile resistance inasmuch as non-load-bearing or weak strata with $q_c \leq 7{,}5$ MN/m² or $c_{u,k} \leq 0{,}06$ MN/m² are taken into consideration. Reference is made also to the application principles and limitations in 5.4.3, in particular with regard to the upper table values (not as a rule). In this example the upper table values are adopted to evaluate the static pile load tests.

The empirical skin friction values in the zone of the load-bearing boulder clay from a depth of 0,80 m above driving depth are given in Tables 5.10 and 5.11 of these Recommendations and, adopting the associated pile skin areas, the ultimate pile shaft resistance $R_{s,k}$ after Table B7.5.

Table B7.5 Ultimate pile shaft resistance for the upper table values

Stratum i [m]	$A_{s,i}$ [m²]	q_c or $c_{u,k,i}$ [MN/m²]	$q_{s,k,i}$ [MN/m²]	$R_{s,k,i}$ [MN]
0,0 to 3,0	5,28	3,60	0,066	0,348
3,0 to 5,0	3,52	11,10	0,121	0,426
5,0 to 7,0	3,52	0,020	0,034	0,120
7,0 to 10,0	5,28	0,050	0,042	0,222
10,0 to 11,0	1,76	0,100	0,056	0,099
11,0 to 14,0	5,28	0,180	0,076	0,401
14,0 to 16,0	3,52	0,110	0,059	0,208
16,0 to 18,2	3,87	0,140	0,067	0,259
				$R_{s,k}$ = 2,083 MN

Using the standard driving work ratio W = 0,71 after B7.3.2 and the pile base volume V = 1,05 m³ from Figure 5.11 the pile base resistance $R_{b,k}$ is:

$$R_{b,k} = 3,890 \text{ MN}.$$

The axial pile resistance based on empirical data is:

$$R_{c,k} = R_{b,k} + R_{s,k} = 3,890 + 2,083 = 5,973 \text{ MN}.$$

B7.4.2 Comparing to the static load test

Extrapolating the resistance-settlement curve from Figure B7.2 using the hyperbola method gives the axial pile resistance for an ultimate settlement $s_{ult} = s_g = 0,1 \cdot D_s$:

$$R_{c,m} = 5,966 \text{ MN}.$$

The deviation in the calculated pile resistance from the measured value is $\Delta R_1 = -0,1\%$.

In this case the upper empirical data thus correspond well to the results of the static load test taken for reference.

B8 Negative Skin Friction for a Displacement Pile as a Result of Fill

B8.1 Objective

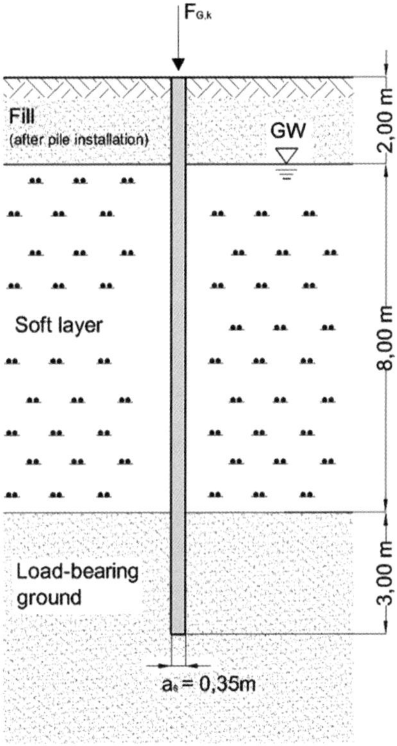

Figure B8.1 System and boundary conditions

Table B8.1 Characteristic soil properties

Stratum	Soil properties
Fill (sand)	φ'_k = 30° γ_k = 16,0 kN/m³
Soft stratum	$c_{u,k}$ = 35,0 kN/m²

The ultimate and serviceability limit states shall be analysed for a prefabricated, square, reinforced concrete pile with an edge length $a_s = 0{,}35$ m and a permanent action $F_{G,k} = 450$ kN imposed by structural loads. The results of a static pile load test are available as shown in Figure B8.2 and Table B8.2. It is known from a settlement analysis that the soft stratum will settle by 5 cm below the assumed infinite fill. The settlements caused by the fill in the load-bearing ground can be ignored. In the serviceability limit state (SLS) a maximum pile head settlement allow. $s_k = 0,5$ cm is permitted, which in this case is compared to a lower bound of the characteristic RSC in the serviceability limit state. The pile is also assumed to be rigid.

B8.2 Determining the characteristic resistance-settlement curve

The characteristic pile resistance in the ultimate limit state (ULS) $R_{c,k}$ is given by the measured value for $R_{c,m}$ from the load test using Eq. (A4.1) as follows:

$$R_{c,k} = \frac{R_{c,m}}{\xi_1}$$

where the correlation factor is $\xi_1 = 1,35$ after Annex A.4. The correlation factors given there refer to "flexible" compression piles and are applied for single pile behaviour, as is assumed here.

For the range of small pile settlements, after 6.4, Eq. (6.13) characteristic boundary lines with values of $\kappa = 0,15$ (based on [59]), were derived to form the basis for the serviceability limit state analysis, here applied in approximation to the measured values in the serviceability limit state.

Note: The method adopted here for the characteristic boundary lines in the service load range is only one possible option. Other reasoned procedures are also possible.

Figure B8.2 and Table B8.2 show the results determined for $R_{c,k}$ (SLS) and $R_{c,k} = R_{c,k}$ (ULS).

In the ultimate limit state (ULS), $s_{ult} = s_g = 0,10 \cdot D_b$ shall be adopted for the pile head settlement, if no other criteria are selected. An equivalent pile diameter $D_{eq} = 39,5$ cm $= 0,395$ m is given for the square pile.

$$s_{ult} = s_g = 0,10 \cdot 39.5 = 3,95 \text{ cm} \approx 4 \text{ cm}.$$

The characteristic pile resistance in the ultimate limit state ULS in Table B8.2 is thus:

$$R_{c,k} = 1,176 \text{ kN}.$$

The characteristic pile resistance $R_{c,k}$(SLS) in the serviceability limit state for allow. s = 0,5 cm from Figure B8.2 and Table B8.2 is:

$$R_{c,k}(SLS) = 850 \text{ kN}.$$

Table B8.2 Pile load test results and derived characteristic pile resistances

s [cm]	$R_{c,m}$ [kN]	κ [–]	$R_{c,k}$ [kN]
0,0	0	–	0
0,5	978	0,15	$R_{c,k}$(SLS) = 850
1,0	1 198	0,15	1 042
1,5	1 320		
2,0	1 410		
3,0	1 532	ξ_1	–
$s_g \approx 4{,}0$	1 587	1,35	$R_{c,k}$ = 1 176

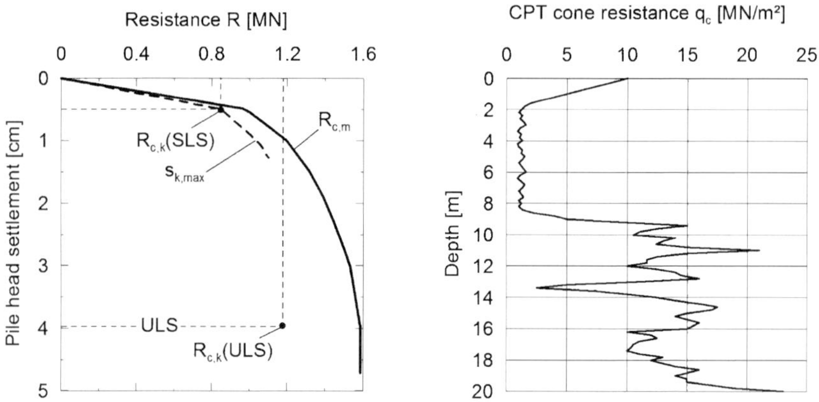

Figure B8.2 Results of a pile load test and of a cone penetration test, and the derived characteristic resistance-settlement curve

A settlement of exist. $s_k \approx 0{,}3$ cm results from the characteristic RSC for the existing characteristic action $F_{G,k} = 450$ kN imposed on the pile by structural loads in the serviceability limit state (SLS).

Note: It is assumed in approximation that the pile resistances in the pile load test only result from the load-bearing soil and the strata above make no contribution.

B8.3 Determining the characteristic actions $F_{n,k}$ from negative skin friction

In Figure B8.3 the pile settlements under the effect $F_{G,k}$ for the ultimate and serviceability limit states are compared to the settlements of the soft stratum.

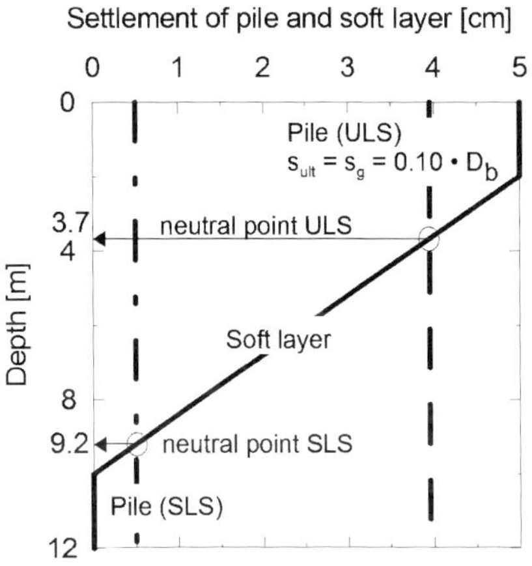

Figure B8.3 Determining the neutral points from the settlements of the pile and the soft stratum for ULS and SLS

In the ultimate limit state (ULS) the neutral point is located 3,7 m below the ground surface and within the soft stratum. The characteristic actions from negative skin friction shown in Table B8.3 are determined using Eq. (4.1) for the soft stratum and Eq. (4.2) for the fill.

Table B8.3 Actions from negative skin friction in the ultimate limit state

Depth [m]	d_i [m]	$A_{s,i}$ [m²]	$\tau_{n,k,i}$ [kN/m²]	$F_{n,k,i}$ [kN]
0,00	2,00	2,80	0,0	12,9
2,00			9,2	
2,00	1,70	2,38	35,0	83,3
3,70				
				$F_{n,k,ULS}$ = 96,2 kN

After determining the existing settlement s ≈ 0,3 cm for the serviceability limit state (SLS), the neutral point is located at a depth of 9,5 m below grade. This deviates from Figure B8.3 where the neutral point is determined for s ≈ 0,5 cm; the difference is however ignored. The characteristic actions from negative skin friction are determined in Table B8.4 for the serviceability limit state.

Table B8.4 Actions from negative skin friction in the serviceability limit state

Depth [m]	d_i [m]	$A_{s,i}$ [m²]	$\tau_{n,k,i}$ [kN/m²]	$F_{n,k,i}$ [kN]
0,00	2,00	2,80	0,0	12,9
2,00			9,2	
2,00	7,50	10,50	35,0	367,5
9,50				
				$F_{n,k,SLS}$ = 380,4 kN

B8.4 Bearing capacity analysis

For analysis of the ultimate limit state the limit state condition must be adhered to:

$$F_{c,d} \leq R_{c,d}$$

The negative skin friction is adopted here as a permanent action analogous to 4.4.3 (1) in the DS-P design situation.

$$F_{c,d} = (F_{G,k} + F_{n,k,ULS}) \cdot \gamma_G = 450{,}0 \cdot 1{,}35 + 96{,}2 \cdot 1{,}35 = 737{,}4 \text{ kN}$$

$$R_{c,d} = R_{c,k}/\gamma_t = 1\,176{,}0/1{,}10 = 1\,069{,}1 \text{ kN}$$

$$F_{c,d} = 737{,}4 \text{ kN} < R_{c,d} = 1\,069{,}1 \text{ kN}$$

B8.5 Serviceability analysis

In the serviceability limit state the limit state condition:

$$F_{c,d}(SLS) \leq R_{c,d}(SLS)$$

must be adhered to.

$$F_{c,d}(SLS) = F_{c,k} = F_{G,k} + F_{n,k,SLS} = 450{,}0 + 380{,}4 = 830{,}4 \text{ kN}$$

$$R_{c,d}(SLS) = R_{c,k}(SLS) = 850{,}0 \text{ kN}$$

$$F_{c,d}(SLS) = 830{,}4 \text{ kN} < R_{c,d}(SLS) = 850{,}0 \text{ kN}$$

Note: The adopted pile settlements s(SLS) and $s_{ult} = s_g$ resulting from structural loads are increased slightly by the characteristic pile effect from negative skin friction, in particular in terms of s(SLS). Iteration can normally be dispensed with, because the procedure shown here adopting s(SLS) ≈ 0,3 cm (without negative skin friction) is conservative and normally leads to an unfavourable characteristic pile effect $F_{c,k}$ as a result of the large depth of the neutral point in the serviceability limit state.

B8.6 Analysis of internal capacity (structural failure)

For analysis of internal capacity (structural failure) the largest effect at the depth of the neutral point in the serviceability limit state is given by:

exist. $F_{c,k}(SLS) = F_{G,k} + F_{n,k,SLS} = F_k$
(for internal pile capacity design).

The internal pile design (structural failure of the pile material) in the ultimate limit state shall be done using this effect $F_k = F_{G,k} + F_{n,k,SLS}$.

B9 Determining the Effect on a Laterally Loaded Pile (Perpendicular to the Pile Axis) and Analysis of Structural Failure

B9.1 Objective

The internal capacity of the pile shown in Figure B9.1 is analysed. The soil resistance and the effect on the pile must first be determined. The pile is then designed, adhering to the corresponding internal pile capacity analyses.

The following characteristic actions are taken into consideration:

Permanent vertical action:	$F_{G,k} = 3{,}333$ MN
Variable vertical action:	$F_{Q,rep} = 2{,}000$ MN
Permanent horizontal action:	$H_{G,k} = 0{,}600$ MN
Variable horizontal action:	$H_{Q,rep} = 0{,}400$ MN.

The system comprising the characteristic actions, the ground profile and the depth-dependent subgrade modulus distribution is shown in Figure B9.1.

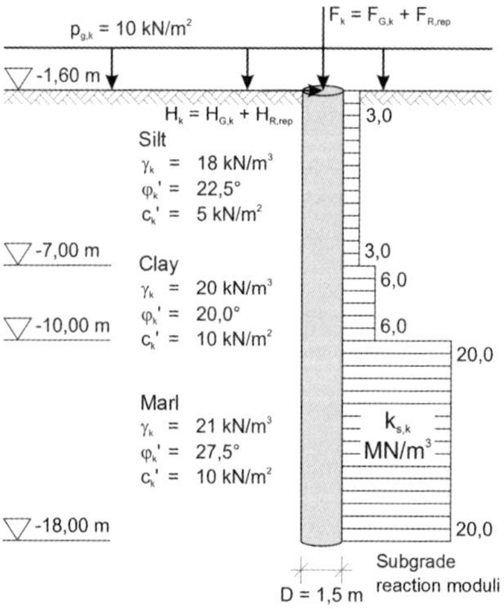

Figure B9.1 System and boundary conditions

According to 6.3.2 the maximum mobilisable characteristic normal stress $\sigma_{h,k}$ between the pile and the soil is limited by the characteristic passive earth pressure $e_{ph,k}$ calculated for the plane state.

$$\sigma_{h,k} \leq e_{ph,k} \tag{B9.1}$$

The design value of the characteristic lateral normal stress resultant, referred to as the lateral soil resistance $B_{h,d}$, may not exceed the design value of the three-dimensional passive earth pressure $E^{r}_{ph,d}$ for the embedment length of the pile as far as the pivot point (displacement zero point).

$$B_{h,d} \leq E^{r}_{ph,d} \tag{B9.2}$$

Adherence to the conditions arising from Eq. (B9.1) and Eq. (B9.2) is to be checked prior to pile design.

B9.2 Determining the characteristic action effects and stresses

Figure B9.2 b) shows the pile's characteristic mobilised subgrade stress from the structural analysis of an elastically supported beam, adopting the subgrade reaction moduli from Figure B9.2 a). The distribution of the characteristic passive earth pressure is shown in Figure B9.2 c). It is clear from the superimposed subgrade stress and passive earth pressure in Figure B9.2 d) that the characteristic mobilised subgrade stresses $\sigma_{h,k}$ in the upper region of the pile exceed the characteristic passive earth pressure $e_{ph,k}$. The approach for the subgrade reaction modulus distribution is modified such that Eq. (B9.1) is adhered to.

Figure B9.3 a) shows the iteratively determined subgrade reaction moduli, reduced in the upper pile region, which are subsequently adopted for further design. The corresponding characteristic mobilised subgrade stress is shown in Figure B9.3 b) and the characteristic passive earth pressure in Figure B9.3 c). The superimposed subgrade stress and passive earth pressure in Figure B9.3 d) show that the values of the characteristic mobilised normal stresses $\sigma_{h,k}$

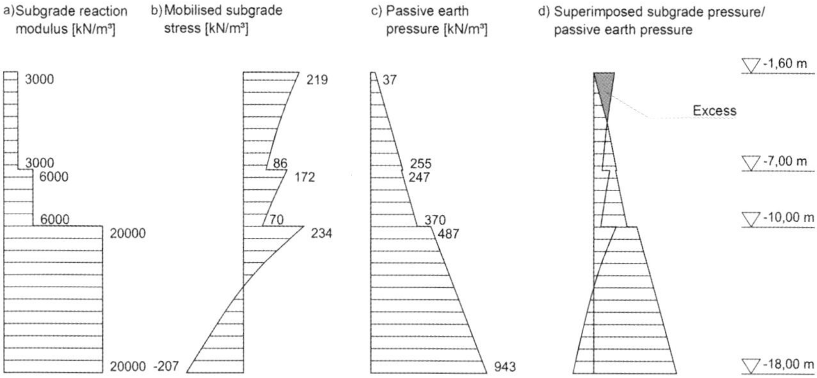

Figure B9.2 Subgrade reaction moduli distribution (a) and characteristic mobilised subgrade stress from structural analysis (b), characteristic plane state passive earth pressure (c) and superimposed subgrade stress and passive earth pressure (d)

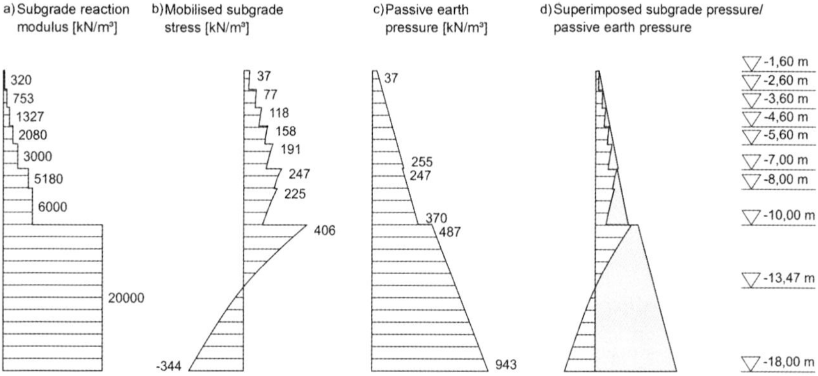

Figure B9.3 Adapted subgrade reaction moduli distribution (a) and characteristic mobilised subgrade stress from structural analysis (b), characteristic plane state passive earth pressure (c) and superimposed subgrade stress and passive earth pressure (d)

are now below the characteristic passive earth pressure $e_{ph,k}$. The condition in Eq. B9.1 is thus met.

While determining the action effects and stresses it must also be demonstrated that the resulting lateral soil resistance is smaller than the design value of the passive earth pressure as far as the pile pivot (zero displacement point).

The lateral soil resistance is given by integrating the mobilised subgrade stress from the pile head to the pivot (displacement zero point), located at an elevation of –13.47 m, using beam analysis. The characteristic soil resistance forces given by integrating the stress areas as shown in Figure B9.3 b) are:

$$B_{hg,k} = 1091 \text{ kN}$$

$$B_{hq,rep} = 727 \text{ kN}$$

The design value of the soil resistance is therefore:

$$B_{h,d} = \gamma_G \cdot B_{hg,k} + \gamma_Q \cdot B_{hq,rep} = 1{,}35 \cdot 1\,091 + 1{,}5 \cdot 727 = 2\,563 \text{ kN} \qquad \text{(B9.3)}$$

The three-dimensional passive earth pressure as a result of the self-weight of the soil is calculated to DIN 4085:2007-10 as follows:

$$E^{r}_{ph,k} = \gamma \cdot \frac{h^2}{2} \cdot K_{pgh} \cdot D^{Er}_{pg} + c \cdot h \cdot D^{Er}_{pc} \qquad \text{(B9.4)}$$

Where:

$D < 0{,}3 \cdot z$: $D^{Er}_{pg} = 0{,}55 \cdot (1 + 2 \cdot \tan\varphi) \cdot \sqrt{D \cdot z}$ $D^{Er}_{pc} = 1{,}1 \cdot (1 + 0{,}75 \cdot \tan\varphi) \cdot \sqrt{D \cdot z}$

$D \geq 0{,}3 \cdot z$: $D^{Er}_{pg} = D + 0{,}6 \cdot z \cdot \tan\varphi$ $D^{Er}_{pc} = D + 0{,}3 \cdot (1 + 1{,}5 \cdot \tan\varphi)$

D: Pile diameter
z: Depth in question.

For the elevation here of –13,47 m:

$$D < 0{,}3 \cdot z \qquad (1{,}5\ \text{m} < 3{,}56\,\text{m} = 0{,}3 \cdot (13{,}47 - 1{,}60)$$

Silt:
$$D_{pg}^{Er} = 0{,}55 \cdot (1 + 2 \cdot \tan 22{,}5°) \cdot \sqrt{1{,}5 \cdot (13{,}47 - 1{,}60)} = 4{,}24$$
$$D_{pc}^{Er} = 1{,}1 \cdot (1 + 0{,}75 \cdot \tan 22{,}5°) \cdot \sqrt{1{,}5 \cdot (13{,}47 - 1{,}60)} = 6{,}08$$

Clay:
$$D_{pg}^{Er} = 0{,}55 \cdot (1 + 2 \cdot \tan 20°) \cdot \sqrt{1{,}5 \cdot (13{,}47 - 1{,}60)} = 4{,}01$$
$$D_{pc}^{Er} = 1{,}1 \cdot (1 + 0{,}75 \cdot \tan 20°) \cdot \sqrt{1{,}5 \cdot (13{,}47 - 1{,}60)} = 5{,}91$$

Marl:
$$D_{pg}^{Er} = 0{,}55 \cdot (1 + 2 \cdot \tan 27{,}5°) \cdot \sqrt{1{,}5 \cdot (13{,}47 - 1{,}60)} = 4{,}74$$
$$D_{pc}^{Er} = 1{,}1 \cdot (1 + 0{,}75 \cdot \tan 27{,}5°) \cdot \sqrt{1{,}5 \cdot (13{,}47 - 1{,}60)} = 6{,}45$$

The three-dimensional passive earth pressure $E_{ph,k}^{r}$ to the elevation of –13.47 m is given by the passive earth pressure distribution in the planar case as follows:

$$\begin{aligned} E_{ph,k}^{r} &= \frac{1}{2} \cdot (10 + 107{,}2) \cdot 5{,}40 \cdot 2{,}240 \cdot 4{,}24 + 5 \cdot 5{,}4 \cdot 6{,}08 \\ &+ \frac{1}{2} \cdot (107{,}2 + 167{,}2) \cdot 3{,}00 \cdot 2{,}040 \cdot 4{,}01 + 10 \cdot 1{,}92 \cdot 5{,}91 \\ &+ \frac{1}{2} (167{,}2 + 240{,}0) \cdot 3{,}47 \cdot 2{,}716 \cdot 4{,}74 + 10 \cdot 1{,}92 \cdot 6{,}45 \\ &= 15\,869\,\text{kN} \end{aligned}$$

Applying the partial factor $\gamma_{R,e}$ the design value of the three-dimensional passive earth pressure is:

$$E_{ph,d}^{r} = \frac{E_{ph,k}^{r}}{\gamma_{R,e}} = \frac{15\,869}{1{,}4} = 11\,335\ \text{kN}\,.$$

For analysis of the lateral soil resistance:

$$B_{h,d} = 2\,563\ \text{kN} < 11\,335\ \text{kN} = E_{ph,d}^{r}\,.$$

The three-dimensional passive earth pressure is not exceeded.

B9.3 Design values of the action effects

The design values of the action effects are given by adopting the partial factors for the limit state in the structure and the ground (STR and GEO-2) after the EC 7-1 Handbook [44], also see Annex A2. The actions are allocated to the permanent design situation DS-P.

The design value of the normal force is:

$$N_{Ed} = N_{Ek,G} \cdot \gamma_G + N_{Ek,Q,rep} \cdot \gamma_Q = 3\,333 \cdot 1{,}35 + 2\,000 \cdot 1{,}50 = 7\,500 \text{ kN} . \quad \text{(B9.6)}$$

If the changeable load acts favourably, the design value is:

$$N_{Ed,fav} = N_{Ek,G} \cdot \gamma_G + N_{Ek,Q,rep} \cdot \gamma_Q = 3\,333 \cdot 1{,}35 + 2\,000 \cdot 0 = 4\,500 \text{ kN} . \quad \text{(B9.7)}$$

The design value of the shear force is:

$$V_{Ed} = V_{Ek,G} \cdot \gamma_G + V_{Ek,Q,rep} \cdot \gamma_Q = 600 \cdot 1{,}35 + 400 \cdot 1{,}50 = 1\,410 \text{ kN} . \quad \text{(B9.8)}$$

The design value of the bending moment is:

$$M_{Ed} = M_{Ek,G} \cdot \gamma_G + M_{Ek,Q,rep} \cdot \gamma_Q = 2\,728 \cdot 1{,}35 + 1\,819 \cdot 1{,}50 = 6\,411 \text{ kNm} . \quad \text{(B9.8)}$$

Figure B9.4 and Figure B9.5 show the distribution of action effect design values with reduced subgrade reaction modulus adopted for structural analyses.

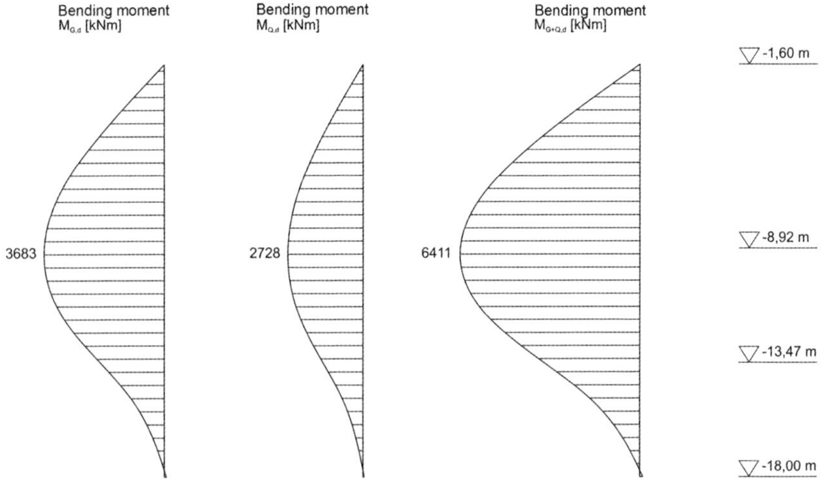

Figure B9.4 Design bending moments for pile design (structural analyses)

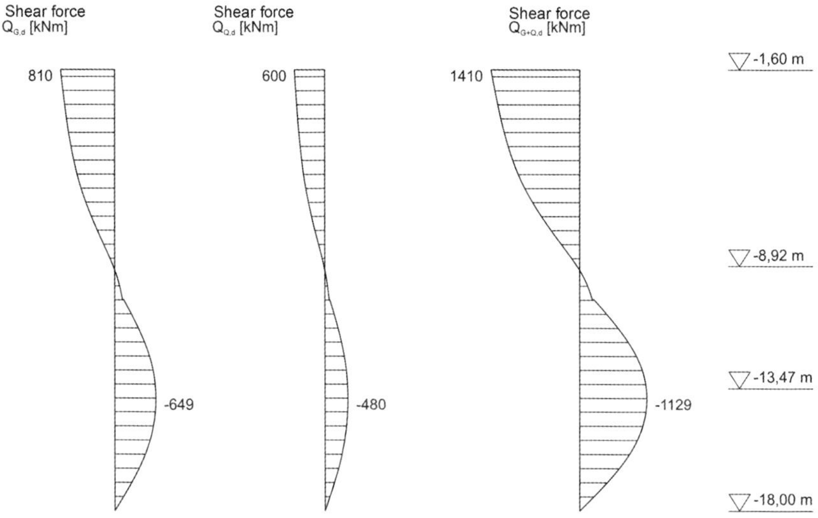

Figure B9.5 Design shear forces for pile design (structural analyses)

B9.4 Minimum strength class of concrete and concrete cover

The minimum concrete strength class to DIN 1045-1:2008-08 for exposure class XC2 with regard to foundation components is C16/20. DIN EN 1536:1999-06 envisages a concrete strength class between C20/25 and C30/37 for bored piles.

Adopted: C25/30 XC2

The nominal concrete cover c_{nom} to DIN 1045-1:2008-08 is given by:

$$c_{nom} = c_{min} + \Delta c \tag{B9.9}$$

Where c_{min} is the minimum concrete cover to DIN EN 1536:1999-06 as a function of D:

$c_{min} = 50$ mm For piles where $D \leq 0{,}6$ m

$c_{min} = 60$ mm For piles where $D > 0{,}6$ m, here $c_{min} = 60$ mm

Δc Allowance value to DIN 1045:2008-08 as a function of the exposure class:

$\Delta c = 10$ mm For exposure class XC1

$\Delta c = 15$ mm In all other cases, here $\Delta c = 15$ mm

Giving the nominal concrete cover:

$c_{nom} = 60 + 15 = 75$ mm.

To guarantee the bond to DIN 1045-1:2008-08 the condition:

$c_{min} \geq d_s$ (d_s = bar diameter)

must also be adhered to.

B9.5 Design values of materials

The partial factors of the materials for the ultimate limit state are adopted from DIN 1045-1:2008-08:

Concrete < C55/67 $\gamma_C = 1{,}50$

Reinforcement steel BSt 500 $\gamma_S = 1{,}15$

The design value of the concrete compressive strength is given in accordance with DIN 1045-1:2008-08 by taking the long-term impact on the compressive strength into consideration using the factor α (α = 0,85 for normal weight concrete):

$$f_{cd} = \alpha \cdot f_{ck} / \gamma_C \tag{B9.10}$$

Where $f_{ck} = 25{,}0$ N/mm².

$$f_{cd} = 0{,}85 \cdot 25{,}0 / 1{,}50 = 14{,}2 \text{ N/mm}^2 .$$

The design value of the yield strength of the reinforcement steel BSt 500 S (A) is given by:

$$f_{yd} = f_{yk} / \gamma_S \tag{B9.11}$$

Where $f_{yk} = 500{,}0$ N/mm².

Thus $f_{yd} = 500{,}0 / 1{,}15 = 435{,}0$ N/mm².

B9.6 Ultimate limit state design

B9.6.1 Design for bending and normal force

The circular section is designed using the ω method with dimensionless coefficients.

Assuming a diameter of 12 mm for the transverse and 32 mm for the longitudinal reinforcement the centres d_1 of the longitudinal reinforcement are at:

$$d_1 = \text{nom } c + d_{s,cross} + d_{s,long} / 2 = 7{,}5 + 1{,}2 + 3{,}2/2 = 10{,}3 \text{ cm} .$$

For the purposes of analysis the bending reinforcement centres are adopted at $d_1 = 0{,}1 \cdot h = 15$ cm, i.e. $d_1/h = 0{,}1$.

The relative bending moment is given by:

$$\mu_{Ed} = \frac{M_{Ed}}{A_c \cdot h \cdot f_{cd}} = \frac{6\,411}{\pi \cdot 1,5^2 / 4 \cdot 1,5 \cdot 14\,00} = 0,171 \quad \text{(B9.12)}$$

The relative normal force is given by:

$$\nu_{Ed} = \frac{N_{Ed,fav}}{A_c \cdot f_{cd}} = \frac{-4\,500}{\pi \cdot 1,5^2 / 4 \cdot 14\,200} = -0,180 \quad \text{(B9.13)}$$

The corresponding, necessary degree of mechanical reinforcement is given by reading the diagram for $d_1/h = 0,1$ as:

$$\omega_{tot} = 0,41$$

The corresponding strains are:

$$\varepsilon_{C2} / \varepsilon_{S1} = -3,5 / 5,177‰$$

The necessary longitudinal reinforcement is given by:

$$\text{erf } A_{S,tot} = \frac{\omega_{tot} \cdot A_c}{f_{yd} / f_{cd}} = \frac{0,41 \cdot \pi \cdot 1,5^2 \cdot 100^2}{4 \cdot 435\,000 / 14\,200} = 236,5\,\text{cm}^2 \quad \text{(B9.14)}$$

The minimum longitudinal reinforcement to DIN EN 1536:1999-06, Table 4 is given as a function of the nominal pile diameter by:

$$A_S \geq 0,25\% \cdot A_C = 0,0025 \cdot \pi \cdot 1,5^2 / 4 \cdot 100^2$$
$$= 44,2\text{ cm}^2 < 236,5\text{ cm}^2 = \text{erf } A_{S,tot} \quad \text{(B9.16)}$$

The minimum reinforcement in this case is not a governing factor.

The adopted longitudinal reinforcement is:

30 dia. 32 = 241,3 cm² > 236,5 cm² = req. A_S

DIN EN 1536:1999-06 requires that a minimum spacing of 100 mm and a maximum spacing of 400 mm is adhered to for the longitudinal bars.

$$10\text{ cm} \leq a = U / n = \pi \cdot (150,0 - 15,0 \cdot 2) / 30 = 12,6\text{ cm} \leq 40\text{ cm} \quad \text{(B9.17)}$$

The necessary minimum spacing is therefore adhered to.

B9.6.2 Design for shear force to DIN 1045-1

The Civil Engineering Standards Committee (*NABau*) interpretation states that the smaller value of the cross-sectional width between the reinforcement centroid (tension flange) and the pressure resultant (corresponds to the smallest width normal to the inner lever arm z) is used as the effective width for shear force design of circular cross-sections.

Bending design gives:

$$\varepsilon_{C2} / \varepsilon_{S1} = -3,50/5,177 .$$

Using the longitudinal reinforcement centres adopted in bending design the height of the compression zone is:

$$x = (150,0 - 15,0) \cdot 3,5 / (5,177 + 3,5) = 54,5 \text{ cm} . \quad \text{(B9.18)}$$

Because the zero strain line lies within the cross-section, a stress block may be adopted in simplification for the distribution of the concrete compressive stress to DIN 1045:2008-08. The height of the stress block is given by:

$$k \cdot x = 0,8 \cdot 54,5 = 43,6 \text{ cm} . \quad \text{(B9.19)}$$

The pressure resultant lies in the centroid of a circle segment with an opening angle α_c of:

$$\begin{aligned} \alpha_c &= 2 \cdot \arccos(1 - k \cdot x \cdot 2/D) \\ &= 2 \cdot \arccos(1 - 43,6 \cdot 2/150,0) = 130,5° . \quad \text{(B9.20)} \end{aligned}$$

The centroid distance from the circle centre is:

$$\begin{aligned} z_C &= 2/3 \cdot D \cdot \left(\sin\left(\alpha_c/2\right)\right)^3 / (\pi \cdot \alpha_c / 180° - \sin \alpha_c) \\ &= 2/3 \cdot 150,0 \cdot \left(\sin\left(130,5°/2\right)\right)^3 / (\pi \cdot 130,5°/180° - \sin 130,5°) \\ &= 49,4 \text{ cm} \quad \text{(B9.21)} \end{aligned}$$

In line with the interpretation of the Civil Engineering Standards Committee the true tensile reinforcement, i.e. the reinforcement within the tension zone, is adopted for shear force design. To determine the reinforcement centroid the reinforcement is assumed to be uniformly distributed along its centroid axis. The radius of the reinforcement centroid axis is:

$$R_S = 150,0/2 - 15,0 = 60,0 \text{ cm} . \quad \text{(B9.22)}$$

The centroid of the ring section with an opening angle α_s of:

$$\begin{aligned} \alpha_S &= 360° - 2 \cdot \arccos(1 - (x - d_1)/R_S) \\ &= 360° - 2 \cdot \arccos(1 - (54,5 - 15,0)/60,0) \\ &= 220,0° \quad \text{(B9.23)} \end{aligned}$$

lies at a distance from the centre of the circle:

$$\begin{aligned} z_S &= 360° \cdot \sin\left(\alpha_S/2\right) \cdot R_S / (\pi \cdot \alpha_S °) \\ &= 360° \cdot \sin\left(220,0°/2\right) \cdot 60,0 / (\pi \cdot 220,0°) = 29,4 \text{ cm} \quad \text{(B9.24)} \end{aligned}$$

This approximation deviates from the true location of the centroid of the steel tensile forces, because stress distribution in the reinforcement is not taken into consideration. However, the approach is conservative due to the larger true lever arm.

The smaller value of the cross-sectional width is therefore at the height of the pressure resultant and is:

$$b_w = 2 \cdot \sqrt{(150,0/2)^2 - 49,4^2} = 112,9 \text{ cm} . \tag{B9.25}$$

The structural height is given by:

$$d = D/2 + z_S = 75,0 + 29,4 = 104,4 \text{cm} . \tag{B9.26}$$

The inner lever arm is:

$$z = z_C + z_S = 49,4 + 29,4 = 78,8 \text{cm} . \tag{B9.27}$$

The design value of the shear capacity without shear reinforcement in the equivalent rectangle to DIN 1045:2008-08 is:

$$V_{Rd,ct} = \left(\frac{0,15}{\gamma_C} \cdot \kappa \cdot \eta_1 \cdot (100 \cdot \rho_l \cdot f_{ck})^{1/3} - 0,12 \cdot \sigma_{cd} \right) \cdot b_w \cdot d . \tag{B9.28}$$

Where $\kappa = 1 + \sqrt{200/d[mm]} = 1 + \sqrt{200/1044} = 1,44 < 2,0$.

$\eta_1 = 1,0$ for normal weight concrete

$$\rho_l = \frac{A_{Sl}}{b_w \cdot d} = \frac{\text{vorh } A_{S,tot} \cdot \alpha_S/360°}{b_w \cdot d} = \frac{241,3 \cdot 220,0°/360°}{112,9 \cdot 104,4} = 0,0125 < 0,02$$

$\sigma_{cd} = N_{Ed}/A_C = -4\,500/(\pi/4 \cdot 1,50^2) = -2\,546 \text{ kN/m}^2$

(negative compressive force)

Giving:

$$V_{Rd,ct} = \left(\frac{0,15}{1,5} \cdot 1,44 \cdot 1,0 \cdot (100 \cdot 0,0125 \cdot 25)^{1/3} - 0,12 \cdot (-2,546) \right) \cdot 1,129 \cdot 1,044$$

$$= 895 \text{ kN} < 1\,410 \text{ kN} = V_{Ed}$$

Shear reinforcement is therefore required.

The limitation in the compression member inclination is:

$$0,58 \le \cot\theta \le (1,2 - 1,4 \cdot \sigma_{cd}/f_{cd}) / (1 - V_{Rd,c}/V_{Ed}) \le 3,0 . \tag{B9.29}$$

Where:

$$V_{Rd,c} = 0,24 \cdot \eta_1 \cdot f_{ck}^{1/3} \cdot (1 + 1,2 \cdot \sigma_{cd} / f_{cd}) \cdot b_w \cdot z$$
$$= 0,24 \cdot 1,0 \cdot 25^{1/3} \cdot (1 + 1,2 \cdot (-2\,546)/14\,200) \cdot 1,129 \cdot 0,788 \qquad \text{(B9.30)}$$
$$= 490 \text{ kN}$$

giving:

$$0,58 \leq \cot\theta \leq (1,2 - 1,4 \cdot (-2\,546)/14\,200) / (1 - 490/1\,410)$$
$$= 2,22 < 3,0 \quad \text{(for standard concrete)}$$

The design value of the maximum shear capacity is:

$$V_{Rd,max} = b_w \cdot z \cdot \alpha_c \cdot f_{cd} / (\cot\theta + \tan\theta)$$
$$= 1,129 \cdot 0,788 \cdot 0,75 \cdot 14\,200 / (2,22 + 1/2,22). \qquad \text{(B9.31)}$$
$$= 3\,548 \text{ kN} > 1\,410 \text{ kN} = V_{Ed}$$

The necessary shear reinforcement is given by:

$$A_{Sw}/s_w = V_{Ed} / (f_{yd} \cdot z \cdot \cot\theta)$$
$$= 1.410 / (435\,000 \cdot 0,788 \cdot 2,22) \cdot 100^2 = 18,5 \text{ cm}^2/\text{m}. \qquad \text{(B9.32)}$$

However, this approach does not take the increased effect of the shear reinforcement due to radial thrust forces caused by the continuous support of the compression strut on the circular link (boiler pressure) into consideration. Therefore, [5] derives the shear resistances from the basic equations in DIN 1045-1, taking the thrust forces of the circular link and the non-uniform distribution of the shear force on the longitudinal reinforcement into consideration for axis-symmetrical, longitudinally reinforced beams with shear reinforcement. If shear reinforcement is in helix format, the helix inclination α must also be taken into consideration. Deriving the tension strut strength for circular links leads to the design value given in DIN 1045-1:2008-08:

$$V_{Rd,sy} = \alpha_k \cdot A_{Sw}/s_w \cdot f_{yd} \cdot z \cdot \cot\theta \cdot \sin\alpha \qquad \text{(B9.33)}$$

supplemented by a factor α_k known as a the *effectiveness factor*, dependent only on dimensionless geometry parameters and a term for the helix inclination α. Due to the small range of values of $0,715 \leq \alpha_k \leq \pi/4 = 0,785$ for practical applications, [5] proposes a simplified estimate of $\alpha_k = 0,75$.

The shear force component $V_{Rd,c}$ to DIN 1045-1:2008-08 is adopted to determine the compression strut inclination, whereby the effective width can be estimated as $b_w = A_{eff}/z \approx 0,9 \cdot D = 0,9 \cdot 1,5 = 1,35$ m. Using Eq. (B9.30):

$$V_{Rd,c} = 0,24 \cdot \eta_1 \cdot f_{ck}^{1/3} \cdot (1 + 1,2 \cdot \sigma_{cd} / f_{cd}) \cdot b_w \cdot z$$
$$= 0,24 \cdot 1,0 \cdot 25^{1/3} \cdot (1 + 1,2 \cdot (-2\,546)/14\,200) \cdot 1,35 \cdot 0,788$$
$$= 586 \text{ kN}$$

Using Eq. (B9.29) the limitation in the compression strut inclination is:

$$0,58 \leq \cot\theta \leq (1,2-1,4\cdot(-2\,546)/14\,200)/(1-0,586/1,410)$$
$$= 2,48 < 3,0 \quad \text{(for standard concrete)}$$

The capacity of the compression strut is limited in accordance with DIN 1045-1:2008-08, whereby the effective width may be adopted as $b_w = 2\cdot R_S \approx D = 1,5$ m using the approach after [5]. Given an angle of inclination of the helix of $\alpha = 85°$ (*cf.* adopted reinforcement) the capacity of the compression strut is limited to:

$$V_{Rd,max} = \alpha_k \cdot b_w \cdot z \cdot \alpha_c \cdot f_{cd} \cdot \cot\theta / \left((\cot\theta + \cot\alpha)^2 + 1\right)$$
$$= 0,75\cdot 1,5\cdot 0,788\cdot 0,75\cdot 14\,200\cdot 2,48/\left((2,48+\cot 85°)^2+1\right) \quad \text{(B9.34)}$$
$$= 3\,084 \text{ kN} > 1\,410 \text{ kN} = V_{Ed}$$

The necessary shear reinforcement is given by Eq. (B9.33):

$$A_{Sw}/s_w = V_{Ed}/\left(\alpha_k \cdot f_{yd} \cdot z \cdot \cot\theta \cdot \sin\alpha\right)$$
$$= 1\,410/(0,75\cdot 435\,000\cdot 0,788\cdot 2,48\cdot \sin 85°)\cdot 100^2 = 22,2 \text{ cm}^2/\text{m}$$

B9.6.3 Design for shear force after [5]

In order to take into consideration the favourable impact of central normal compression forces, [5] divides the bearing resistance into individual components in a truss model with an additive concrete contribution, whereby the shear force transfer as a result of crack formation is ignored, such that by adopting $V_{Rd,c} = 0$

$$0,58 \leq \cot\theta = 1,2 - 1,4\cdot\sigma_{cd}/f_{cd}$$
$$= 1,2-1,4\cdot(-2,546)/14,2 = 1,45 < 3,0 \quad \text{(B9.35)}$$

In addition to the effect of the truss model, the resistance in the compression and failure process zone is adopted:

$$V_{Rd,sy} = V_{Rd,sy}(V_{Rd,c} = 0) + V_{Rd,ct} \quad \text{(B9.36)}$$

The semi-empirical approach was adopted by [5] for shear force resistance:

$$V_{Rd,ct} = 0,216/\gamma_c \cdot \kappa \cdot (100\cdot\rho_l)^{0,42} \cdot f_{ck}^{\ 1/3} \cdot b_{eff} \cdot z - \lambda \cdot N_{Ed} \quad \text{(B9.37)}$$

where the term $\lambda \cdot N_{Ed}$ considers the favourable influence of central normal compression forces via a strutted frame. The inclination of the strutted frame $\lambda = z_C/a$ corresponds to the change in the distance z_N between the centroid axis and the centroid of the compression zone over the length a. In simplification, the length a may be estimated as the distance between the point of acting of the shear force and the location of the maximum moment where

$a = 892 - 160 = 732\ \text{cm}$. Hence: $\lambda = z_C / a = 49,4/732 = 0,067$. A normal compression force is adopted by applying the appropriate partial factor for favourable actions.

In contrast to the NABau interpretation, [5] adopts the reinforcement of the half cross-section in the tension zone in simplification for the degree of longitudinal reinforcement.

$$\rho_l = \frac{1}{2} \cdot A_{s,tot} / A_C = \frac{1}{2} \cdot 241,3 / \left(\pi \cdot 150^2 / 4\right) = 0,0068 < 0,02 \qquad \text{(B9.38)}$$

Using $b_{eff} = 0,9 \cdot D = 1,35\ \text{m}$ and the values previously determined in Equation (B9.37):

$$\begin{aligned} V_{Rd,ct} &= 0,216/1,5 \cdot 1,44 \cdot (100 \cdot 0,0068)^{0,42} \cdot 25^{1/3} \cdot 1,35 \cdot 0,788 - 0,067 \cdot (-3,333) \\ &= 772\ \text{kN} \end{aligned}$$

The required reinforcement in accordance with Equation (B9.36) is therefore:

$$\begin{aligned} A_{Sw}/s_w &= \left(V_{Ed} - V_{Rd,ct}\right) / \left(\alpha_k \cdot f_{yd} \cdot z \cdot \cot\theta \cdot \sin\alpha\right) \\ &= (1\,410 - 772)/(0,75 \cdot 435\,000 \cdot 0,788 \cdot 1,45 \cdot \sin 85°) \cdot 100^2 = 17,2\ \text{cm}^2/\text{m} \end{aligned}$$

In contrast to the NABau interpretation, the approach in accordance with [5] gives an approx. 16% lower shear force reinforcement. This is caused by the increase in shear capacity due to the strutted frame approach for the normal forces. If no, or only minor, compressive forces act on the structural element, the design approach in accordance with [5] returns more shear reinforcement than the NABau interpretation, because in this case the additional effect imposed by the circular reinforcement's boiler pressure not covered by the DIN 1045-1:2008-08 design concept becomes governing. The approach after [5] is therefore recommended for design.

B9.6.4 Minimum reinforcement for shear force to DIN 1045-1

The minimum reinforcement to DIN 1045-1 is given by the base value for determining the minimum reinforcement $\rho = 0,068\,\%$ as:

$$A_{Sw}/s_w = \rho_w \cdot b_w \cdot \sin\alpha = 0,068 \cdot 135,0 \cdot \sin 90,0° = 9,2\ \text{cm}^2/\text{m}. \qquad \text{(B9.39)}$$

The minimum diameter of the transverse reinforcement in accordance with DIN EN 1536:1999-06 is:

$$d_{sw,min} \geq \begin{cases} 6\ \text{mm} \\ \frac{1}{4} \cdot d_{sl} = \frac{1}{4} \cdot 32,0 = 8\ \text{mm} \end{cases} \qquad \text{(B9.40)}$$

The minimum clear bar spacing in accordance with DIN EN 1536:1999-06 may not be smaller than for longitudinal bars at:

$$s_w \geq 10 \text{ cm} .$$

The minimum longitudinal spacing in accordance with DIN 1045-1:2008-08 for

$$0,30 \cdot V_{Rd,max} < V_{Ed} < 0,60 \cdot V_{Rd,max}$$
$$0,30 \cdot 2\,899 = 870 \text{ kN} < 1\,410 \text{ kN} < 0,60 \cdot 2\,899 = 1\,739 \text{ kN} \quad \text{(B9.41)}$$

is:

$$s_{max} \leq 0,5 \cdot h = 0,5 \cdot 150,0 = 75,0 \text{ cm} . \quad \text{(B9.42)}$$

The following shear reinforcement is used:

Helix 2-core dia. 14/15 cm = 20,5 cm^2 > 17,2 cm^2 = req. A_{Sw}.

B10 Laterally Loaded Piles

B10.1 Objective and systems

Figure B10.1 shows a bridge abutment founded on 8 tubular steel piles. The lateral pressure on the piles as a result of the backfill must be determined, based on [35].

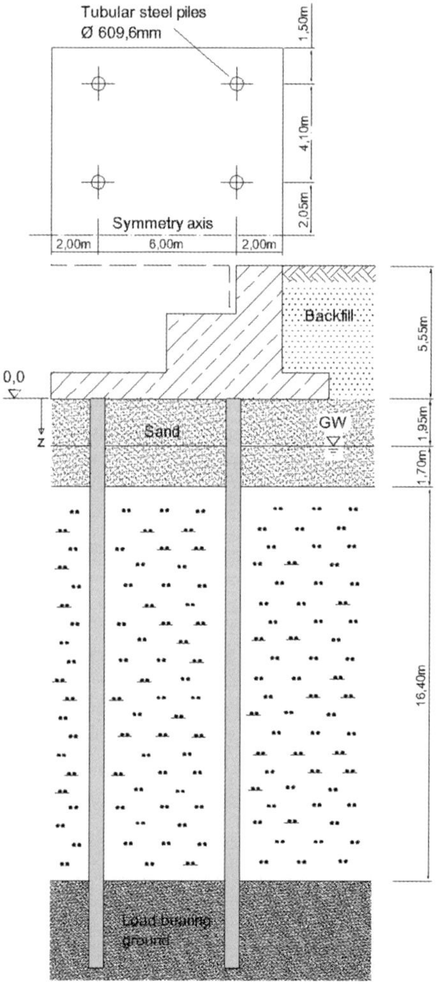

Tubular steel piles:

$A = 0{,}018837\ m^2$
$I = 8{,}47 \cdot 10^{-4}\ m^4$
$E = 210\,000\ N/mm^2$

Soil properties:

Backfill:

$\varphi'_k = 30°$
$c'_k = 0\ kN/m^2$
$\gamma = 19\ kN/m^3$

Sand:

$\varphi'_k = 30° \Rightarrow K_{agh} = 0.33$
$c'_k = 0\ kN/m^2$
$\gamma/\gamma' = 19/11\ kN/m^3$

Tidal mud:

$\varphi'_k = 15° \Rightarrow K_{agh} = 0{,}59$
$c'_k = 10\ kN/m^2$
$c_{u,k} = 25\ kN/m^2$
$\gamma' = 6{,}5\ kN/m^3$

Boundary conditions:

- Surcharge from sand layer in tidal mud, consolidated
- Surcharge from backfill in tidal mud, unconsolidated (initial state) ($\rightarrow U = 0$)

Figure B10.1 System sketch and soil properties

B10.2 Determining the characteristic actions and effects

The characteristic actions from lateral pressure are determined in line with 4.5, whereby the characteristic flow pressure $p_{f,k}$ is determined using Eq. (4.5) and adopting the geometry of the pile group as shown in Figure 4.7.

$$p_{f,k} = \eta_a \cdot 7 \cdot c_{u,k} \cdot D_s = 1{,}20 \cdot 7 \cdot 25 \cdot 0{,}61 = 128{,}1 \text{ kN/m} .$$

The characteristic effect of the flow pressure $P_{f,k}$ on the individual group pile is calculated for the entire action height h_w as:

$$P_{f,k} = p_{f,k} \cdot h_w = 128{,}1 \cdot 16{,}40 = 2\,100{,}8 \text{ kN} .$$

The flow pressure $P_{f,k}$ is compared to the resultant earth pressure ΔE_k over the entire action height. For partial consolidation Eq. (4.9) gives for $e_{a,k}$

$$e_{a,k} = (\gamma \cdot z + U_c \cdot \Delta p_k) \cdot K_{agh} + (1 - U_c) \cdot \Delta p_k - 2 \cdot c'_k \cdot \sqrt{K_{agh}}$$

and for the passive earth pressure $e_{p,k}$ using Eq. (4.10):

$$e_{p,k} = \gamma \cdot z \cdot K_{pgh} \quad \text{where} \quad K_{pgh} = 1{,}0$$

The resultant earth pressure Δe_k using Eq. (4.6) is determined in Table B10.1.

Table B10.1 Determining ΔE_k for partial consolidation in tidal mud (surcharge as a result of consolidated sand layer; as a result of unconsolidated backfill)

Depth z [m]	$e_{a,k}$ [kN/m²]	$e_{p,k}$ [kN/m²]	Δe_k [kN/m²]	ΔE_k [kN]
3,65	(1,95 · 19 + 1,70 · 11) · 0,59 + 5,55 · 19 – 2 · 10 · 0,77 = 122,9	1,95 · 19 + 1,70 · 11 = 55,8	67,1	
20,05	(1,95 · 19 + 1,70 · 11+ 16,4 · 6,5) · 0,59 + 5,55 · 19 – 2 · 10 · 0,77 = 185,8	1,95 · 19 + 1,70 · 11 + 16,4 · 6,5 = 162,4	23,4	742,1

The resultant undrained earth pressure is calculated for comparison. For the active earth pressure using Eq. (4.7):

$$e_{a,k} = \gamma \cdot z + \Delta p_k - 2 \cdot c_{u,k}$$

The governing resultant earth pressure determined from the partially consolidated and undrained condition is given by a comparison of the resultant earth pressures ΔE_k. In this case the resultant earth pressure in the undrained condition is higher. The governing resultant earth pressure Δe_k. is therefore that in the undrained condition.

Table B10.2 Determining ΔE_k in the undrained condition

Depth z [m]	$e_{a,k}$ [kN/m²]	$e_{p,k}$ [kN/m²]	Δe_k [kN/m²]	ΔE_k [kN]
3,65	1,95 · 19 + 1,70 · 11 + 5,55 · 19 − 2 · 25 = 111,20	1,95 · 19 + 1,70 · 11 = 55,8	55,4	908,6
20,05	1,95 · 19 + 1,70 · 11 + 16,4 · 6,5 + 5,55 · 19 − 2 · 25 = 217,8	1,95 · 19 + 1,70 · 11 + 16,4 · 6,5 = 162,4	55,4	

Analysis of the characteristic action from earth pressure on the individual pile $p_{e,k}$, taking the group effect into consideration, is based on Section 4.5.4 (6) for each elevation for the governing resultant earth pressure Δe_k. The minimum of the following conditions a) to d) is governing.

a) Mean pile centres perpendicular to the direction of force:

$$p_{e,k} = b \cdot \Delta e_k = 4{,}10 \cdot 55{,}4 = 227{,}1 \text{ kN/m}$$

b) Triple pile diameter Ds:

$$p_{e,k} = b \cdot D_s \cdot \Delta e_k = 3 \cdot 0{,}61 \cdot 55{,}4 = 101{,}4 \text{ kN/m}$$

c) Thickness of the stratum generating the lateral pressure:

$$p_{e,k} = b \cdot \Delta e_k = 16{,}4 \cdot 55{,}4 = 908{,}6 \text{ kN/m}$$

d) Overall width of the pile group divided by the total number of piles:

$$p_{e,k} = b \cdot \Delta e_k = 1/8 \cdot 12{,}3 \cdot 55{,}4 = 85{,}2 \text{ kN/m} .$$

The provision of Section 4.5.4 (8) is not governing. Taking the pile centres and pile diameter into consideration (pile centres > 4 · D_s), investigation of the approach after Section 4.5.4 (7) can also be dispensed with.

The governing resultant earth pressure ΔE_k is calculated for the entire action height h_w as:

$$\Delta E_k = p_{e,k} \cdot h_w = 85{,}2 \cdot 16{,}40 = 1\,397{,}3 \text{ kN} .$$

The governing lateral pressure is given by the smaller action resulting from flow pressure $P_{f,k}$ and the resultant earth pressure ΔE_k. In this case, the characteristic earth pressure is therefore adopted as an action from lateral pressure. A linear load of $p_{e,k} = 85{,}2$ kN/m acts on the individual pile perpendicular to the pile axis over the entire action height of the soft stratum.

B11 Pillar Foundation on 9 Piles – Ultimate and Serviceability Limit State Analyses Taking the Group Effect into Consideration

B11.1 Objective and system

The following example is taken from [125]. The geometry of a pillar foundation on 9 piles, the soil parameters and the resistance-settlement curve acquired in a pile load test on a similar individual pile are given in Figure B11.1. A maximum settlement $s = 0{,}03 \cdot D = 0{,}03 \cdot 130 = 3{,}9$ cm is defined for the serviceability limit state.

Load test evaluation returned a mean $q_{s,k} = 28$ kN/m² and $q_{b,k} = 692$ kN/m².

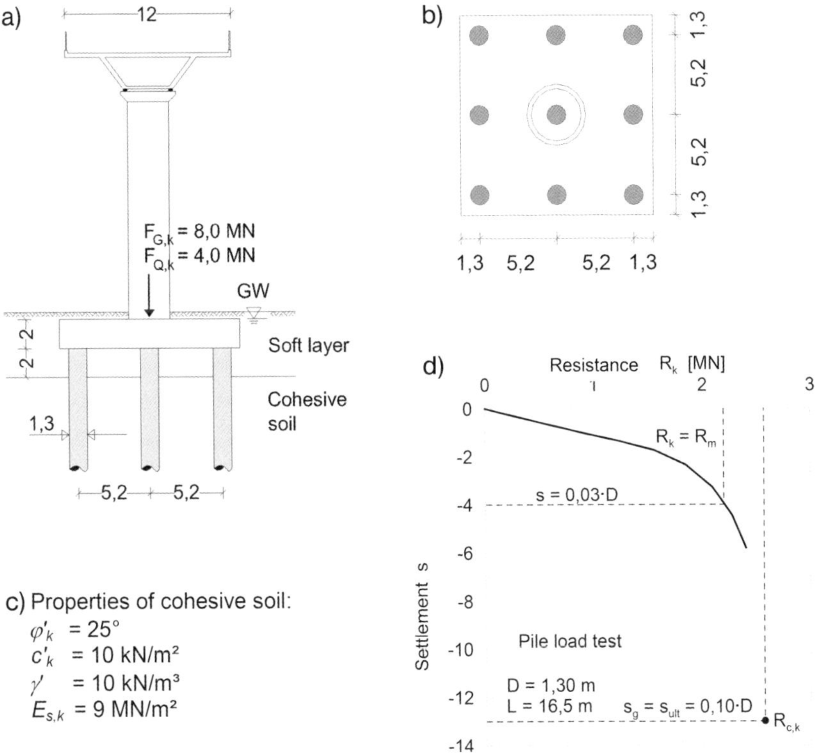

Figure B11.1 Pillar foundation on 9 piles, a) elevation, b) plan, c) soil parameters, d) characteristic resistance-settlement curve for a similar individual pile from a pile load test; lengths in [m]

All serviceability and ultimate limit state analyses must be performed for the pile group, and the characteristic bending moments in the pile capping slab must be determined.

a) Serviceability limit state analysis

A maximum settlement $s = 0{,}03 \cdot D$ is defined for the serviceability limit state. The nomogram reproduced in Figure 8.8 b is used to determine the pile resistances.

The input variable is the pile centre-embedment depth ratio:

$$a/d = 5{,}2/14{,}5 = 0{,}359$$

Using these variables the first influence factor λ_1 is given from Figure 8.8 b as a function of the pile position at $\lambda_1 = 0{,}76$ (corner pile), $\lambda_1 = 0{,}59$ (edge pile) and $\lambda_1 = 0{,}35$ (internal pile).

From Figure 8.11 a the second influence factor λ_2 used to consider the group size for the internal piles is $\lambda_2 = 1{,}0$. Additional influence factors need not be taken into consideration for this example.

Using the characteristic pile resistance of an individual pile $R_{E,k} = 2{,}21$ MN for a settlement of $s = 0{,}03 \cdot D$ in accordance with Figure B11.1 d, and the previously determined influence factors, the following pile resistances in the group are given:

$$R_{\text{corner, k}}\,(s = 0{,}03 \cdot D) = 2{,}21 \cdot 0{,}76 = 1{,}68 \text{ MN}$$

$$R_{\text{edge, k}}\,(s = 0{,}03 \cdot D) = 2{,}21 \cdot 0{,}59 = 1{,}30 \text{ MN}$$

$$R_{\text{internal, k}}\,(s = 0{,}03 \cdot D) = 2{,}21 \cdot 0{,}35 \cdot 1{,}00 = 0{,}77 \text{ MN}$$

Given a settlement of $s = 0{,}03 \cdot D$ the characteristic total resistance of the pile group is:

$$R_{G,k}\,(s = 0{,}03 \cdot D) = 4 \cdot 1{,}68 + 4 \cdot 1{,}30 + 0{,}77 = 12{,}69 \text{ MN}\,.$$

The limit state condition for the serviceability limit state demands that the effects remain smaller than the resistances. All partial safety factors are adopted at $\gamma_F = \gamma_R = 1{,}0$.

$$F_{c,d}\,(SLS) = F_{c,k}\,(SLS) \le R_{c,d}\,(SLS) = R_{c,k}\,(SLS)$$
$$\rightarrow 12{,}0 \text{ MN} < 12{,}69 \text{ MN}$$

Under d) the true characteristic group settlements anticipated under the characteristic action are determined in an additional iteration.

b) Ultimate limit state analysis after 8.3.1 (2) to (4)

In the ultimate limit state the characteristic total resistance of the group is given analogous to the procedure under a), but using a settlement of $s_g = s_{ult} = 0{,}1 \cdot D$. With the characteristic ultimate resistance of the individual pile $R_{c,k} = 2{,}57$ MN from Figure B11.1d and the influence factors from Figure 8.8d and Figure 8.11a, the characteristic pile resistances are:

$$R_{\text{corner, k}}\,(s = 0{,}1 \cdot D) = 2{,}57 \cdot 1{,}0 = 2{,}57 \text{ MN}$$

$$R_{\text{edge, k}}\,(s = 0{,}1 \cdot D) = 2{,}57 \cdot 0{,}99 = 2{,}54 \text{ MN}$$

$$R_{\text{internal, k}}\,(s = 0{,}1 \cdot D) = 2{,}57 \cdot 0{,}97 \cdot 1{,}0 = 2{,}49 \text{ MN}$$

This gives the characteristic total resistance of the pile group in the ultimate limit state as:

$$R_G\,(s = 0{,}1 \cdot D) = 4 \cdot 2{,}57 + 4 \cdot 2{,}54 + 2{,}49 = 22{,}93 \text{ MN}$$

The characteristic values of the actions and resistances are converted to design values using the appropriate partial factors given in the EC 7-1 Handbook [44], see Annexes A2 and A3. The limit state equation is therefore:

$$F_{c,d} = F_{G,k} \cdot \gamma_G + F_{Q,k} \cdot \gamma_Q \leq R_{c,d} = R_{c,k}/\gamma_t$$

$$F_{c,d} = 8 \cdot 1{,}35 + 4 \cdot 1{,}50 = 16{,}8 \text{ MN} < R_{c,d} = 22{,}93/1{,}10 = 20{,}85 \text{ MN}\,.$$

Stability is therefore given for both the serviceability limit state after a) and the ultimate limit state after b).

c) Ultimate limit state analysis after 8.3.1.1 (5) and (6) – Pile group as large equivalent pile

Individual pile base area

$$A_{b,i} = \frac{\pi \cdot D^2}{4} = \frac{\pi \cdot 1{,}3^2}{4} = 1{,}33 \text{ m}^2$$

Equivalent shaft area

$$A^*_{s,j} = 4 \cdot (2 \cdot 5{,}2 + 1{,}3) \cdot (16{,}5 - 2) = 678{,}60 \text{m}^2$$

$$R_{c,k,G} = q_{b,k} \cdot \sum A_{b,i} + \sum q_{s,k,j} \cdot A^*_{s,j}$$

$$= (692 \cdot 9 \cdot 1{,}33 + 28 \cdot 678{,}60)/1000$$

$$= 27{,}28 \text{ MN}$$

$$F_{c,d} = 16{,}8 \text{ MN} < 27{,}28/1{,}10 = 24{,}80 \text{ MN} = R_{c,d,G}$$

This alternative bearing capacity analysis for the pile group is also adhered to.

d) Mean characteristic settlements of the pile group and corresponding pile resistances

The serviceability limit state analysis can also be performed indirectly by comparing the forecast mean settlement to the specified maximum settlement. The mean settlement is estimated using the nomograms in 8.2.1.2. Input variables are the pile centre-embedment depth ratio a/d and the relative action, giving:

$$a/d = 5,2/14,5 = 0,359$$

$$F_{G,k}/(n_G \cdot R_{E,s=0,1\cdot D}) = 12/(9 \cdot 2,57) = 0,52$$

The first influence factor is given by Figure 8.2a as $S_1 = 2,2$. The second influence factor for taking the group size into consideration is given as $S_2 = 0,8$ in Figure 8.4e. With a mean action on the group pile of $F_{G,k}/n_G = 12$ MN/9 = 1,33 MN, a settlement $s_E = 1,5$ cm is acquired from Figure B11.1d for a similar isolated pile.

The mean settlement s_G of the pile group can be determined thus:

$$s_G = s_E \cdot S_1 \cdot S_2 = 1,5 \cdot 2,2 \cdot 0,8 = 2,6 \text{ cm}$$

The settlement of the group relative to the pile diameter is:

$$s_G/D = 2,6/130 = 0,020$$

The following characteristic pile resistances are acquired from Figure B11.1d, Figure 8.8a and Figure 8.11a for the determined settlement of s = 2,6 cm = 0,02 · D:

$$R_{\text{corner, k}}(s = 0,02 \cdot D) = 2,05 \cdot 0,64 = 1,31 \text{ MN}$$

$$R_{\text{edge, k}}(s = 0,02 \cdot D) = 2,05 \cdot 0,46 = 0,94 \text{ MN}$$

$$R_{\text{internal, k}}(s = 0,02 \cdot D) = 2,05 \cdot 0,30 \cdot 1,0 = 0,62 \text{ MN}$$

The characteristic total resistance of the pile group is therefore:

$$R_{G,k}(s = 0,02 \cdot D) = 4 \cdot 1,31 + 4 \cdot 0,94 + 0,62 = 9,62 \text{ MN}.$$

With the calculated settlement $s_G = 2,6$ cm the required total resistance of the pile group is fallen short of by 20%. The settlement is therefore corrected. With an assumed settlement of $s_G = 3,6$ cm the pile resistances correspond to the existing total action. This leads to the following pile resistances, whereby the influence factors are interpolated between Figures 8.8a and b or are taken from Figure 8.11a:

$$R_{\text{corner, k}}(s = 0,028 \cdot D) = 2,18 \cdot 0,73 = 1,59 \text{ MN}$$

$$R_{edge,k}(s=0,028\cdot D)=2,18\cdot 0,56=1,22 \text{ MN}$$

$$R_{internal,k}(s=0,028\cdot D)=2,18\cdot 0,34\cdot 1,0=0,74 \text{ MN}$$

The characteristic total resistance of the pile group is therefore:

$$R_{G,k}(s=0,028\cdot D)=4\cdot 1,59+4\cdot 1,22+0,74=11,98 \text{ MN}.$$

The settlement $s_G = 3,6$ cm is smaller than the maximum proposed settlement of $s_{G,max} = 3,9$ cm ($0,03 \cdot D$). This also meets the serviceability limit state conditions.

e) Effect on the rising structure as a result of group effect

Under d) the characteristic pile resistances and the mean settlement of the pile group under the given action were determined. Spring stiffnesses for the governing section of the characteristic resistance-settlement curve are derived from these values in order to determine the effects resulting from the group effect in a structural analysis. A structural model is used for this purpose, in which the piles are modelled as linear springs.

With a mean settlement of $s = 0,028 \cdot D = 0,036$ m the following spring stiffnesses are adopted:

$$R_{corner,k}/s=1,59/0,036=44 \text{ MN/m}$$

$$R_{edge,k}/s=1,22/0,036=34 \text{ MN/m}$$

$$R_{internal,k}/s=0,74/0,036=21 \text{ MN/m}.$$

All piles were modelled with a uniform stiffness without taking the group effect into consideration. As stated in d) the settlement of the isolated pile is $s_E = 0,015$ m for an isolated pile resistance $R_{E,k} = 1,33$ MN.

$$R_{E,k}/s=1,33/0,015=89 \text{ MN/m}$$

To evaluate the influence of the group effect on the rising structure, the characteristic action effects in the pile capping slab from the structural analyses are compared to the equivalent spring stiffnesses. Figure B11.2 shows the characteristic bending moments $M_{x,k}$. By way of symmetry these bending moments also correspond to $M_{y,k}$. The maximum value of the moment effect increases under the group effect by 9 % from $M_{x,k} = 1\,728$ kNm/m to $M_{x,k} = 1\,880$ kNm/m.

The characteristic shear forces in the pile capping slab $Q_{x,k}$ are shown in Figure 11.3. Because of the symmetry of the pile group they also correspond to the shear forces $Q_{y,k}$. Taking the group effect into consideration, the maximum shear force effect increases by 4 % from $Q_{x,k} = 562$ kN/m to $Q_{x,k} = 587$ kN/m.

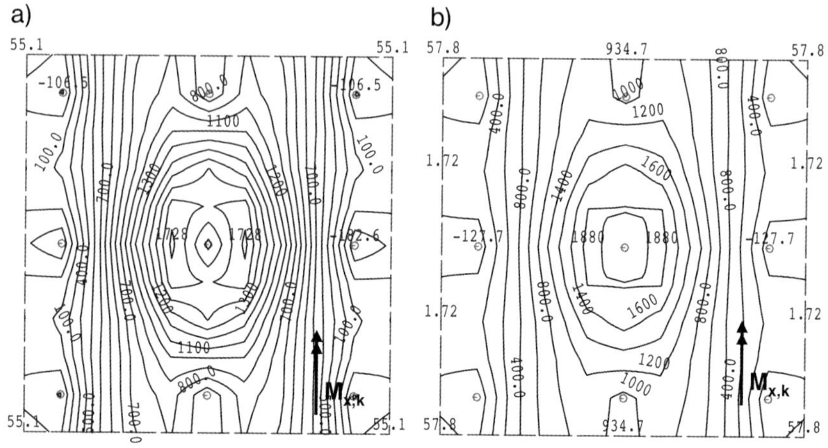

Figure B11.2 Characteristic bending moments $M_{x,k}$ [kNm/m] in the pile capping slab a) without group effect; b) with group effect

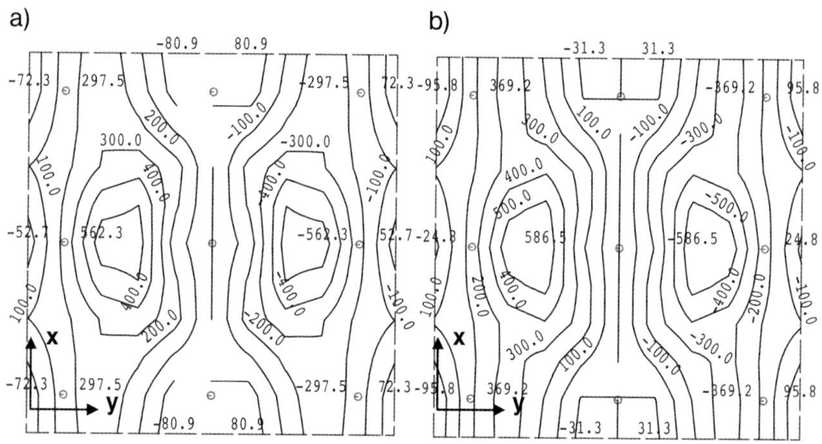

Figure B11.3 Characteristic shear forces $Q_{x,k}$ [kN/m] in the pile capping slab a) without group effect; b) with group effect

B12 Tension Pile Group Analyses in the Ultimate Limit State

B12.1 Objective

A square pile group containing four piles serves to anchor the suspension cable of a suspension bridge. The pile diameter is D = 0,90 m. The pile length necessary to comply with the analysis of the ultimate limit state of the pile foundation with regard to the vertical component of the suspension cable anchor is determined. The horizontal component is accepted through pile subgrade and passive earth pressure and does not form part of the analysis.

A mean characteristic pile skin friction $q_{s,k}$ = 60 kN/m² was derived from a similar tension pile load test, confirmed for this case as an empirical value by the geotechnical expert.

Cone penetration tests in the cohesionless soil returned a mean cone resistance q_c = 7,5 MN/m². The following soil parameters were determined from this:

$\varphi'_k = 32{,}5°$; $c'_k = 0$; $\gamma_k = 18$ kN/m³; $\gamma'_k = 10{,}5$ kN/m³

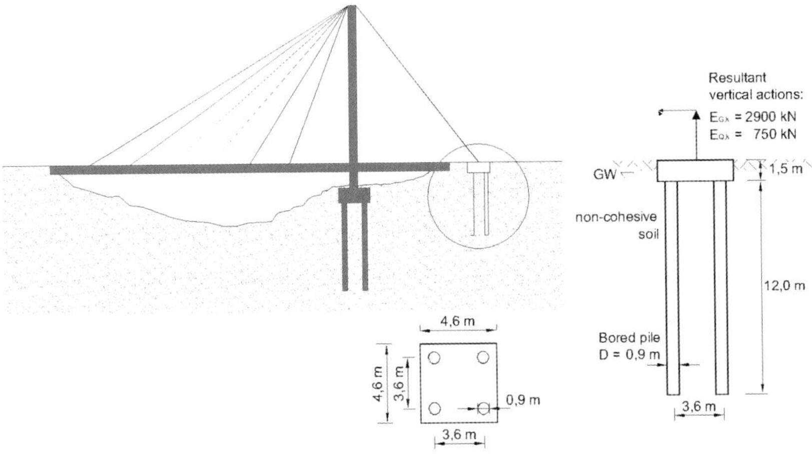

Figure B12.1 a) Suspension bridge, b) suspension cable anchored to a pile foundation, c) plan view

B12.2 Isolated pile analysis

In order to guarantee sufficient safety against pull-out of a foundation structure anchored using tension piles, the following limit state condition must be met for an individual pile within a group in the ultimate limit state (ULS, GEO-2) in accordance with the EC 7-1 Handbook [44]:

$$F_{t,d} \leq R_{t,d}$$

The characteristic value $R_{t,k}$ of the pull-out resistance of an individual pile with the dimensions given in Figure B12.1 and $q_{s,k} = 60$ kN/m² is given by:

$$R_{t,k} = A_s \cdot q_{s,k} = \pi \cdot D \cdot L \cdot q_{s,k}$$

$$= \pi \cdot 0{,}9 \cdot L \cdot 60 = 169{,}6 \cdot L \quad [\text{kN}]$$

The design value $R_{t,d}$ of the pull-out resistance of a pile is then:

$$R_{t,d} = \frac{R_{t,k}}{\gamma_{s,t}} = \frac{169{,}6 \cdot L}{1{,}50} = 113{,}1 \cdot L \qquad [\text{kN}]$$

The design value of the tensile effect is obtained from:

$$F_{t,d} = F_{t,G,k} \cdot \gamma_G + F_{t,Q,rep} \cdot \gamma_Q - F_{c,G,k} \cdot \gamma_{G,inf}$$

The self-weight of the pile capping slab is adopted as a favourable permanent action at $4{,}6^2 \cdot 1{,}5 \cdot 24 = 762$ kN and the self-weight of the piles ($\gamma_{concrete} =$ 24 kN/m³) at $4 \cdot \pi \cdot 0{,}45^2 \cdot L \cdot (24 - 10) = 35{,}6 \cdot L$ [kN]. For four piles the tensile effect per pile is:

$$\gamma_{concrete} = 24 \text{ kN/m}^3$$

$$F_{t,d} = \tfrac{1}{4}(2900 \cdot 1{,}35 + 750 \cdot 1{,}50 - (762 + L \cdot 35{,}6) \cdot 1{,}00)$$
$$= (1069{,}5 - 8{,}9 \cdot L) \quad [\text{kN}].$$

The limit state equation gives:

$$F_{t,d} \le R_{t,d}$$

$$1069{,}5 - 8.9 \cdot L = 113{,}1 \cdot L$$

$$\rightarrow L = \frac{1069{,}5}{113{,}1 + 8{,}9} = 8{,}77 \text{ m}.$$

B12.3 Analysis of the pile group effect (attached soil monolith)

The UPL limit case is analysed using the following equation for the group effect in closely spaced piles:

$$G_{dst,k} \cdot \gamma_{G,dst} + Q_{dst,rep} \cdot \gamma_{Q,dst} \le G_{stb,k} \cdot \gamma_{G,stb} + G_{E,k} \cdot \gamma_{G,stb}$$

The characteristic value $G_{E,k}$ of the self-weight of the attached soil is given by the formula:

$$G_{E,k} = n_Z \cdot \left[l_a \cdot l_b \left(L - \frac{1}{3} \cdot \sqrt{l_a^{\ 2} + l_b^{\ 2}} \cdot \cot\phi \right) \right] \cdot \eta_Z \cdot \gamma'$$

$$= 4 \cdot \left[3{,}6 \cdot 3{,}6 \left(L - \frac{1}{3} \cdot \sqrt{3{,}6^2 + 3{,}6^2} \cdot \cot 32{,}5 \right) \right] \cdot 0{,}8 \cdot 10{,}5$$

$$= 435{,}5 \cdot L - 1\,160{,}0 \quad [\text{kN}].$$

If this is adopted in the limit state equation with the partial factors $\gamma_{G,dst} = 1{,}05$ and $\gamma_{Q,dst} = 1{,}50$ the necessary pile length is:

$$2\,900 \cdot 1{,}05 + 750 \cdot 1{,}50 \leq 762 \cdot 0{,}95 + (435{,}5 \cdot L - 1160{,}0) \cdot 0{,}95$$

$$\rightarrow L = \frac{2\,900 \cdot 1{,}05 + 750 \cdot 1{,}50 - 762 \cdot 0{,}95 + 1160{,}0 \cdot 0{,}95}{435{,}5 \cdot 0{,}95} = 10{,}99 \text{ m}.$$

The group effect analysis is governing here and numerically the piles must be at least 10,99 m long (rounded to 11 m).

B13 Laterally Loaded Pile Groups: Determining the Distribution of Horizontal Subgrade Moduli

Employing the procedure described in 8.2.3 for laterally loaded pile groups, the horizontal subgrade modulus distributions in the pile group are derived for the generalised task shown in Figure B13.1 (from DIN 4014:1990-03).

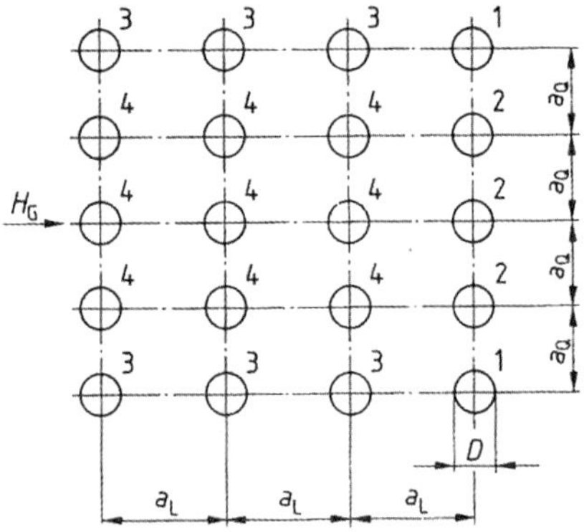

Figure B13.1 Pile group geometry

Reduction factors:

From Figure 8.12: $a_L / D = 3{,}5$ $\alpha_L = 0{,}69$

From Figure 8.13: $a_Q / D = 2{,}5$ $\alpha_{QA} = 0{,}95$

$\alpha_{QZ} = 0{,}88$

From Figure 8.14: $\alpha_1 = 1 \cdot 0{,}95 = 0{,}95$

$\alpha_2 = 1 \cdot 0{,}88 = 0{,}88$

$\alpha_3 = 0{,}69 \cdot 0{,}95 = 0{,}66$

$\alpha_4 = 0{,}69 \cdot 0{,}88 = 0{,}61$

Subgrade reaction moduli of the piles in the group $l / L = 3$ by interpolation between $l / L = 4$ and $l / L = 2$:

a) For linearly increasing subgrade reaction moduli:

$$n_{h1} = \frac{1}{2} \cdot \left(0{,}95^{1{,}67} + 0{,}95\right) \cdot n_{hE} = 0{,}93 \cdot n_{hE}$$

$$n_{h2} = \frac{1}{2} \cdot \left(0{,}88^{1{,}67} + 0{,}88\right) \cdot n_{hE} = 0{,}84 \cdot n_{hE}$$

$$n_{h3} = \frac{1}{2} \cdot \left(0,66^{1,67} + 0,66\right) \cdot n_{hE} = 0,58 \cdot n_{hE}$$

$$n_{h1} = \frac{1}{2} \cdot \left(0,61^{1,67} + 0,61\right) \cdot n_{hE} = 0,52 \cdot n_{hE}$$

b) For constant subgrade reaction moduli:

$$k_{s1} = \frac{1}{2} \cdot \left(0,95^{1,33} + 0,95\right) \cdot k_{sE} = 0,94 \cdot k_{sE}$$

$$k_{s2} = \frac{1}{2} \cdot \left(0,88^{1,33} + 0,88\right) \cdot k_{sE} = 0,86 \cdot k_{sE}$$

$$k_{s3} = \frac{1}{2} \cdot \left(0,66^{1,33} + 0,66\right) \cdot k_{sE} = 0,62 \cdot k_{sE}$$

$$k_{s4} = \frac{1}{2} \cdot \left(0,61^{1,33} + 0,61\right) \cdot k_{sE} = 0,56 \cdot k_{sE}$$

Annex C
Examples of Dynamic Pile Load Testing and Integrity Testing

C 1 Dynamic Pile Load Test Evaluation: Example using the Direct Method

C 1.1 Objectives and test data

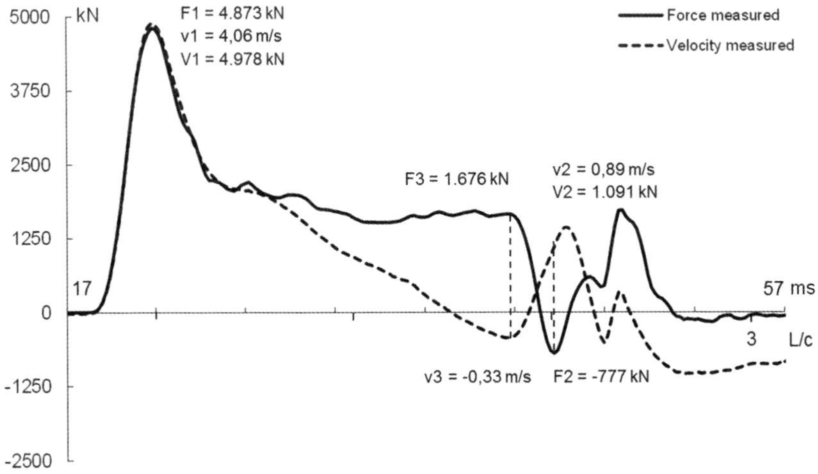

Figure C1.1 Measured force and velocity time history and individual values for the direct method

The following measurement plots and pile data are available:

Steel tube cross-section D = 762 mm (30″), t = 12,7 mm (1/2″):
Area A = 298,96 cm²

Length below transducer: 56,5 m

Soil: Sandy silt, sand along the lower shaft area,

Impedance

$$Z = \frac{E \cdot A}{c} = \frac{210\,000 \cdot 0,0299}{5\,123} = 1,226 \quad [\text{MN/(m/s)}]$$

$$(Z = F/v = 4\,873/4,06 = 1\,200) \quad [\text{kN/(m/s)}]$$

Recommendations on Piling – EA Pfähle.
Editors: German Geotechnical Society (*Deutsche Gesellschaft für Geotechnik e.V.*)

Using the data (taken from the time history):

$$R_{tot} = \frac{1}{2}(F_1 + Z \cdot v_1) + \frac{1}{2}(F_2 - Z \cdot v_2) = \frac{1}{2}(4\,873 + 4\,978) + \frac{1}{2}(-777 - 1\,091)$$

$$R_{tot} = 3\,991 \quad [kN].$$

For evaluation both the Case and the TNO method are used.

C 1.2 Case method

Equations (10.5) and (10.6) and $J_c = 0{,}3$ for sand/silt from Table 10.1 give:

$$R_{dyn} = J_c \cdot Z \cdot \left(v_1 + \frac{(F_1 - R_{tot})}{Z} \right) = 0{,}3 \cdot 1\,226 \cdot \left(4{,}06 + \frac{(4\,873 - 3\,991)}{1\,225} \right)$$

$$R_{dyn} = 0{,}3 \cdot 1\,226 \cdot 4{,}78 = 1\,758 \quad [kN].$$

And using Eq. (10.12)

$$R_{m,i} = R_{stat} = R_{tot} - R_{dyn} = 3\,991 - 1\,758 = 2\,233 \; [kN] \qquad (C1.4)$$

C1.3 TNO method

Eq. (10.8) and C_b = 1,0 [MN/m²/(m/s)] for sand in the area of the pile base embedment give:

$$R_{bdyn} = 0{,}0299 \cdot 1{,}0 \cdot 4{,}78 = 143 \; [kN]. \qquad (C1.5)$$

Eq. (10.9) and F_3, v_3 from the time history, as well as $C_s = 0{,}006$ [MN/m²/(m/s)] from Table 10.2 for sandy silt in the embedment zone over 30 m give:

$$v_s = \frac{1}{2}(4{,}06 + \frac{4\,873}{1\,226}) - \frac{1}{2}(\frac{1\,676}{1\,226} - (-0{,}33)) = 3{,}17 \; [m/s] \qquad (C1.6)$$

$$R_{sdyn} = 0{,}006 \cdot (30 \cdot 0{,}762 * 3{,}14) \cdot 3{,}17 = 1\,366 \; [kN] \qquad (C1.7)$$

$$R_{m,i} = R_{stat} = R_{tot} - R_{bdyn} - R_{sdyn}$$
$$= 3\,991 - 143 - 1\,366 = 2\,482 \; [kN]. \qquad (C1.8)$$

C2 Dynamic Pile Load Test Evaluation Example Using the Extended Method with Complete Modelling

C2.1 Objectives and test data

An open steel tube was driven using a Menck MHF3-10 hydraulic hammer. Toward the end of the driving process (blow 847) the graph shown in Figure C1.1 was recorded and is subsequently evaluated using the extended method.

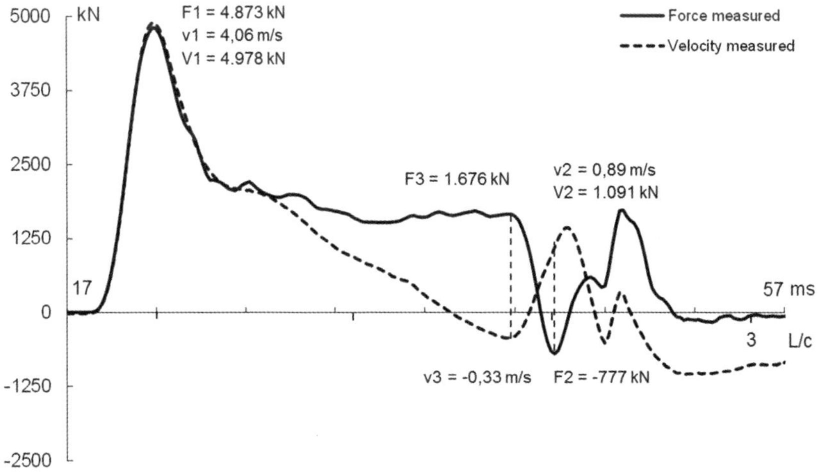

Figure C1.1 Recorded force and velocity histories and individual values for the direct method

The following pile data are available:

Steel tube cross-section D = 762 mm (36″), t = 12 mm (5/8″):
Area A = 298,96 cm²

Length below transducer: 56,5 m

Soil: Sandy silt, sand along the lower shaft area,

Impedance

$$Z = \frac{E \cdot A}{c} = \frac{210\,000 \cdot 0{,}0299}{5\,123} = 1{,}226 \text{ [MN/(m/s)]}$$

$$(Z = F/v = 4\,873/4{,}06 = 1\,200) \text{ [kN/(m/s)]}$$

The result of the extended method is a resistance-settlement curve from a dynamic pile load test. Because it is computed from the complete model of the pile in the ground, the resistance-settlement behaviour can also be calculated and plotted for the pile base resistance, see Figure C2.1.

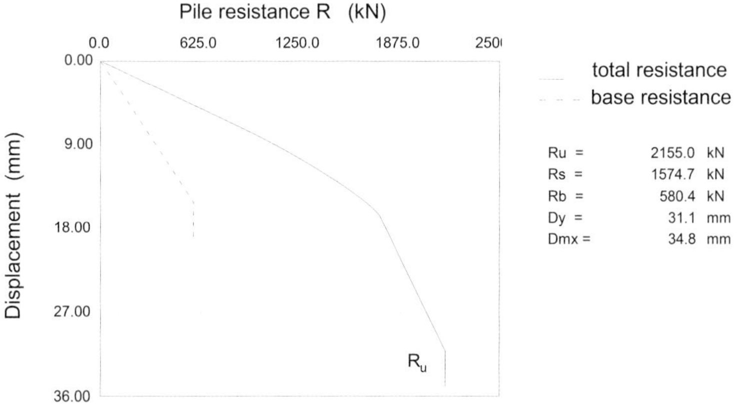

Figure C2.1 Resistance-settlement curve from a dynamic load tests using complete modelling

The static component of the activated resistance is given as R_u (ultimate resistance), the corresponding shaft resistance as R_s and the base resistance as R_b.

It can be seen from the resistance-settlement curve that, after exceeding the skin friction, further forces are taken by increasing base resistances only. Due to the bilinear approximation of the base resistance, the load-settlement curve ends as a vertical line. Normally, for an unlimited load, a small increase would be anticipated for increasing settlements.

The static resistance R_u = 2 155 kN achieved is the recorded or test value $R_{m,i}$ forming the basis for deriving the characteristic pile resistance.

The analysis using complete modelling and explicit determination of the dynamic components of the total resistance displays a slightly lower limit load than the direct method (CASE, TNOWAVE) using empirical damping coefficients.

Another result of complete modelling is the distribution of skin friction over depth. This corresponds to the deviations between the velocity and the force curves. The force and velocity curves are proportional along pile lengths in water or in very soft soils at shallow depth, as there will be no action of skin friction.

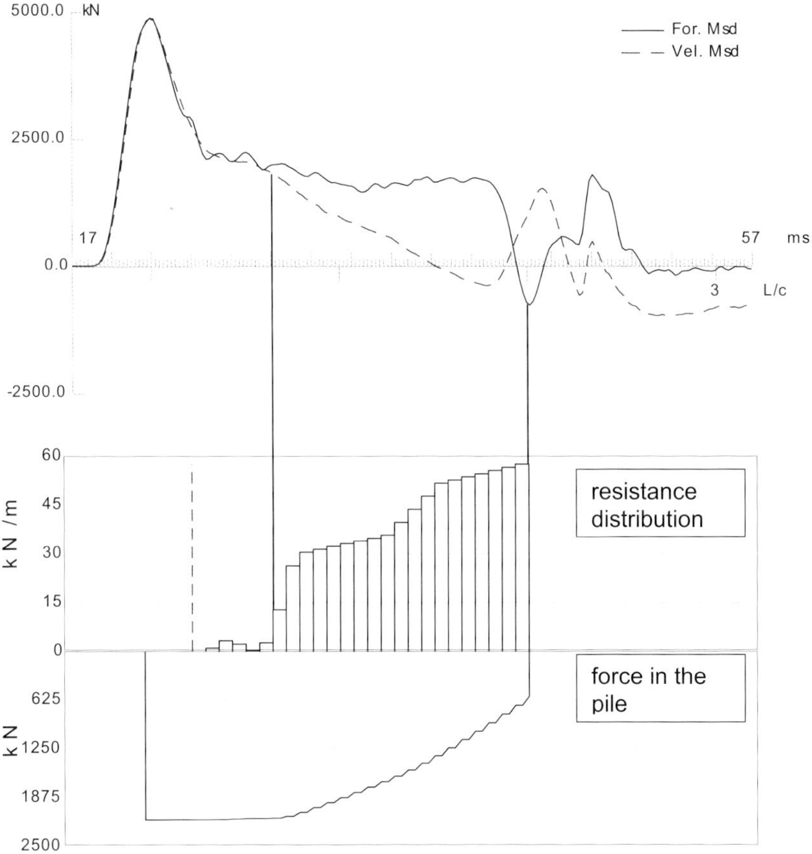

Figure C2.2 Relationship between the recorded time history and the skin friction from dynamic load tests with complete modelling

The accuracy of a CAPWAP solution can be controlled by comparing the recorded curve to the curves calculated using the model. Modelling should not only be performed using the velocity as input variable and the force at the pile head as the controlling variable (Figure C2.3), but also vice versa using the force as the input variable and the velocity as the controlling variable (Figure C2.4).

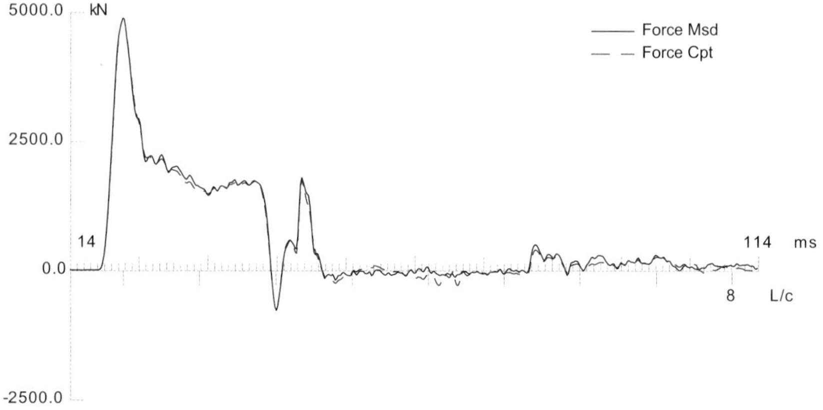

Figure C2.3 Force-time history– comparison of the recorded curve and the curve calculated using the model

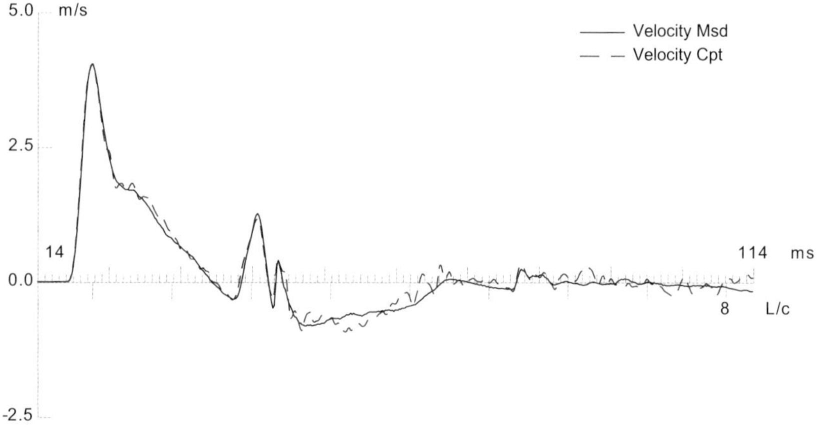

Figure C2.4 Velocity-time history– comparison of recorded curve and the curve calculated using the model

C3 Rapid Load Tests Evaluation Example Using the Unloading Point Method

Step A: Apply the load and record the resulting settlements as an initial indication of system stiffness. Perform the rapid load test and record the force, displacement and acceleration data, see Figures C3.1 and C3.2.

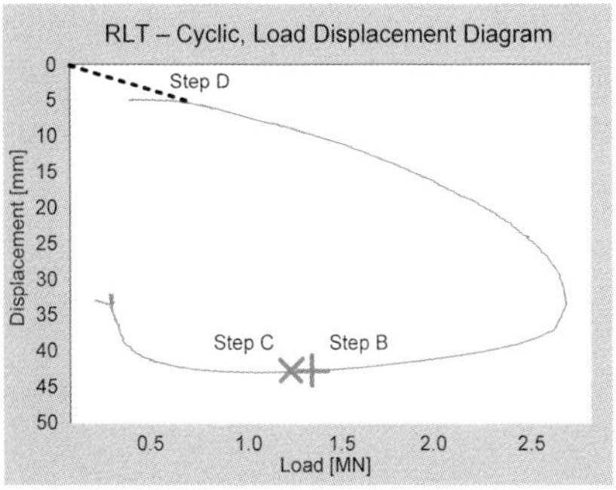

Figure C3.1 Rapid load test data and results using the unloading point method

Step B: Analyse the mobilised capacity (UP+) derived from the measurement and determine the corresponding displacement, see Figure C3.2.

(B1): Calculate the pile mass based on the installation data.

Example:

A = pile cross-section $= 0{,}126\ m^2$

L = pile length $= 12{,}0\ m$

ρ = density $= 2\,400\ kg/m^3$

$m = \text{mass} = A \cdot L \cdot \rho = 3\,619\ kg$.

(B2): Determine the force, displacement, velocity and acceleration time histories from the rapid load test measurement data.

(B3): Determine the time $t_{w\text{-}max}$ of maximum displacement and the velocity zero point.

(B4): Determine the force $F(t_{w\text{-}max})$ at the time $t_{w\text{-}max}$.

(B5): Determine the acceleration $a(t_{w\text{-}max})$ at the time $t_{w\text{-}max}$.

(B6): Adopt the determined value in the UPM analysis and calculate the derived mobilised resistance $F_{statder}(t_{w\text{-}max})$:

$$F_{statder}(t_{w\text{-}max}) = F(t_{w\text{-}max}) - m \cdot a(t_{w\text{-}max})$$

Example: (cf. Figure C 3.2):

$t_{w\text{-}max}$ = 112 ms

$F(t_{w\text{-}max})$ = 1 083 · 10^6 N

m = 3 619 kg

$a(t_{w\text{-}max})$ = –70,3 m/s²

$F_{statder}(t_{w\text{-}max}) = 1{,}083 \cdot 10^6 - (-70{,}3 \cdot 3\,619)$

$F_{statder}(t_{w\text{-}max}) = 1{,}34 \cdot 10^6\ N = 1{,}34\ MN.$

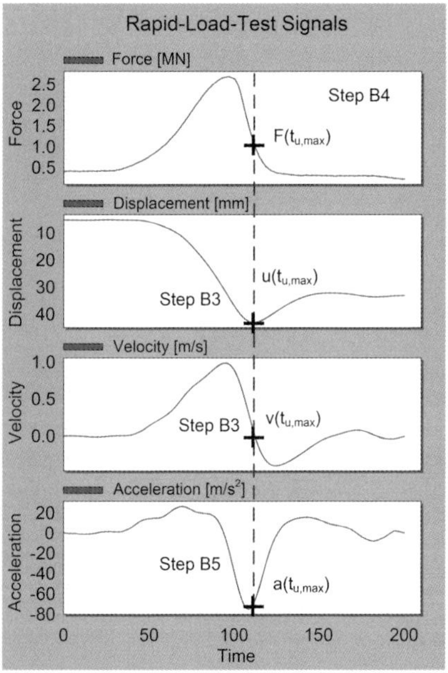

Figure C3.2 Data recorded at time $t_{w\text{-}max}$

Step C: Adapt the calculated unloading point (UP+) using the correction factor η (UP**x**) to take the dynamic soil response into consideration (loading rate effect).

Example:

Correction factor $\eta_{driven\ pile}$ for sand = 0,94

$F_{statder}(\eta_{driven\ pile} \cdot F_{statder}(t_{w\text{-}max}))$ = 0,94 · 1,34 MN

$F_{statder}$ = 1,260 MN

Maximum displacement $w(t_{w\text{-}max})$ at $t_{w\text{-}max}$ = 43,0 mm.

Step D: Determine the initial system stiffness from the measured force-displacement data when applying the reaction mass, and the analysis result (cf. Step B) at time $F(t_c)$, when the measured force corresponds to the characteristic value of the action F_c of the pile.

Step E: Draft the resistance-settlement curve based on the initial system stiffness and the mobilised capacity with the aid of a hyperbolic approximation, see Figure C3.3.

Step F: Draft the resistance-settlement curve based on the hyperbolic approximation for the mobilised static capacity modified by the load duration-dependent effects, see Figure C3.3.

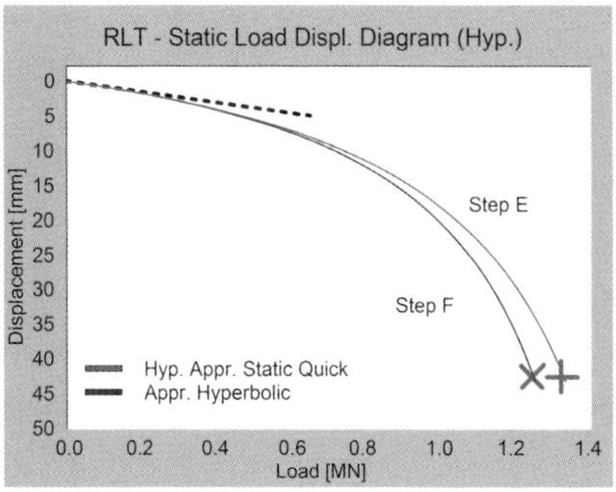

Figure C3.3 Resistance-settlement curve evaluation using the unloading point method

C4 Low Strain Integrity Test Case Studies

C4.1 Example: pile in accordance with specification – Class A1

The base signal in the form of a tension wave (a compression wave occurs only in piles standing directly on rock) is pronounced and the depth corresponds to the specified pile length. Reflections between the input and the base signal cannot be recognised, see Figure C4.1.

A clear base signal (depending on pile length, diameter, concrete age) is normally received in uncoupled, short, prefabricated concrete piles and bored piles with lengths between 5 m and 25 m in non-cohesive soils.

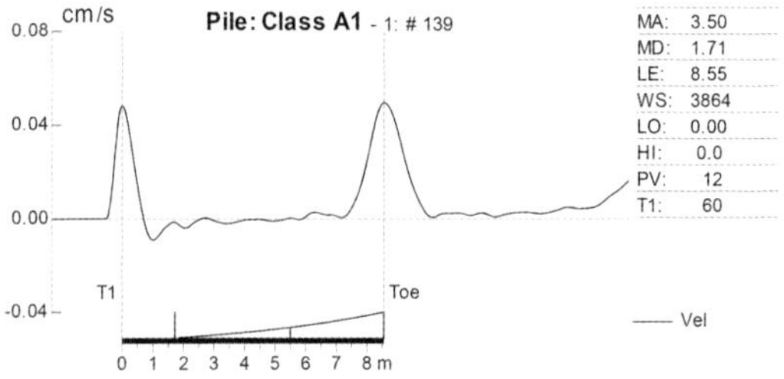

Figure C4.1 Pile without change in impedance – Class A1

C4.2 Example: pile in accordance with specification – Class A2

A pile displays a planned or unplanned change in impedance (increased cross-section).

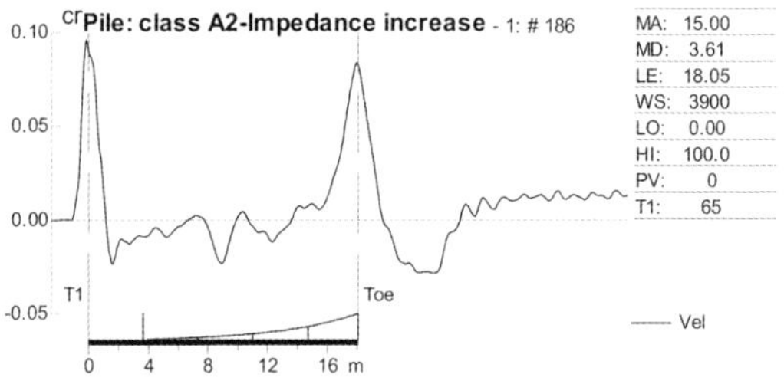

Figure C4.2 Pile with increased impedance – Class A2

A pile embeds in very firm soil (rock) or very dense sand (glacially compacted), such that no base reflection occurs, see Figure C4.3.

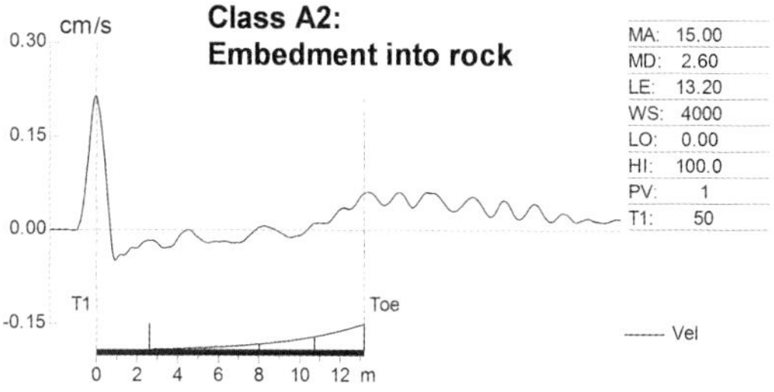

Figure C4.3 Pile showing no base reflection due to rock embedment – Class A2; no conclusion on wave velocity or pile length

Because of the large pile length the introduced wave does not reach the pile base or the reflected wave does not reach the pile head. In long, small diameter piles (around L/D > 30) and/or piles with high skin friction values, irregular cross-section or higher internal damping, the impact wave energy can be too small and the reflected wave signal very weak. Conclusions on the integrity of the lower section of the pile are therefore only possible if a reflection can be clearly identified.

If in doubt Class 0 should be assigned.

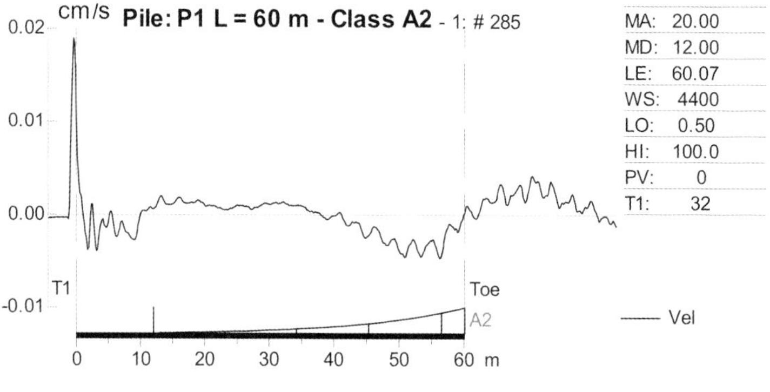

Figure C4.4 Long pile with no reduction in impedance to a determinable depth – Class A2

C4.3 Example: pile with minor deviations – Class A3

A wave reflection can be identified in the pile shaft region, and can be clearly interpreted as a deviation from the normal cross-section or the planned material quality, see Figure C4.5.

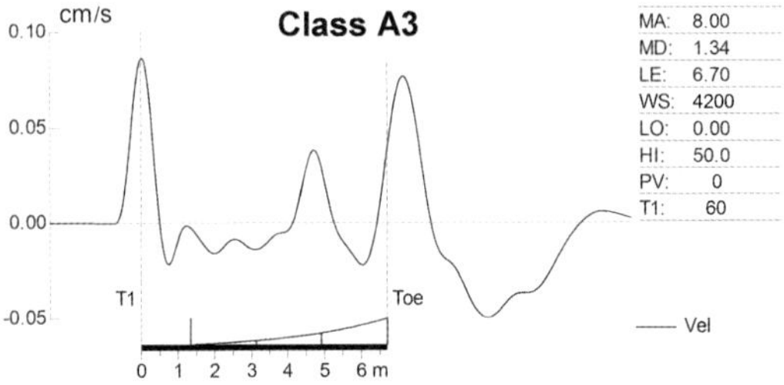

Figure C4.5 Class A3 – initial increase of impedance, thereafter decrease to 81% residual impedance

An impedance decrease to 81% was determined using a analysis with the wave equation method. A clear base signal can be recognised, so the pile is classified as Class A3.

C4.4 Example: pile with substantial impedance reduction – Class B

Pile with clear impedance reduction, see Figure C4.6.

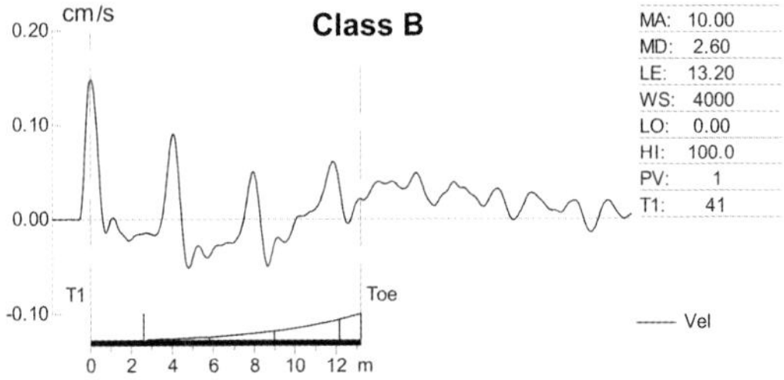

Figure C4.6 Substantial impedance reduction or even interruption in the concrete column at approx. 4 m – Class B

If a pile displays multiple changes in impedance, the lower deviations can often not be clearly determined, or can even not be determined at all, due to the wave reflection caused by the upper defect. If in doubt, assessment remains limited to the uppermost impedance reduction.

C4.5 Example: measurement can not be-evaluated – Class 0

If a defect near the pile head – thinning/necking – is indicated, the following changes in impedance shall not be evaluated.

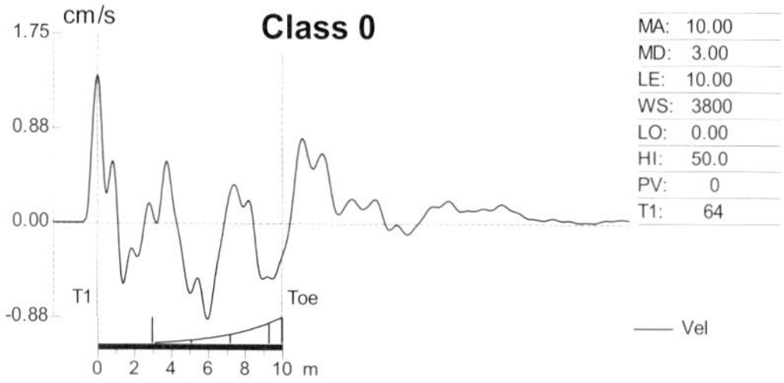

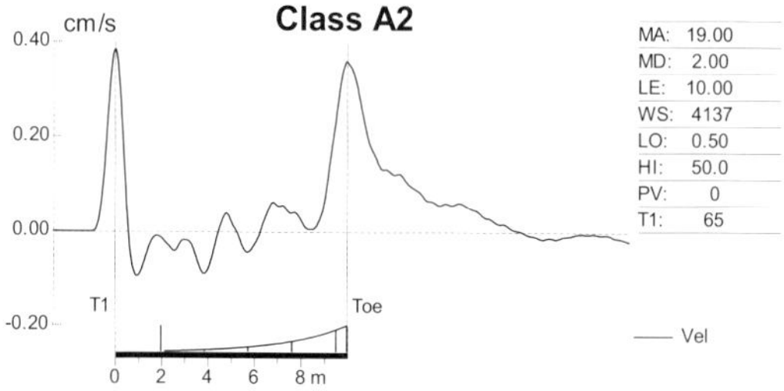

Figure C4.7 Because of the defect near the pile head, the pile cannot be evaluated – Class 0

C5 Integrity Tests during Driving and/or High Strain Integrity Tests

C5.1 Introduction

Dynamic pile load test with impact application after Section 10 can be used for integrity testing. Rapid load test can however not be used for this purpose. If the measured force and velocity show a deviation which obviously does not correspond to a change in skin friction caused by known stratification of the ground, a conclusion with regard to the pile material and cross section can be drawn.

This test using a driving rig produces a signal as shown in Figure C5.1.

When carrying out high strain integrity tests, the test report shall contain the basis on which the tested piles were assessed. This applies to both, piles judged serviceable and deficient.

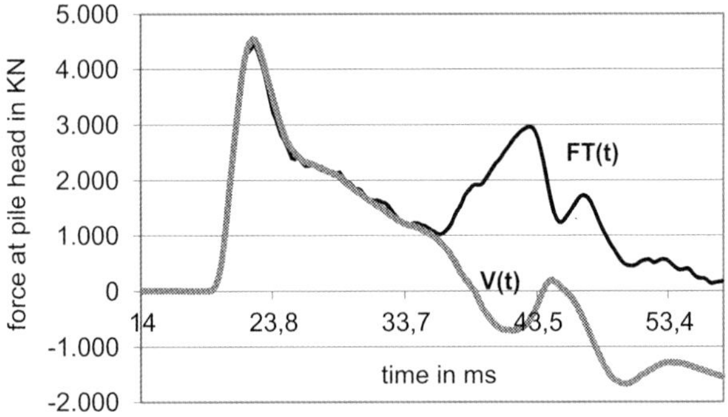

b)

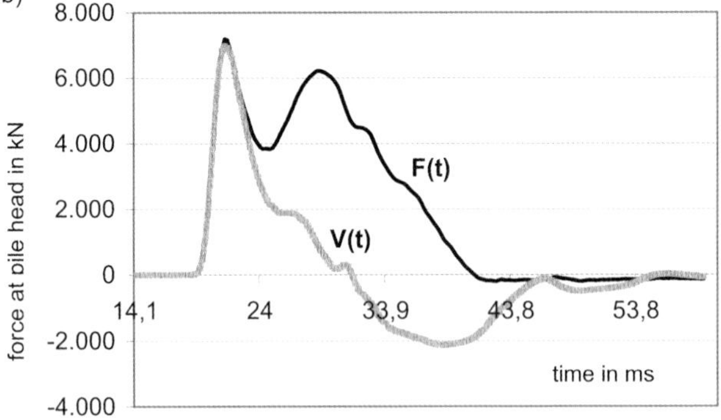

Figure C5.1 Pile without impedance reduction

C5.2 Example: pile in accordance with the specification

The results of two "high strain" tests are shown in Figure C5.1. For pile a), which is standing in water (three-quarters of its length) and where skin friction resistance is low, a clear reflection from the pile base (like in a "low strain" test) can be seen in the velocity profile (lower curve).. For pile b) the pile base reflection is not much distinct due to high skin friction.

C5.3 Example: defective pile

Figure C5.2 shows the development of pile failure on a 70 m long steel pile. The measuring results shown are for the still intact pile during driving (blow 403) and after an impedance decrease occurred approx. 10 m above the pile base (blow 863). The yield stress in a part of the cross-section is obviously exceeded, such that the Young's modulus for the plasticised zone wanes. It should be noted that the tension wave caused by the impedance reduction vir-

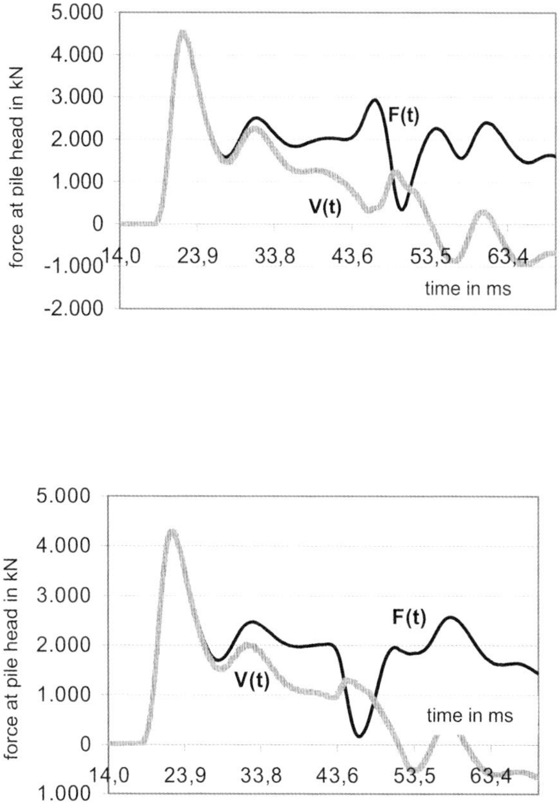

Figure C5.2 Pile damaged during driving

tually pulls the pile from below the hammer, such that the defect is recognised by a kink in the force profile.

C5.4 Example: coupled pile

Figure C5.3 shows measurements taken at the beginning and at the end of restriking. While coupling is barely noticeable when restriking begins because of the high skin friction of the consolidated pile, it can be easily recognised after approx. 200 blows. In both cases the coupling must be taken into consideration in the CAPWAP model.

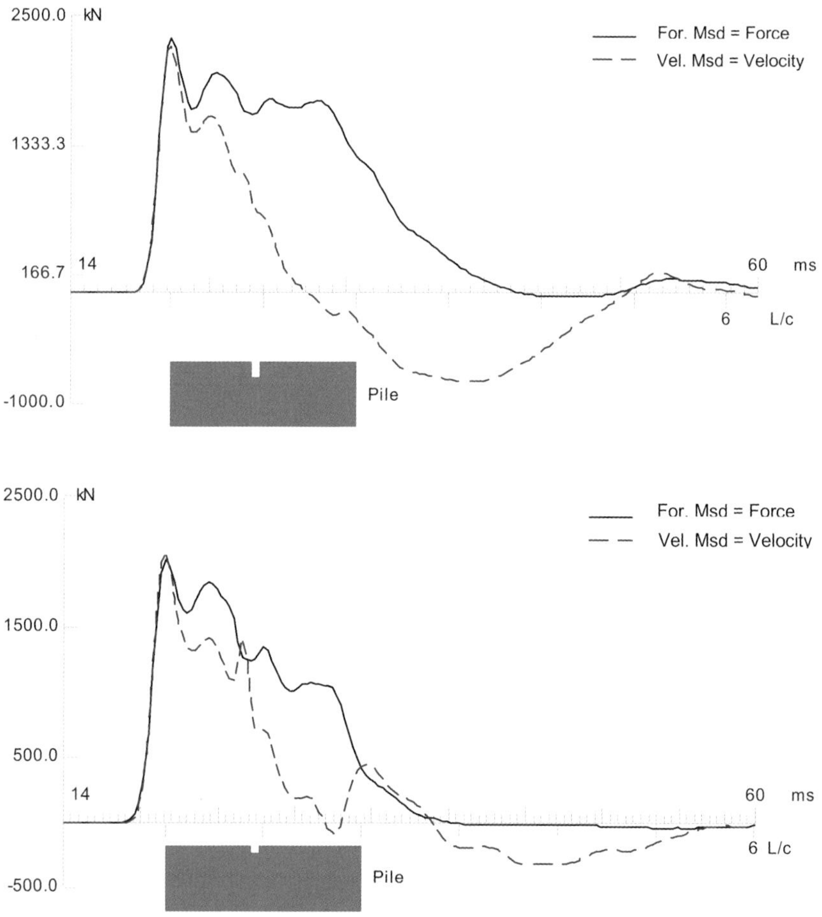

Figure C5.3 Coupled prefabricated concrete pile

C6 Example: Ultrasonic Integrity Testing

In a cast-in-place concrete pile, diagonal and perimeter cross-hole surveys were performed for an ultrasonic investigation on all installed measuring tubes and in all ten orientations, see Figures C6.1 C6.2. The length of the pile was checked and the measured length compared to the target length of 35 m. At depths of 24 m to 25 m and 32 m to 33 m the ultrasonic signal displayed both a delay in arrival time and an energy loss in all ten measured orientations. These signal deviations are accounted for by Osterberg cells and their steel frames installed at the corresponding depths and do not represent an anomaly in the concrete.

The pile was constructed with an empty bore; wave propagation in the concrete therefore starts approximately 1 m below the reference level.

At a depth of 14 m the signal displays a delay in arrival time between measuring tubes 4–5, 1–5, 3–5, 2–5 and 2–4 and a drop in energy, see Table C6.2. This is the result of a minor anomaly in the region of measuring tube 5. Because the anomaly was only found in a single cross-section it is approximately

Table C6.1 Arrival time and wave velocity in sound pile concrete and at the height of the anomaly

Orientation between measuring tube nos.	Measuring tube distances [m]	Mean arrival time in zones with sound pile concrete [ms]	Maximum arrival time at a depth of 14 m [ms]	Deviation [%]	Wave velocity in zones with sound pile concrete [m/s]	Wave velocity at anomaly depth [m/s]
1–2	0,96	0,22	0,22	0%	4 364	4 364
1–3	1,36	0,34	0,34	0%	4 000	4 000
1–4	1,26	0,30	0,30	0%	4 200	4 200
1–5	0,54	0,15	0,22	47%	3 600	2 455
4–5	0,89	0,22	0,25	14%	4 045	3 560
3–4	0,79	0,23	0,23	0%	3 435	3 435
2–3	0,92	0,22	0,22	0%	4 182	4 182
3–5	1,32	0,35	0,45	29%	3 771	2 933
2–5	1,29	0,32	0,44	38%	4 031	2 932
4–2	1,38	0.35	0.48	37%	3 943	2 875
				Mean wave velocity:	3 957	3 493

10 cm high. Table C6.1 shows the mean arrival time in zones with sound pile concrete, the maximum arrival time at a depth of the 14 m, the deviation in arrival time at a depth of 14 m, the corresponding wave velocity in zones with sound concrete and the minimum wave velocity at a depth of 14 m between the measuring tubes.

The possible causes for the defect are summarised below:

a) An air-filled void is located at 14 m depth in the zone of the threaded sleeve on measuring tube 5. The air layer is approximately 0,5 cm to 1 cm thick. Sound concrete is located around this air layer.
b) Located at 14 m depth and around measuring tube 5 lies a zone with inclusions, segregated concrete or with insufficient cement paste. The extent of this encountered defect depends on the deviation of the concrete quality from the sound concrete. A smaller extent results for a large deviation, for a small deviation the extent is correspondingly larger. The exact wave velocity in the zone of the anomaly is unknown, any prediction of the extent of the defect is therefore always an estimate, see Figure C6.2.
c) An alternative option for estimating the extent and influence of a defect consists of adopting a lower bound of the wave velocity in the defective zone. For example, this can be the wave propagation velocity in air. However, it must be assumed that wave propagation in this zone is still possible (in fact, only very little energy is transported in air and the wave is completely reflected at the concrete surface). The size of the defective zone can then be determined directly from the magnitude of the deviation in the wave velocity.

   For example, measurement run 3–5:

   - Extent of sound concrete: L_1 [m];
   - Extent of air L_2 [m] where $c_{air} = 340$ m/s;
   - Geometry: $L_1 + L_2 = 1{,}32$ m;
   - Travel time: $L_1/3\,957 + L_2/340 = 0{,}45$ s;
   - Solution: Concrete thickness $L_1 = 1{,}28$ m, air thickness $L_2 = 0{,}04$ m.

   The air thicknesses for the remaining measurement runs (tube pairs) are calculated accordingly. If the air thicknesses are drawn along straight lines between the tubes, emanating from pipe 5, the geometry of the defect can be seen. If it is also assumed that the defect extends outside of the reinforcement, the total defective area within the cross-section is 0,01 m^2 as shown in Figure C6.3.

   Even if the defect is limited to the inner pile area – within the reinforcement – the reduction is far less than 1%. If, on the other hand, qualitatively poor concrete with a wave velocity of 2 000 m/s is assumed instead of air, the extent of the defective zone can be very much larger, at $L_2 = 0{,}53$ m.

d) If the zone of reduced quality is calculated directly from the ratio of true to average wave velocity, a greater total reduction of 12,1% results for the cross-section, see Table C6.2.

Table C6.2 Determining a quality indicator using Eq. (12.2) for the cross-section in the example in Figure C6.2 and Table C6.1 – the measuring runs with anomalies are indicated by bold/italic type

Measuring runs	Length of measuring runs a_{ij} [m]	Average arrival time in sound concrete ΔT_{good} [ms]	Arrival time in cross-section at depth 14 m $\Delta T_{anomaly}$ [ms]	Wave velocity in sound concrete $c_a = \Delta T_{good}/a_{ij}$ [m/s]	Wave velocity in cross-section at depth 14 m $c_{ij} = \Delta T_{anomaly}/a_{ij}$ [m/s]	Ratio $\chi_{ij} = c_{ij}/ca_{ij}$	Relative to measuring run $\chi_{ij}\, a_{ij}$
1–2	0,96	0,22	0,22	4 364	4 364	1,00	0,96
1–3	1,36	0,34	0,34	4 000	4 000	1,00	1,36
1–4	1,26	0,30	0,30	4 200	4 200	1,00	1,26
1–5	0,54	0,15	0,22	3 600	2 455	***0,68***	0,37
4–5	0,89	0,22	0,25	4 045	3 560	0,88	0,78
3–4	0,79	0,23	0,23	3 435	3 435	1,00	0,79
2–3	0,92	0,22	0,22	4 182	4 182	1,00	0,92
3–5	1,32	0,35	0,45	3 771	2 933	***0,78***	1,03
2–5	1,29	0,32	0,44	4 031	2 932	***0,73***	0,94
4–2	1,38	0,35	0,48	3 943	2 875	***0,73***	1,01
			Mean:	3 957	3 493		
Σ	10,71					Σ	9,42
					Reduction:	$1 - \eta_Q$	12,1%

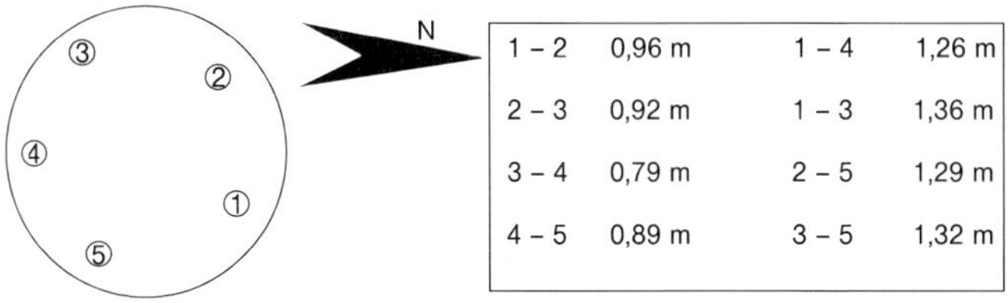

Figure C6.1 Measuring tubes and measuring runs

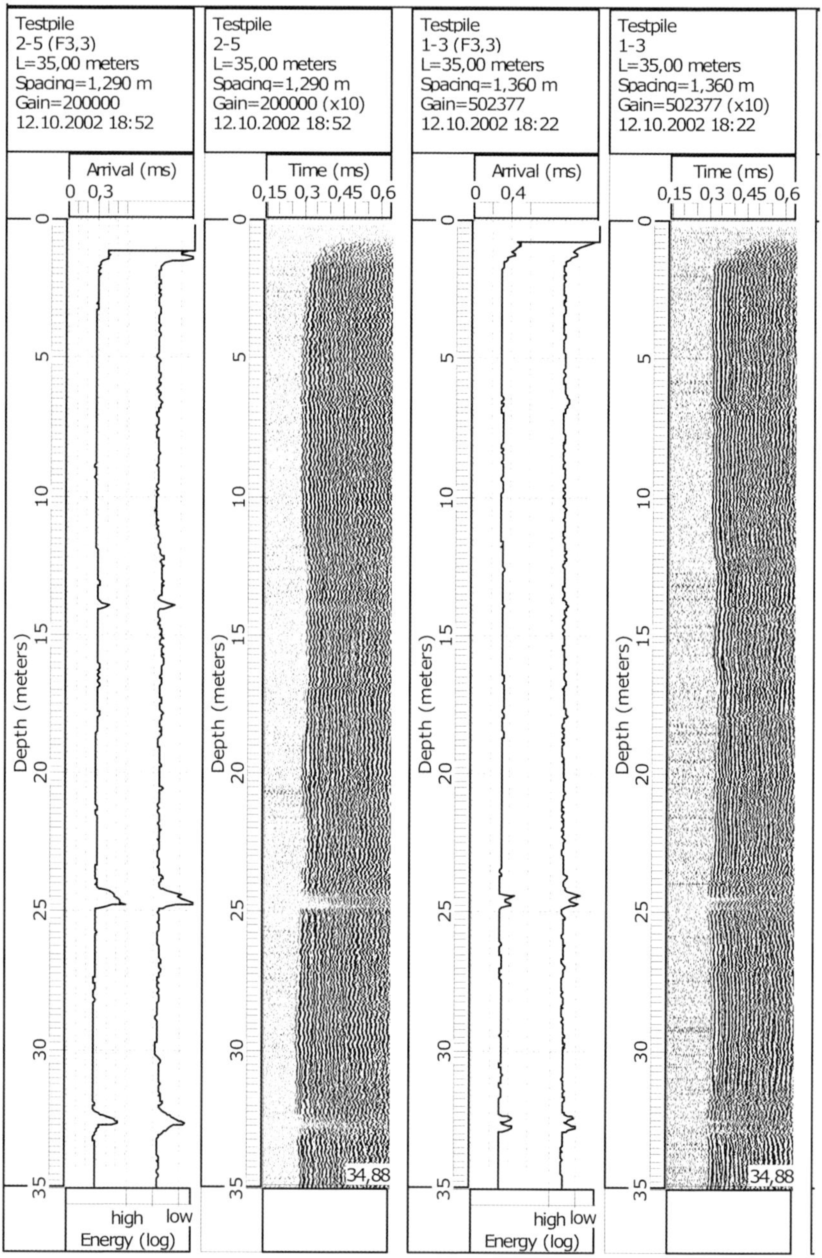

Figure C6.2 Measuring results; only two of ten measuring runs are shown as examples

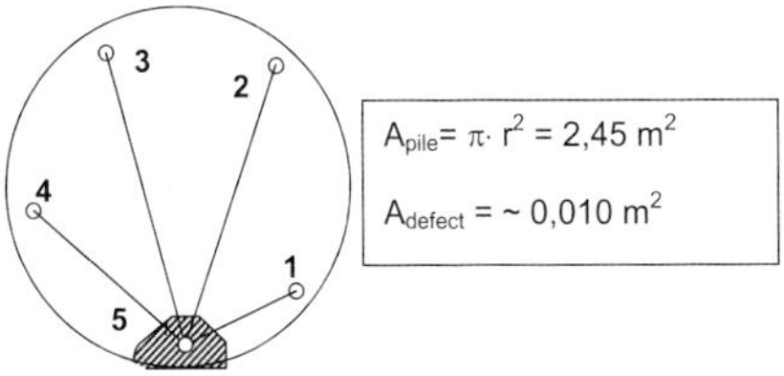

Figure C6.3 Direct defect identification

Annex D
Analysis Methods and Examples for Cyclically Loaded Piles (Informative)

D1 Guidance notes

(1) When using the methods described below to determine pile bearing behaviour under cyclic loads it should be noted that they have only been tested in research projects. Partially, the methods were calibrated against model tests and large-scale tests, but that extensive, project-specific experience is not available at present. The methods therefore do not represent generally recognised best practice, which should be noted by the user in the course of appropriate construction projects. This also applies to the model factors below. Other analysis approaches suited to specific applications can be employed. The applicability of the methods must be confirmed by a geotechnical expert.

(2) The methods are presented here for information only.

(3) The following calculation methods and examples frequently refer to the pile resistance in the ultimate limit state or the failure state under static actions $R_{ult} = R_g$. Hereby, the pile resistance is to be understood as outlined in 13.2.2 (4) and taking the EC 7-1 Handbook [44] into consideration as follows:

a) If based on static or dynamic pile load tests, then $R_{ult} = (R_{c,m})_{mean}$ and $R_{ult} = (R_{t,m})_{mean}$, respectively, or, if the minimum value is governing in line with the provisions of the EC 7-1 Handbook [44] and Appendix A4, then $R_{ult} = (R_{c,m})_{min}$ and $R_{ult} = (R_{t,m})_{min}$.
b) If determining from empirical data, e.g. after 5.4, then $R_{ult} = R_{c,k}$ and $R_{ult} = R_{t,k}$.

(4) Determination of the design values of the static pile resistances $R_{c,d}$ and $R_{t,d}$ in Eq. (13.9a) and (13.9b), as well as Eq. (13.10a) and (13.10b) is carried out in line with the provisions of the EC 7-1 Handbook [44] and 6.3.1 (1), Eq. (6.5) with Appendices A3.2 and A4.

(5) In the following calculation examples for cyclically loaded piles the partial factors from the EC 7-1 Handbook [44] are adopted. However, other partial factors can apply for offshore applications.

Recommendations on Piling – EA Pfähle.
Editors: German Geotechnical Society (*Deutsche Gesellschaft für Geotechnik e.V.*)

D2 Piles Subjected to Cyclic Axial Loads

D2.1 Analysis methods

D2.1.1 Pile resistance in the ultimate limit state based on interaction diagrams

(1) Bearing capacity analysis can be performed in approximation with the aid of interaction diagrams, see Figures D2.1 and D2.2. By use of a utilisation factor μ, a reduced pile capacity is determined.

(2) The following procedure can be followed: the load levels X_{cyc} and X_{mean} are marked in an interaction diagram and, by modifying the pile resistance in the ultimate limit state R_{ult}, the point is found at which the existing action from F_{mean} and F'_{cyc} reaches the corresponding boundary line for the existing load cycles. This then gives the corresponding value R^*_{ult}. The system's utilisation factor is then defined as follows:

$$\mu = R^*_{ult} / R_{ult} . \tag{D2.1}$$

where:

R_{ult} $R_{c,k}$ and $R_{t,k}$ or $R_{(c,m)mean}$ and $R_{(t,m)mean}$ after 13.2.2 (4) and D1 (3);
R^*_{ult} static pile resistance leading to failure under the same action after N load cycles.

In practice, this means that μ is determined using the following equation adopting the terms in Figure D2.1:

$$\mu = \frac{l_1}{l_1 + l_2} . \tag{D2.2}$$

In cases where $\mu = 1{,}0$, the considered point lies on the boundary line for the considered number of load cycles (then $l_2 = 0$) and the maximum cyclic capacity reduction (cases where $\mu > 1$ are not allowed), relative to the static pile resistance represented by the total length $l_1 + l_2 + l_3$, is given by the following equation:

$$\Delta R_{cyc,max} = \frac{l_3}{l_1 + l_2 + l_3} \cdot R_{ult} . \tag{D2.3}$$

where:

l_1, l_2, l_3 lengths derived from the interaction diagram in Figure D2.1.

In cases where $\mu < 1{,}0$ the cyclic capacity reduction is smaller. No reduction occurs for $\mu = 0$. In a first approximation it is assumed that ΔR_{cyc} develops linearly with μ. The cyclic capacity reduction is then given by Eq. (D2.4).

$$\Delta R_{cyc} = \mu \cdot \frac{l_3}{l_1 + l_2 + l_3} \cdot R_{ult} . \tag{D2.4}$$

A stability analysis for the ultimate limit state (ULS) in accordance with 13.7.1 must therefore be performed. A model factor $\eta_{cyc} = 1{,}2$ may be adopted for the interaction diagram in Figure D2.2.

(3) Interaction diagrams compiled on the basis of model and field tests, and cyclic pile load tests, can be found in several literature sources, see e.g. [98], [56], [59] and [67].

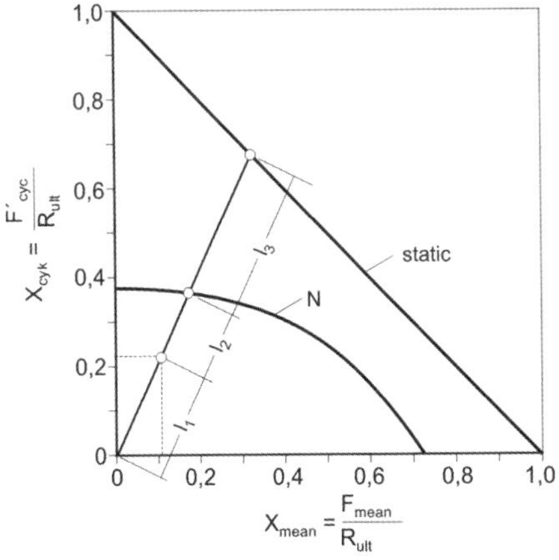

Figure D2.1 Evaluation of an interaction diagram

(4) The interaction diagram in Figure D2.2 is recommended for non-cohesive soils.

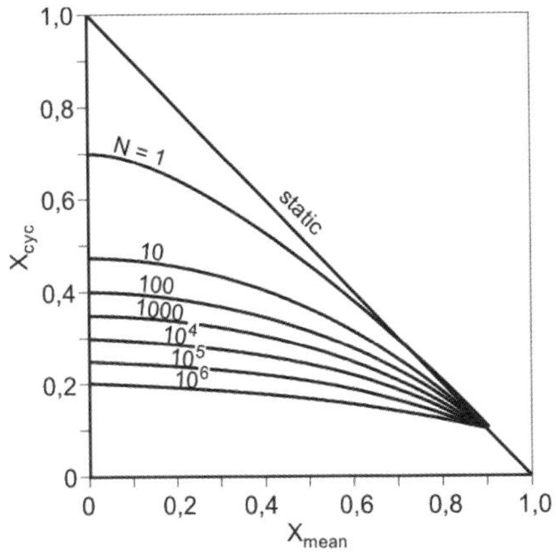

Analytical description of interaction curves:

$$X_{cyc} = \kappa\,[1 - 1{,}11^{EXP} \cdot X_{mitt}^{EXP}] + 0{,}1235 \cdot X_{mean}^{EXP}$$

where $\kappa = 0{,}5 + 0{,}67[\log(N+1) - 1{,}0746 \log(N)]$

$EXP = 2 - 1{,}5[\log(N+1) - \log(N)]$

Figure D2.2 Interaction diagram, from [67]

(5) For cohesive soils, the value κ in Figure D2.2 may be increased by the factor 1,3 for $N \geq 10$. For $N = 1$ the increase factor is 1,0. Values of N between 1 and 10 may be linearly interpolated.

D2.1.2 Displacement accumulation using an empirical approach

(1) Cyclic pile settlement and heave behaviour as a result of characteristic cyclic actions can be described in approximation by Eq. (D2.5), e.g. see [135],. The displacement and the displacement rate after the first load cycle, and the inclination coefficient, should be determined in a cyclic pile load test using at least a few load cycles. The further course of cyclic displacement can then be estimated using Eq. (D2.5).

(2) The inclination coefficient λ in Eq. (D2.5) depends on the soil, the load and the pile type. Corresponding values basing on model and field tests, and cyclic pile load tests are given e.g. in [59]. It should be noted that these values are subjected to strong scatter as a result of the different boundary conditions. For a ratio of cyclic to mean load X_{cyc}/X_{mean} between 0,15 and 0,40, a value for λ of 0,7 to 0,9 can be used as a first approximation for piles in non-cohesiove soils under pulsating loads. However, the inclination coefficient should preferably be determined on a project-specific basis in a static pile load test with unloading and reloading stages.

$$s_{cyc,k} = s^1 + \frac{\dot{s}_{pl}^{\,1}}{1-\lambda} \cdot (N^{1-\lambda} - 1) \quad \text{for } \lambda \neq 1. \qquad \text{(D2.5)}$$

where:

$s_{cyc,k}$ axial displacement after N load cycles;

s^1 displacement after the first load cycle;

$\dot{s}_{pl}^{\,1}$ plastic displacement rate after the first load cycle;

λ inclination coefficient.

(3) The serviceability limit state (SLS) analysis in accordance with 13.7.2 can be approximately performed using Equations (13.12) and (13.13) in conjunction with Equation (D2.5).

D2.1.3 Approximation methods for calculating pile bearing behaviour under cyclic loads after [66]

D2.1.3.1 Introduction

(1) The method presented hereafter for determining the bearing capacity and deformations of piles subjected to cyclic axial loads is described in detail in [66]. Comparisons of this method with in-situ test results are provided in [122].

(2) Based on a simple engineering model, the method returns the value of the cyclic bearing capacity reduction ΔR_{cyc} and the axial displacements exist. s_k of

cyclically axially loaded piles in non-cohesive soils above and below the groundwater table and in overconsolidated, cohesive soils.

(3) The method takes into consideration the decrease of the soil's shear strength at the pile shaft as a result of compaction of the soil surrounding the pile under cyclic shear loading.

(4) Accumulated displacements, consisting of a displacement component resulting from cyclic compaction and an additional component resulting from cyclic creep, occur as a result of the cyclic load amplitude F'_{cyc} of the axial cyclic load.

(5) It is assumed that the pile is subjected to a load from an equivalent, single-stage load spectrum as outlined in D4.1.1, whereby the mean action may not be zero.

(6) Shear strains γ on the pile shaft as a result of the mean load F_{mean} and the cyclic load amplitude F'_{cyc} are added to calculate the total displacements.

(7) The pile displacements are calculated from the total shear strains γ_{ges} adopting the range estimate in accordance with [15].

(8) The distribution of pile resistance resulting from cyclic effects to different sections of the pile shaft and of the pile base is iteratively carried out via the condition of equal, separately determined displacements, by initially estimating the respective components and calculating the displacements in accordance with D2.1.3.3. The distribution of the actions is iteratively adapted until all shaft elements and the pile base element experience the same displacement after the cyclic loading.

D2.1.3.2 Pile resistance in the ultimate limit state

(1) The reduction in the pile's ultimate skin friction as a result of cyclic compaction of non-cohesive soils can be determined using the following equation:

$$\Delta\tau(N) = 2 \cdot G_W \cdot \tan\delta \cdot \Delta D^* \cdot \left(\gamma_{zyk} \cdot \left(\frac{\gamma_{cyc}}{\gamma_{ult}} - 1 \right) - \frac{1}{2} \cdot \alpha \cdot \gamma_{ult} \left[\left(\frac{\gamma_{cyc}}{\gamma_{ult}} \right)^2 - 1 \right] \right) \tag{D2.6}$$

where:

$$\gamma_{cyc} = \tau_{cyc} / G_{cyc}$$

and

$$\Delta D^* = \Delta D \cdot \log(N+1) = 0{,}5 \cdot I_d^{-2{,}32} \cdot \log(N+1) \tag{D2.7}$$

where:

N	number of cycles;
G_W	shear modulus on reloading;
δ	activated wall friction angle;
I_D	initial density of the soil;
γ_{cyc}	cyclic shear strain;
τ_{cyc}	cyclic shear stress;
G_{cyc}	shear modulus for cyclic loading, depending on γ_{cyc}, see [18];
γ_{ult}	ultimate shear strain;
α	dilatation parameter.

(2) Either empirical data or the results of laboratory tests can be adopted for the individual parameters, see D2.2.3, for example.

(3) The cyclic capacity reduction is given by the reduction in the ultimate skin friction $\Delta\tau(N)$ multiplied by the pile shaft area A_m as:

$$\Delta R_{cyc} = \Delta\tau(N) \cdot A_m \tag{D2.8}$$

(4) For analysing the bearing capacity using Eq. (13.8) or (13.9), in line with current knowledge, the model factor $\eta_{cyc} = 1{,}20$ shall be used.

D2.1.3.3 Displacement accumulation in the serviceability limit state

(1) The shear strain resulting from cyclic compaction, i.e. from the reduction in the ultimate shear stress and thus the increase in the utilisation factor, is given by:

$$_1\Delta\gamma = \gamma_2 - \gamma_1 \tag{D2.9}$$

where:

$$\gamma_2 = \frac{\kappa_2}{1 - \kappa_2 / c} \cdot \gamma_r \tag{D2.10}$$

$$\gamma_1 = \frac{\kappa_1}{1 - \kappa_1 / c} \cdot \gamma_r \tag{D2.11}$$

and the load levels κ_1 and κ_2

$$\kappa_1 = (\tau_{mean} + \tau_{cyc}) / \tau_{ult} \tag{D2.12}$$

$$\kappa_2 = (\tau_{mean} + \tau_{cyc}) / \tau_{ult}(N) = (\tau_{mean} + \tau_{cyc}) / (\tau_{ult} - \Delta\tau(N)) \tag{D2.13}$$

and

τ_{mean} mean shear stress;
τ_{cyc} cyclic shear stress amplitude;
τ_{ult} ultimate shear stress (corresponds to $q_{s,k}$);
c curvature of the hysteresis loop.

The meaning of indeces 1 and 2 are described in the Example D2.2.3.

(2) The shear strain from cyclic creep is given by:

$$_2\Delta\gamma = {_2\Delta\gamma_1} \cdot (1 + \zeta \cdot \ln(N)) \tag{D2.14}$$

where:

$$_2\Delta\gamma_1 = \gamma_r \left[\frac{\kappa_1}{1 - \kappa_1/c} - \frac{\kappa_1}{1 + \kappa_1/c} \right]$$

and

$$\gamma_r = \tau_{ult}/G_{max}$$

where:

ζ abatement constant.

(3) The pile displacements resulting from characteristic cyclic actions are therefore given by:

$$s_{cyc,k} = ({_1\Delta\gamma} + {_2\Delta\gamma}) \cdot r_0 \cdot \ln(r_m/r_0) \tag{D2.15}$$

where:

$$r_m = 2{,}5 \cdot L \cdot (1 - \nu) \tag{D2.16}$$

where:

r_0 pile radius;
L embedded length;
ν Poisson's ratio of the soil.

(4) Using Eq. (D2.17) the total displacements are given by:

$$\text{exist. } s_k = s_{G,k} + s_{Q,k} + s_{cyc,k} \tag{D2.17}$$

Analysis of the serviceability limit state can then be performed using Eq. (13.12).

D2.1.4 Approximation method for analysing pile bearing behaviour under cyclic loads after [142]

(1) The ZYKLAX [142], [144] analysis model described below serves to approximately determine the static, cyclic and post-cyclic pile bearing behav-

iour. The calculation model returns both the value of the change (increase or decrease) in pile capacity ΔR_{cyc} and the plastic deflection $s_{pl} = s_{cyc}$ after N load cycles, as well as a complete static and cyclic resistance-settlement curve analogous to Figure 13.7 or a resistance-heave curve. These can then be used to analyse the ultimate and serviceability limit states in accordance with 13.7.1 and 13.7.2.

(2) It is assumed that the pile is subjected to a load from an equivalent, single-stage load spectrum as explained in D4.1.1.

(3) The adopted calculation approaches initially consider the pile shaft and the pile base separately. Static load-bearing behaviour is modelled using analytical approaches after [32] and [71]. The cyclic calculation approaches extend the static approaches by using modified *Masing* rules [89]. A calculation program based on the load transfer approach [16] was developed to allow several soil strata to be taken into consideration. The procedure described below can be adopted for calculation of pile load-bearing behaviour in homogeneous soil. Otherwise, the source code described in [142], where the calculation model is also dealt with in more detail, can be used.

(4) The approach after [32] is used to model the displacement of the pile shaft s_s on initial loading and for post-cyclic loading:

$$s_s = \frac{\tau_0 \cdot r_0}{G_0 \cdot g_s} \ln\left(\frac{\left(\frac{r_m}{r_0}\right)^{g_s} - R_{fs}\left(\frac{\tau_0}{\tau_{ult}}\right)^{g_s}}{1 - R_{fs}\left(\frac{\tau_0}{\tau_{ult}}\right)^{g_s}}\right) \tag{D2.18}$$

where:

τ_0	shear stress at the pile shaft;
τ_{ult}	pile skin friction in the failure state (q_s);
G_0	shear modulus for small strains;
r_0	pile radius;
r_m	influence radius after Eq. (D2.20);
R_{fs}	empirical model parameter;
g_s	empirical model parameter.

(5) The displacement of the pile base under static, cyclic and post-cyclic loads is modelled on the basis of the approach after [71], whereby the approach is extended by the parameter g_b:

$$s_b = \frac{R_b \cdot (1-\nu)}{4 \cdot G_0 \cdot r_0 \cdot \left(1 - R_{fb} \cdot \left(\frac{R_b}{R_{b,ult}}\right)^{g_b}\right)} \tag{D2.19}$$

where:

R_b pile base resistance;
R_{fb} empirical model parameter;
g_b empirical model parameter;
$R_{b,ult}$ pile base resistance in the failure state;
ν Poisson's ratio.

(6) The following approach is adopted for the influence radius r_m on initial loading, in deviation from the proposal in [120]:

$$r_m = D \tag{D2.20}$$

where:

D pile diameter.

(7) Cyclic loading is described by an initial and reloading curve, which are modelled on the basis of the rules in [89]. The rules are extended by the model parameters κ, δ_N, β_N and r_m. The pile shaft load-displacement relationship when the pile is unloaded can then be described by:

$$s_{s,E}(N) = s_{max} + \frac{(\tau_0 - \tau_{max}) \cdot r_0}{G_0 \cdot g_s} \cdot \ln \left(\frac{\left(\frac{r_m}{r_0}\right)^{g_s} - R_{fs} \cdot \left(\frac{|\tau_0 - \tau_{max}|}{\kappa \cdot \beta_N \cdot \tau_{ult}} \right)^{g_s}}{1 - R_{fs} \cdot \left(\frac{|\tau_0 - \tau_{max}|}{\kappa \cdot \beta_N \cdot \tau_{ult}} \right)^{g_s}} \right) \tag{D2.21}$$

where:

$s_{s,E}(N)$ pile shaft displacement at unloading in load cycle N;
s_{max} pile shaft displacement at the beginning of unloading at $F = F_{max}$;
τ_{max} shear stress at the beginning of unloading at $F = F_{max}$;
κ model parameter describing the form of the hysteresis;
β_N model parameter describing the change in pile capacity;

and on pile reloading by:

$$s_{s,W}(N) = s_{min} + \frac{(\tau_0 - \tau_{min}) \cdot r_0}{G_0 \cdot g_s} \cdot \ln \left(\frac{\left(\frac{r_m}{r_0}\right)^{g_s} - R_{fs} \cdot \left(\frac{|\tau_0 - \tau_{min}|}{\kappa \cdot \delta_N \cdot \beta_N \cdot \tau_{ult}} \right)^{g_s}}{1 - R_{fs} \cdot \left(\frac{|\tau_0 - \tau_{min}|}{\kappa \cdot \delta_N \cdot \beta_N \cdot \tau_{ult}} \right)^{g_s}} \right) \tag{D2.22}$$

where:

$s_{s,W}(N)$ pile shaft displacement at reloading in load cycle N;
s_{min} pile element displacement at the beginning of reloading at $F = F_{max}$;
τ_{min} Shear stress at the beginning of reloading at $F = F_{min}$;
δ_N model parameter describing the accumulation of plastic displacement.

The decrease or increase in pile capacity is assessed with the aid of the parameter β_N in Eq. (D2.22), which modifies the shear strength failure value of the soil or the pile skin friction. This parameter is determined on the basis of the approach in [91] as follows:

$$\beta_N = (1-\dot{\beta})\cdot(\beta_{N-1} - \beta_{lim}) + \beta_{lim} \quad \text{for} \quad N > 1 \qquad \text{(D2.23)}$$

$$\beta_N = 1{,}0 \quad \text{for} \quad N = 1 \qquad \text{(D2.24)}$$

where:

β_{N-1} modification factor in the previous load cycle;
β_{lim} upper or lower bound for the modification factor;
$\dot{\beta}$ reduction rate.

The change in the shear strength failure value or the pile skin friction failure value during cyclic loading is specified by the parameter β_{lim}. The parameter $\dot{\beta}$ determines within how many load cycles the limit value β_{lim} is reached.

The accumulation of the plastic pile shaft displacement for a decreasing displacement rate is described in Eq. (D2.22) by the parameter δ_N, which modifies the secant modulus on pile reloading curve as follows:

$$\delta_N = 1 - \frac{1}{a \cdot N^b} \qquad \text{(D2.25)}$$

where:

a, b empirical model parameters.

If a progressive increase in the displacement rate is to be modelled, this behaviour can be taken into consideration using the following approach:

$$\delta_N = \delta_{N-1} \cdot \left(1 + \frac{1}{c}\right)^N \quad \text{for} \quad N > 1 \qquad \text{(D2.26)}$$

where:

δ_{N-1} value of δ_N in previous load cycle;
c empirical model parameter.

(8) A reversal in the displacement rate after N load cycles can be taken into consideration by the limit criterion for the maximum load level X_{ult} in accordance with Eq. (D2.27).

$$X_{max}(N) = \frac{F_{max}}{\beta_N \cdot R_{ult}} > X_{ult} \qquad \text{(D2.27)}$$

where:

β_N model parameter after eq. (D2.23)

(9) The calculation model also allows modelling of a load cycle-dependent influence radius. This can be useful in non-cohesive soils, as demonstrated by the soil behaviour model tests in [142].

$$r_{m,N} = (1 - \dot{r}_m) \cdot (r_{m,N-1} - r_{m,lim}) + r_{m,lim} \quad \text{for} \quad N > 1 \tag{D2.28}$$

$$r_{m,N} = D \quad \text{for} \quad N = 1 \tag{D2.29}$$

where:

$r_{m,N-1}$ influence radius in the previous load cycle;
$r_{m,lim}$ upper bound of influence radius;
$\dot{r}_m$ increase rate.

The static approach using Eq. (D2.20) can be used for the influence rate in the first load cycle. The static approach should also be adopted for cohesive and mixed-grained soils.

(10) Optimally, the model parameters required for the approaches are derived from static and cyclic pile load tests. The static model parameters should be calibrated against static pile load tests and then be adopted unaltered in the cyclic calculation.

Table D2.1 provides guide values for the cyclic model parameters. Static model parameter guide values derived from model tests and pile load tests are given below: In the investigations in [142] values between 0,97 and 1,00 have proven suitable for the parameters R_{fs} and R_{fb}. The parameters g_s and g_b were adopted in the range 0,01 to 0,30. Further guidance on determining the model parameter can be found in [142].

(11) It is anticipated that, given increasing experience from applications, it will be possible to drive the model parameters in approximation also from suitable cyclic shear tests.

Table D2.1 Guidance values for the cyclic model parameters, derived from model tests in [142]

a [–]	b [–]	c [–]	κ [–]	β_{lim} [–]	$\dot{\beta}$ [–]	$r_{m,lim}$ [m]	$\dot{r}_m$ [–]
1,40 to 15	0,40 to 2,00	500 to 5 000	0,70 to 2,00	0,50 to 1,30	0,001 to 0,010	D to 3 • D	0,001 to 0,010

D2.2 Calculation examples

D2.2.1 Ultimate limit state analysis based on interaction diagrams after D2.1.1

A reinforced concrete driven pile (∅ 0,46 m) with a length of 19,0 m subjected to cyclic tensile loads serves as example. Based on load tests the static ultimate capacity of the pile embedded in non-cohesive soil can be adopted as $R_{ult} = (R_{t,m})_{mean} = 2\,500$ kN. Three static pile load tests were performed, such that the characteristic tension pile resistance is $R_{t,k} = (R_{t,m})_{mean}/\xi_1 = 2\,500/1{,}15 = 2\,174$ kN in accordance with the EC 7-1 Handbook [44]. $(R_{t,m})_{min}$ is not governing here.

A pulsating tensile load is applied to the pile. This load is $F_{mean} = 700$ kN (mean action adopted as a permanent action) and $F'_{cyc} = 700$ kN (cyclic load amplitude). The pile is loaded with N = 200 load cycles.

It must first be examined whether the prevalent cyclic load necessitates a special investigation:

$$\frac{F'_{cyc}}{R_{ult}} = \frac{700}{2\,500} = 0{,}28 > 0{,}10$$

It is therefore necessary to take the cyclic action into consideration after 13.4.2 (1).

In this example the ultimate limit state analysis is based on the interaction diagram in Figure D2.2, which was introduced by [67]. With the aid of Eq. (D2.2) and Figure D2.1 a utilisation factor $\mu = 0{,}81$ is given for $X_{mean} = X_{cyc} = 0{,}28$. The maximum capacity reduction associated with $\mu = 1{,}0$ is given by Eq (D2.3) as 0,31 R_{ult}. The true capacity reduction ΔR_{cyc} in accordance with Eq. (D2.4) is thus 628 kN.

The limit state condition (ULS) is as follows in accordance with 13.7.1:

$$F_{t,d} = F_{mitt} \cdot \gamma_G + F'_{cyc} \cdot \gamma_Q \le R_{t,d}(N) = R_{t,d} - \eta_{cyc} \cdot \Delta R_{cyc} = \frac{R_{t,k}}{\gamma_{s,t}} - \eta_{cyc} \cdot \Delta R_{cyc}$$

Adopting the partial factors $\gamma_G = 1{,}35$, $\gamma_Q = 1{,}50$, $\gamma_{s,t} = 1{,}15$ and $\eta_{cyc} = 1{,}20$ returns the following results:

$$F_{t,d} = 700 \cdot 1{,}35 + 700 \cdot 1{,}50 = 1\,995{,}0 \text{ kN}$$

$$R_{t,d} = \frac{2\,174}{1{,}15} - 1{,}20 \cdot 628 = 1136{,}8 \text{ kN.}$$

The design value of the resistance is smaller than the design value of the actions of 1 995 kN, the effect is therefore classified as not allowable.

D2.2.2 Serviceability limit state analysis with an empirical displacement approach after D2.1.2

In this example the stability in the serviceability limit state of a reinforced concrete driven pile (diameter: 0,66 m, length: 15,0 m) in non-cohesive soil and subjected to tensile pulsating loads is analysed. It is assumed that the equivalent, single-stage load spectrum as discussed in D4.1.1 has already been determined. The pile is thus subjected to a mean action $F_{mean} = 800$ kN and a cyclic load amplitude $F'_{cyc} = 300$ kN, with an equivalent load cycle number $N_{eq} = 1\,000$. The allowable pile displacement is allow. $s_k = 3$ cm.

For the serviceability limit state analysis in accordance with Eq. (13.12) the pile deflection after N_{eq} load cycles must be known and can be determined using Eq. (D2.5). The displacement and the displacement rate after the first load cycle and the inclination coefficient λ are required. These values can either be determined from a cyclic pile load test with only a few load cycles or, alternatively, can be estimated from a static pile load test, in which the pile is unloaded at F_{max} and reloaded at F_{min}.

In this example a tensile static pile load test was performed, in which the pile was unloaded, then reloaded (Figure D2.3). The inclination coefficient λ was adopted in accordance with D2.1.2 (2) at 0,75. The cyclic pile heave after 1 000 load cycles was estimated in accordance with Eq. (D2.5) with the aid of the pile's resistance-heave curve (Figure D2.3) as follows:

$$s_{cyc,k} = 0,45 + \frac{0,10}{1-0,75} \cdot (1\,000^{1-0,75} - 1) = 2,30 \text{ cm} .$$

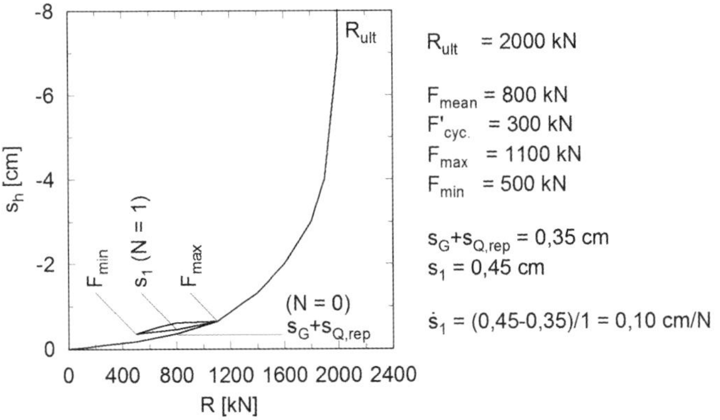

Figure D2.3 Determining parameters from the resistance-heave curve with an unloading cycle in a static pile load test

The total pile heave is given from Eq. (13.13) by the heave components from permanent and usual variable actions as:

$$\text{exist.}\, s_{h,k} = s_G + s_{Q,rep} + s_{cyc,k} = 0{,}35 + 2{,}30 = 2{,}65 \text{ cm} .$$

Stability in the serviceability limit state can now be analysed using Eq. (13.12):

$$\text{exist.}\, s_k = 2{,}65 \text{ cm} < 3{,}00 \text{ cm} = \text{allow.}\, s_k .$$

Stability in the serviceability limit state after 1 000 load cycles has thus been demonstrated.

D2.2.3 Calculation example for the ultimate and the serviceability limit states using the method after D.2.1.3

D2.2.3.1 Objective

A 20 m long pile is investigated with regard to its tensile capacity and displacements under asymmetric, tensile alternating load using the analytical approach after [66]. In simplification and preliminary calculation, a single soil layer is adopted for the entire length of the pile. For practical purposes the ground is to be divided into several strata and the capacity reduction to be determined per stratum applying the condition of equal displacement under the cyclic load. The allowable displacement is allow. $s_k = 2$ cm.

D2.2.3.2 Ultimate limit state analysis

The table below contains the input values for the calculation.

Table D2.2 Adopted and derived input values

Pile length	$L = 20$	[m]
Pile radius	$r_0 = 0{,}4$	[m]
Axial pile resistance at the shaft from empirical data, such that $R_{ult} = R_{t,k}$	$R_{ult} = 4\,500$	[kN]
Axial pile resistance at the base	$R_{b,k} = 0$	[kN]
Wall-ground interface friction angle	$\varphi = 29$	[°]
Relative soil density	$I_D = 0{,}75$	[–]
Soil void ratio	$e = 0{,}45$	[–]
Ratio of reloading stiffness to maximum stiffness of the soil (from laboratory tests, e.g. ResCol test)	$\frac{G_W}{G_{max}} = 0{,}2$	[–]

Stiffness ratio for associated γ_{zyk} and maximum shear modulus of soil (from laboratory tests, e.g. ResCol test)	$\frac{G_{cyc}}{G_{max}} = 0{,}4$	[–]
Poisson's ratio of soil	$\nu = 0{,}35$	[–]
Ultimate shear strain at which no further change in volume occurs (from laboratory tests, e.g. ResCol test)	$\gamma_{ult} = 2 \cdot 10^{-4}$	[–]
Cyclic dilatation parameter (from laboratory tests, e.g. simple shear test)	$\alpha = 0{,}5$	[–]
Constant used to describe the hysteresis curvature for initial loading (from laboratory tests, e.g. simple shear tests or $c_1 = 1{,}0$ after [25])	$c_1 = 1{,}0$	[–]
Constant used to describe the hysteresis curvature in the following load cycles (from laboratory tests, e.g. simple shear tests or $c_2 = 2{,}0$ after [150])	$c_2 = 2{,}0$	[–]
Reference shear strain (from laboratory tests, e.g. simple shear tests)	$\gamma_r = 1 \cdot 10^{-3}$	[–]
Factor for cyclic creep (from laboratory tests, e.g. simple shear tests)	$\zeta = 2{,}0$	[–]
Mean axial tensile load, assumed here: permanent actions only	$F_{mean} = 600$	[kN]
Cyclic load amplitude	$F'_{cyc} = 1\,200$	[kN]
Cycle number	$N = 1\,000$	[–]

Analysis of the capacity reduction at the pile shaft resulting from the cyclic load is performed by first separating the total skin friction into a cyclic and a static component.

$$\tau_{zyk} = \frac{F'_{cyc,s}}{2 \cdot \pi \cdot r_0 \cdot L} = \frac{1\,200}{(2 \cdot \pi \cdot 0{,}4 \cdot 20)} = 23{,}9\ \text{kN/m}^2 \quad \text{and}$$

$$\tau_{mitt} = \frac{F_{mean,s}}{2 \cdot \pi \cdot r_0 \cdot L} = \frac{600}{(2 \cdot \pi \cdot 0{,}4 \cdot 20)} = 11{,}9\ \text{kN/m}^2$$

The ultimate skin friction is given by:

$$\tau_{ult,k} = \frac{R_{s,k}}{2 \cdot \pi \cdot r_0 \cdot L} = \frac{4\,500}{(2 \cdot \pi \cdot 0{,}4 \cdot 20)} = 89{,}5\ \text{kN/m}^2.$$

Determining a normal stress state on the pile shaft representative of the cyclic load:

$$\sigma_0 = \frac{\tau_{ult,k} + \tau_{s,k}(N)}{2 \cdot \tan\varphi} = \frac{89{,}5 + (89{,}5 - 6{,}8)}{2 \cdot \tan 29°} = 155{,}4 \text{ kN/m}^2$$

Determining the maximum shear modulus G_{max} (formula from [18] – not true in dimensions):

$$G_{max} = 6{,}9 \cdot \frac{(2{,}17 - e)^2}{1 + e} \cdot \sigma_0^{0{,}5} = 6{,}9 \frac{(2{,}17 - 0{,}45)^2}{1 + 0{,}45} \cdot 155{,}4^{0{,}5}$$

$$= 175\,494 \text{ kN/m}^2$$

Calculating the cyclic shear strain from:

$$\gamma_{cyc} = \frac{\tau_{cyc}}{G_{cyc}} = \frac{23{,}9}{(0{,}4 \cdot 175\,494)} = 3{,}4 \cdot 10^{-4}$$

Calculating the activated wall-ground friction from:

$$\tan\delta = \tan\varphi \cdot \frac{\left(F_{mean} + F'_{cyc}\right)}{\left(R_{s,k} - \Delta\tau_k(N) \cdot \pi \cdot D \cdot L\right)}$$

$$= \tan 29° \cdot \frac{(600 + 1\,200)}{4\,500 - 6{,}8 \cdot \pi \cdot 0{,}8 \cdot 20} = 0{,}24$$

Calculating the cyclic reduction in the ultimate skin friction:

$$\Delta\tau_k(N) = 2 \cdot G_w \cdot \tan\delta \cdot \Delta D^* \cdot \left[\gamma_{cyc} \cdot \left(\frac{\gamma_{cyc}}{\gamma_{ult}} - 1\right) - \frac{1}{2} \cdot \alpha \cdot \gamma_{ult} \left[\left(\frac{\gamma_{cyc}}{\gamma_{ult}}\right)^2 - 1\right]\right]$$

$$= \left[2 \cdot (0{,}2 \cdot 175\,494) \cdot 0{,}24 \cdot 2{,}92 \cdot \left[\begin{array}{l} 3{,}4 \cdot 10^{-4} \cdot \left(\dfrac{3{,}4 \cdot 10^{-4}}{2 \cdot 10^{-4}} - 1\right) \\ -\dfrac{1}{2} \cdot 0{,}5 \cdot 2 \cdot 10^{-4} \cdot \left[\left(\dfrac{3{,}4 \cdot 10^{-4}}{2 \cdot 10^{-4}}\right)^2 - 1\right] \end{array}\right]\right]$$

$$= 6{,}8 \text{ kN/m}^2$$

where:

$$\Delta D^* = \Delta D \cdot \log(N + 1) = 0{,}5 \cdot I_D^{-2{,}32} \cdot \log(N + 1) = 0{,}5 \cdot 0{,}75^{-2{,}32} \cdot \log 1\,001 = 2{,}92$$

The characteristic utilisation factor is given by:

$$\mu_k = \frac{(F'_{cyc} + F_{mean})}{(R_{s,k} - \Delta\tau_k (N) \cdot \pi \cdot D \cdot L)} = \frac{(1\,200 + 600)}{(4\,500 - 6{,}8 \cdot \pi \cdot 0{,}8 \cdot 20)} = 0{,}43$$

The design value of the pile resistance is given by Eq. (13.9b)

$$R_{t,d}(N) = R_{t,d} - \eta_{cyc} \cdot \Delta R_{cyc}$$

$$= R_{t,k}/\gamma_{s,t} - \eta_{cyc} \cdot \Delta\tau\,(N) \cdot \pi \cdot D \cdot L$$

$$= 4\,500/1{,}50 - 1{,}20 \cdot 6{,}8 \cdot \pi \cdot 0{,}8 \cdot 20 = 2\,590\ \text{kN}$$

where $\gamma_{s,t} = 1{,}50$, $\eta_{cyc} = 1{,}20$

The design value of the action is:

$$F_d = F_{mean} \cdot \gamma_G + F'_{cyc} \cdot \gamma_Q = 600 \cdot 1{,}35 + 1\,200 \cdot 1{,}50 = 2\,610\ \text{kN}$$

where $\gamma_G = 1{,}35$, $\gamma_Q = 1{,}50$

The load is just no more allowable and a small extension of the pile would be necessary.

D2.2.3.3 Serviceability limit state analysis

The displacements at the pile shaft are calculated. To keep the example simple it is not divided into individual sections. The displacements comprise cyclic compaction and cyclic creep components. The shear strains as a result of cyclic compaction are calculated by first determining the reduced ultimate capacity:

$$\tau_{s,k}(N) = \tau_{ult} - \Delta\tau(N) = 89{,}5 - 6{,}8 = 82{,}7\ \text{kN/m}^2$$

where:

τ_{ult} characteristic failure shear stress at the pile shaft (corresponds to $q_s = q_{s,k}$, here from empirical data).

The load level κ_1 and κ_2 before or after cyclic loading is calculated using:

$$\kappa_1 = \frac{(\tau_{cyc} + \tau_{mean})}{\tau_{ult}} = \frac{(23{,}9 + 11{,}9)}{89{,}5} = 0{,}4$$

$$\kappa_2 = \frac{(\tau_{cyc} + \tau_{mean})}{\tau_{s,k}(N)} = \frac{(23{,}9 + 11{,}9)}{82{,}7} = 0{,}433$$

where:

τ_{mean} mean shear load;
τ_{cyc} cyclic shear load.

Calculating the shear strain from compaction:

$$_1\Delta\gamma = \gamma_2 - \gamma_1 = (7{,}64 - 6{,}67) \cdot 10^{-4} = 9{,}7 \cdot 10^{-5}$$

where:

$$\gamma_2 = \frac{\kappa_2}{1-\kappa_2/c_1} \cdot \gamma_r = \left(\frac{0,433}{1-\frac{0,433}{1}}\right) \cdot 10^{-3} = 7,64 \cdot 10^{-4}$$

and

$$\gamma_1 = \frac{\kappa_1}{1-\kappa_1/c_1} \cdot \gamma_r = \left(\frac{0,4}{1-\frac{0,4}{1}}\right) \cdot 10^{-3} = 6,67 \cdot 10^{-4}$$

Calculating the shear strains resulting from cyclic creep is performed separately for the first and the following cycles.

Cyclic creep in the first cycle:

$${}_2\Delta\gamma = \gamma_r \left[\frac{\kappa}{1-\kappa/c_2} - \frac{\kappa}{1+\kappa/c_2}\right] = 10^{-3}\left[\left(\frac{0,4}{1-\frac{0,4}{2}} - \frac{0,4}{1+\frac{0,4}{2}}\right)\right] = 1,7 \cdot 10^{-4}$$

where:

$$\kappa = \frac{\left(F_{mean} + F'_{cyc}\right)}{R_{s,k}} = \frac{(600+1\,200)}{4\,500} = 0,4$$

Cyclic creep in the following cycles:

$${}_2\Delta\gamma_{cyc} = {}_2\Delta\gamma\,(1+\zeta \cdot \ln N) = 1,7 \cdot 10^{-4}\,(1+2,0 \cdot \ln 1\,000) = 2,5 \cdot 10^{-3}$$

In order to calculate the total displacements the shear strains under the total static load are also calculated. Alternatively, the results of load tests can also be adopted here to determine s_{stat}.

$$\gamma_{stat} = \frac{\left(\tau_{cyc} + \tau_{mean}\right)}{G_{stat}} = \frac{(23,9+11,9)}{21\,057} = 1,7 \cdot 10^{-3}$$

Where:

$G_{static} = 21\,057$ kN/m² from the hyperbolic function

and

$r_m = 2{,}5 \cdot L\,(1 - \nu) = 2{,}5 \cdot 20\,(1 - 0{,}35) = 32{,}5$ m

The total displacements result when adding the individual displacement components:

$$s_{ges} = \left({}_1\,\Delta\gamma + {}_2\,\Delta\gamma_{cyc} + \gamma_{stat}\right) \cdot r_0 \cdot \ln\left(\frac{r_m}{r_0}\right)$$

$$= \left(9{,}7 \cdot 10^{-5} + 2{,}5 \cdot 10^{-3} + 1{,}7 \cdot 10^{-3}\right) \cdot 0{,}4 \ln\left(\frac{32{,}5}{0{,}4}\right)$$

$$s_{ges} = 7{,}5 \cdot 10^{-3}\ \text{m} = 7{,}5\ \text{mm}$$

The serviceability limit state is analysed using Eq. (13.12).

$$\text{exist.}\, s_k = 0{,}8\ \text{cm} < 2{,}0\ \text{cm} = \text{allow.}\, s_k\ .$$

Stability in the serviceability limit state after 1 000 load cycles has therefore been demonstrated.

D2.2.4 Calculation example for the ultimate and the serviceability limit states using the method after D.2.1.4

D2.2.4.1 Objective

The subsequent examples serve the analysis after [142] of the stability in the ultimate and serviceability limit states of a bored pile subjected to cyclic axial loads (diameter: 1,20 m, length: 8,50 m) in non-cohesive soil.

It is assumed that the equivalent, single-stage load spectrum as discussed in D4.1.1 has already been determined and the maximum action F_{max} is not greater than $(F_{mean} + F'_{cyc})_{eq}$. The following load situations are considered:

a) The pile is subjected to a basic static compressive load of $F_G = 400$ kN and a cyclic load range of $F_{cyc,eq} = 1\,200$ kN corresponding to load situation 3 in Figure 13.3, but without representative variable actions. The equivalent load cycle number is $N_{eq} = 500$.

a) The pile is subjected to a basic static tensile load of $F_G = 200$ kN and a cyclic load range of $F_{cyc,eq} = 600$ kN corresponding to load situation 1 in Figure 13.3, also without representative variable actions. The equivalent load cycle number is $N_{eq} = 100$.

The allowable displacement of the pile is 3 cm in both load situations.

D2.2.4.2 Calculating subtask a) (compressive pulsating load)

Static pile behaviour was determined in a static pile load test [154] (Figure D2.4). The ultimate capacity determined is around $R_{ult} = R_{c,m} = 4\,600$ kN. For the static analysis using ZYKLAX the model parameter $q_{s,ult}$ was determined from the pile shaft resistance and the parameter $q_{b,ult}$ using the hyperbolic method [123]. The shear modulus G_0 was estimated based on information in [154]. The parameters g_s and g_b are calibrated against the static pile load test, i.e. adopted such that the calculated and measured resistance-settlement curves approximately coincide.

Cyclic model parameters, preferably determined from cyclic pile load tests, are required to determine cyclic pile bearing behaviour using the ZYKLAX model. The static model parameters are adopted unaltered from the static analysis. Because no cyclic pile load test results are available for this example, the required model parameters are estimated. The values given in Table D2.1, determined in [142], are adopted as guide values. Tables D2.3 and D2.4 contain the model parameters adopted for use in this example.

Table D2.3 Model parameters for static analysis

$q_{s,ult}$ [kN/m²]	$q_{b,ult}$ [kN/m²]	G_0 [kN/m²]	g_s [–]	g_b [–]	r_0 [m]	L [m]
62,4	7 500	120 000	0,020	0,058	0,60	8,5

Table D2.4 Model parameters for cyclic analysis

a [–]	b [–]	κ [–]	β_{lim} [–]	$\dot{\beta}$ [–]
1,4	0,75	1,5	0,86	0,010

The plastic displacement after 500 load cycles calculated using ZYKLAX is $s_{pl} = 2,1$ cm (Figure D2.4a). The ultimate capacity after cyclic loading is $R_{ult}(N) = 4.470$ kN (Figure D2.4b). The reduction in pile capacity as a result of cyclic loading is therefore $\Delta R_{cyc} = 130$ kN.

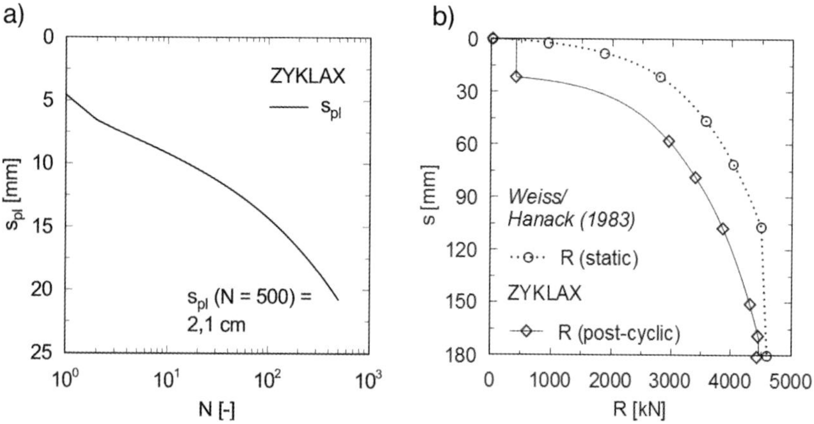

Figure D2.4 Pile bearing behaviour determined using ZYKLAX, a) plastic pile deflection, b) post-cyclic pile bearing behaviour

Stability analyses in accordance with 13.7.1 and 13.7.2 can be performed using these values (D2.2.4.4 and D2.2.4.5).

D2.2.4.3 Calculating subtask b) (tensile pulsating load)

The static ultimate capacity was determined at $R_{ult} = R_{t,m} = 1\,800$ kN in a tensile pile load test, corresponding to a skin friction failure value of $\tau_{ult} = 56{,}2\ \text{kN/m}^2$. Cyclic pile bearing behaviour is now determined with the aid of the formulas in D2.1.4.

The static model parameters in Table D2.4 are adopted for this purpose. The pile displacement as a result of the static load ($F_G = 200$ kN, i.e. $\tau_0 = 6{,}2\ \text{kN/m}^2$) is calculated using Eq. (D2.18):

$$s_s = \frac{6{,}2 \cdot 0{,}6}{120\,000 \cdot 0{,}02} \cdot \ln\left(\frac{\left(\frac{1{,}2}{0{,}6}\right)^{0{,}02} - 0{,}99 \cdot \left(\frac{6{,}2}{56{,}2}\right)^{0{,}02}}{1 - 0{,}99 \cdot \left(\frac{6{,}2}{56{,}2}\right)^{0{,}02}} \right) = 0{,}00036\ \text{m}$$

Parameter δ_N determined using Eq. (D2.25) and parameter β_N determined using Eq. (D2.24) are adopted to analyse the first load cycle. The required cyclic model parameters can be adopted from Table D2.4. The displacement for $F_{max} = 800$ kN, i.e. $\tau_0 = 25{,}0\ \text{kN/m}^2$, is then calculated by applying Eq. (D2.22):

$$s_{s,W}(1) = 0{,}00036 + \frac{(25{,}0 - 6{,}2) \cdot 0{,}6}{120\,000 \cdot 0{,}02}$$

$$\times \ln\left(\frac{\left(\frac{1{,}2}{0{,}6}\right)^{0{,}02} - 0{,}99 \cdot \left(\frac{|25{,}0 - 6{,}2|}{1{,}5 \cdot 0{,}29 \cdot 1{,}0 \cdot 56{,}2}\right)^{0{,}02}}{1 - 0{,}99 \cdot \left(\frac{|25{,}0 - 6{,}2|}{1{,}5 \cdot 0{,}29 \cdot 1{,}0 \cdot 56{,}2}\right)^{0{,}02}} \right)$$

$$s_{s,W}(1) = 0{,}0034\ \text{m}\,.$$

By applying Eq. (D2.21) the pile displacement (pile shaft) after the first load cycle at $F_{min} = F_G = 200$ kN can then be determined:

$$s_{s,E}(1) = 0{,}0034 + \frac{(6{,}2 - 25{,}0) \cdot 0{,}6}{120\,000 \cdot 0{,}02}$$

$$\times \ln\left(\frac{\left(\frac{1{,}2}{0{,}6}\right)^{0{,}02} - 0{,}99 \cdot \left(\frac{|6{,}2 - 25{,}0|}{1{,}5 \cdot 1{,}0 \cdot 56{,}2}\right)^{0{,}02}}{1 - 0{,}99 \cdot \left(\frac{|6{,}2 - 25{,}0|}{1{,}5 \cdot 1{,}0 \cdot 56{,}2}\right)^{0{,}02}} \right)$$

$$s_{s,E}(1) = 0{,}0020\ \text{m}\,.$$

The pile displacement after the first load cycle is therefore 2,0 mm.

By repeatedly applying Eq. (D2.21) and Eq. (D2.22) in conjunction with Eq (D2.23) and Eq. (D2.25), the pile displacement can be determined in analogy for each additional load cycle. A displacement of around 5,0 mm after 100 load cycles was determined in this way.

The reduction in the pile capacity after 100 load cycles is calculated using the parameter β_N and Eq. (D2.23):

$$\beta_{100} = (1-0{,}010)\cdot(0{,}912-0{,}860)+0{,}860 = 0{,}91\,.$$

The pile skin friction failure values reduced as a result of cyclic loading is then:

$$\tau_{ult}(100) = \tau_{ult}\cdot\beta_{100} = 56{,}2\cdot 0{,}91 = 51{,}1\ \text{kN/m}^2.$$

This gives a pile resistance after 100 load cycles of $R_{ult}(100) = 1.637{,}5$ kN and thus a reduction in pile capacity of:

$$\Delta R_{cyc} = R_{ult} - R_{ult}(100) = 1\,800 - 1\,637{,}5 = 162{,}5\ \text{kN}.$$

Knowing the reduction in pile capacity and the pile deflection after 100 load cycles, the stability analyses in accordance with 13.7.1 and 13.7.2 can subsequently be performed.

D2.2.4.4 Ultimate limit state analysis

The design value of the actions is calculated by applying the partial factors γ_G and γ_Q from the EC 7-1 Handbook [44] using Eq. (13.7) for subtask a) (compressive pulsating load) as:

$$F_d = F_G\cdot\gamma_G + \lambda\cdot F'_{cyc}\cdot\gamma_Q = 400\cdot 1{,}35 + 2\cdot 600\cdot 1{,}50 = 2\,340{,}0\ \text{kN}$$

and for subtask b) (tensile pulsating load) as:

$$F_t = F_G\cdot\gamma_G + \lambda\cdot F'_{cyc}\cdot\gamma_Q = 200\cdot 1{,}35 + 2\cdot 300\cdot 1{,}50 = 1170{,}0\ \text{kN}.$$

The characteristic pile resistance $R_{c,k}$ or $R_{t,k}$ under static load can be determined using the correlation factor ξ_2 in accordance with EC 7-1 Handbook [44]:

$$R_{c,k} = R_{ult}/\xi_2 = R_{c,m}/\xi_2 = 4\,600/1{,}35 = 3\,407{,}4\ \text{kN}$$

or

$$R_{t,k} = R_{ult}/\xi_2 = R_{t,m}/\xi_2 = 1\,800/1{,}35 = 1\,333{,}3\ \text{kN}\,.$$

The design value of the pile resistance in the ultimate limit state $R_{c,d}(N)$ or $R_{t,d}(N)$ can thus be calculated using Eq. (13.9a) or Eq. (13.9b) and the partial factors γ_t and $\gamma_{s,t}$, and the adopted model factor $\eta_{cyc} = 1,20$:

$$R_{c,d}(N) = \frac{R_{c,k}}{\gamma_t} - \eta_{cyc} \cdot \Delta R_{cyc} = \frac{3\,407,4}{1,10} - 1,20 \cdot 130 = 2\,941,6 \text{ kN}$$

or

$$R_{t,d}(N) = \frac{R_{t,k}}{\gamma_{s,t}} - \eta_{cyc} \cdot \Delta R_{cyc} = \frac{1\,333,3}{1,15} - 1,20 \cdot 162,5 = 964,4 \text{ kN}.$$

Analysis of stability in the ultimate limit state can now be performed in accordance with Eq. (13.11a) or Eq. (13.11b) using the design values of the action and the pile resistance:

$$F_{c,d} = 2\,340,0 \text{ kN} < 2\,941,6 \text{ kN} = R_{c,d}(N)$$

or

$$F_{t,d} = 1170,0 \text{ kN} > 964,4 \text{ kN} = R_{t,d}(N).$$

Stability is therefore adhered to for the ultimate limit state for the load situation in subtask a). In addition to the analysis using an equivalent, single-stage load spectrum, analysis of the capacity under the maximum load is required. The ultimate limit state for the load in subtask b), on the other hand, is not adhered to. In order to demonstrate stability in this case, either the pile length or the pile diameter could be increased.

D2.2.4.5 Serviceability limit state analysis

In D2.2.4.2 and D2.2.4.3 the plastic pile displacement after 500 and 100 load cycles was determined as $s_k = s_{pl,k} = 2,1$ cm and $s_{h,k} = s_{pl,k} = 0,5$ cm. The allowable displacement in accordance with the specified objective was $\text{allow.}\, s_k = 3,0$ cm respectively.

Stability in the serviceability limit state can now be calculated using Eq. (13.12):

$$\text{exist.}\, s_k = 2,1 \text{ cm} < 3,0 \text{ cm} = \text{allow.}\, s_k \quad \text{for subtask a)}$$

or

$$\text{exist.}\, s_{h,k} = 0,5 \text{ cm} < 3,0 \text{ cm} = \text{allow.}\, s_{h,k} \quad \text{for subtask b).}$$

Stability in the serviceability limit state for the two load situations has therefore been demonstrated.

D3 Piles Subjected to Cyclic Lateral Loads

D3.1 Calculation methods

D3.1.1 Empirical method for estimating the accumulated deflections

(1) The horizontal pile head deflections under the cycle number resulting from a pulsating load can be approximately estimated by applying the logarithmic approach:

$$y_{cyc} = y_{N=1} \cdot (1 + t \cdot \ln N) \qquad \text{(D3.1)}$$

where:

y_{cyc} pile head deflection after N load cycles;
$y_{N=1}$ pile head deflection after the first load cycle (N = 1);
t parameter for system behaviour as a result of cyclic loads;

Or an exponential approach:

$$y_{cyc} = y_{N=1} \cdot N^{m} \qquad \text{(D3.2)}$$

where:

m parameter for system behaviour as a result of cyclic loads;

See e.g. [47], [74] and [114].

(2) In line with test results for piles in sand under a pulsating load, the parameter t lies in the range $0{,}16 \leq t \leq 0{,}22$ [47], [1] and [114]. According to [78] the parameter depends on the relative pile stiffness, the pile construction method and the kind of the load. An approach taking these influencing factos into consideration was proposed.

(3) Alternatively, the decrease in the subgrade stiffness using the modulus of subgrade reaction method can be described as a function of the load cycle number using an exponential approach via a factor $N^{-\alpha}$, see e.g. [82] and [84]. The parameters t and m or α are generally dependent on the pile system, the pile geometry, the soil properties and the load level.

(4) An analysis approach was proposed for the factor α by [84]. According to this it is dependent on the pile construction method, the kind of the load (pulsating or alternating loads) and the density of the sand. For pulsating loads it has an order of magnitude between 0,10 and 0,25.

(5) For short, practically rigid piles in sand the parameter m is identical to the parameter α. For non-rigid piles it is smaller. Theoretically, for very long, flexible piles $m = 0{,}6 \cdot \alpha$ for pure horizontal loading and $m = 0{,}4 \cdot \alpha$ for pure moment actions on the pile. Values for m between around 0,04 and 0,09 were derived in [82] from pile tests on relatively long piles under pulsating loads.

D3.1.2 Calculation approaches for estimating deflection accumulation taking to consideration non-linear soil behaviour

(1) The calculation methods discussed e.g. in [48], [76] and [141], can be adopted to forecast the deflection accumulation using element tests. The methods in [48] and [76] are based on the strain wedge model after [104] and [3], and on finite element analyses. The calculation method after [141] describes the deflection accumulation using a modification of the strain wedge model. The method is described in summary below using an example.

(2) The analysis method proposed in [141] for forecasting the deflection accumulation in piles subjected to cyclic lateral loads using static and cyclic element tests is a modification and extension of the strain wedge model after [104]. The relationship between the subgrade stress and the pile deflection is derived from the three-dimensional stress-strain behaviour of the soil.

(3) The approach described in [25] is adopted to describe the non-linear stress-strain behaviour of the soil, taking loading and unloading processes into consideration.

(4) The input parameters required to describe the analysis model's soil properties are determined from static and cyclic triaxial tests.

(5) The pile deflections and action effects for pulsating loads with variable load amplitudes are calculated in several iterative calculation steps using the solution for an elastically supported beam.

(6) In the proposed model the calculation result are p-y curves, in contrast to the p-y method described in [2]. The curves are not only dependent on the prevalent soil, but also on the pile properties. The p-y curves are therefore not of a generally applicable character for the same soil type.

D3.1.3 Calculation approach with subgrade reaction reduction using the p-y method

(1) The approach using non-linear subgrade springs (p-y curves) to analyse laterally loaded piles is often used in practice. The characteristic spring curves are generally described on the basis of the relationships described in [2]. The curves are defined for static and cyclic loading in the API Recommendations. The p-y curves were derived from field tests using approximately 100 load cycles.

(2) The reduction in the subgrade reaction using the p-y method is demonstrated below in an example for sand. The ultimate resistance p_u is determined as a function of depth as follows:

$$p_u = \min\begin{cases}(C_1 \cdot z + C_2 \cdot D) \cdot \gamma \cdot z \\ C_3 \cdot \gamma \cdot z \cdot D\end{cases} \tag{D3.3}$$

where:

C_1, C_2, C_3 dimensionless factors in accordance with Figure D3.1 (left);
D pile diameter;
γ unit weight;
z depth below grade.

The resistance-deflection relationship is defined as follows:

$$p = A \cdot p_u \cdot \tan h\left(\frac{k \cdot z}{A \cdot p_u} \cdot y\right) \tag{D3.4}$$

where:

k initial stiffness in accordance with Figure D3.1 (right);
A empirical factor, where:

$A = (3{,}0 - 0{,}8 \cdot z/D) \geq 0{,}9$ for a static load;
$A = 0{,}9$ for a cyclic load.

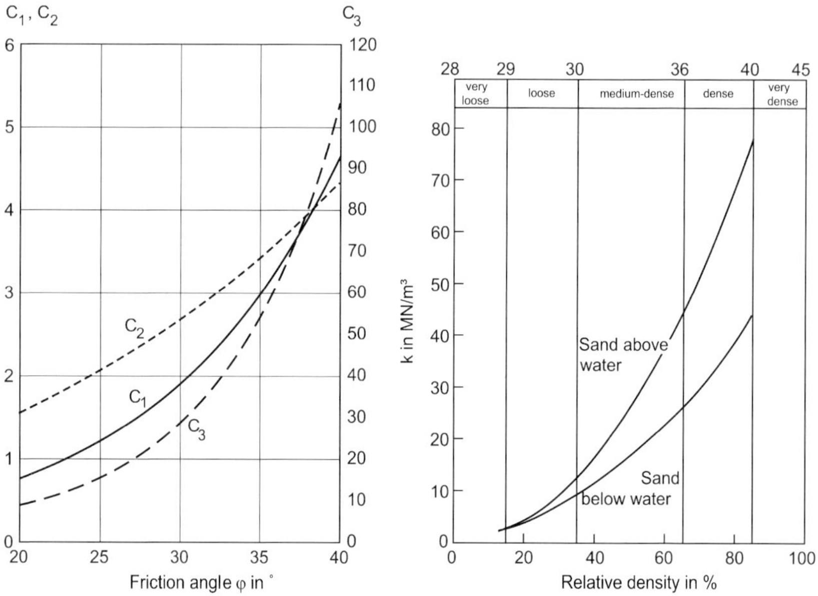

Figure D3.1 Factors C1, C2, C3 (left) and initial stiffness k (right) after [2]

(3) The approach described in (2) comprises a depth-related reduction in the subgrade reaction as a result of cyclic loads of up to 70%. However, it is not dependent on the number of cycles. Investigations in [19] confirm the validity

of the approach up to around 100 cycles. The following modification is proposed for cycle numbers over and above this:

$$p = A_{cyc} \cdot p_u \cdot \tan h\left(\frac{k_i \cdot z}{0{,}9 \cdot p_u} \cdot y\right) \qquad \text{(D3.5)}$$

where:

A_{cyc} empirical factor, where:

$A_{cyc} = r_A (3 - 1\,143 \cdot z/D) + 0{,}343 \cdot z/D \le 0{,}9$

r_A degradation factor between 0 and 0,3.

(4) The degradation factor r_A can be determined in model tests or by numerical investigations. The degradation factor generally decreases with increasing cycle numbers. For an initial estimate of deformation accumulation at more than 10^5 load cycles determining the subgrade reaction using Equation (D3.5) for $r_A = 0$ using:

$$A_{cyc} = 0{,}343 \cdot z/D \le 0{,}9 \qquad \text{(D3.6)}$$

is proposed.

D3.2 Examples

D3.2.1 Estimating deflection accumulation after D3.1.1

A driven tubular steel pile (D = 2 m, L = 30 m) in medium-dense sand is considered (Figure D3.2). Calculation was carried out using the modulus of subgrade reaction method with linear with depth increasing subgrade reaction modulus $k_s(z) = n_h \cdot z/D$. Adopting $n_h = 6$ MN/m^3 the deflection curve shown in Figure D3.2 with a pile head deflection of 1,31 cm under static

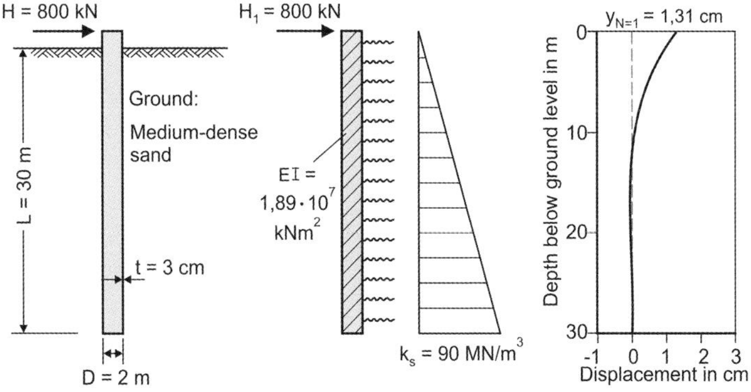

Figure D3.2 Tubular steel pile in sand; system (left) and deflection curve under static load or N = 1 (right) calculated using the modulus of subgrade reaction method

loading (H = 800 kN) was calculated, equated here with the deflection in the 1st load cycle.

The pile head deflections after N = 1 000 load cycles under pulsating loads are now estimated.

a) Using Eq. (D3.1):

The parameter t is estimated here as t = 0,20, see D3.1.1 (2). The deflection accumulation is then obtained from:

$$y_{cyc} = 1{,}31 \cdot (1 + 0{,}20 \cdot \ln 1\,000) = 3{,}12 \text{ cm} .$$

b) After D2.1.1 (3):

The parameter α is estimated here as $\alpha = 0{,}17$ for a driven pile under a pulsating load in medium-dense sand. The subgrade stiffness in the modulus of subgrade reaction method is correspondingly reduced by a factor $1\,000^{-0.17} = 0{,}31$. Recalculating using $n_h = 0{,}31 \cdot 6 = 1{,}86$ MN/m^3 returns a pile head deflection of 2,64 cm (Figure D3.3).

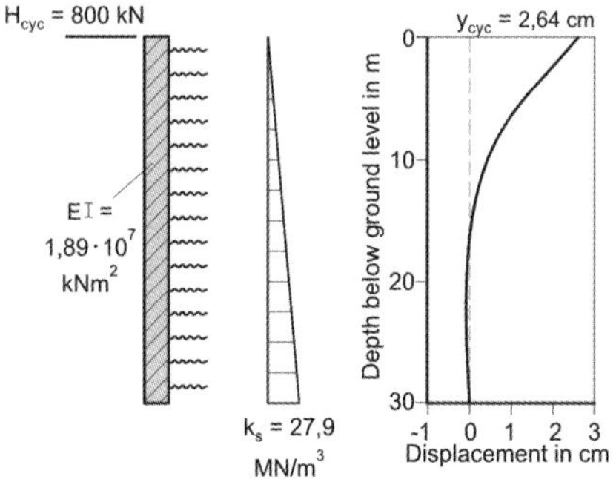

Figure D3.3 Deflection curve after the modulus of subgrade reaction method with reduced subgrade stiffness for H_{cyc} where N = 1 000

c) After D3.1.1 (1), Eq. (D3.2):

Because this case deals with a long, flexible pile (recognised by the deflection curve), parameter m in Eq. (D3.2) can also be determined as $m = 0{,}6 \cdot \alpha = 0{,}6 \cdot 0{,}17 = 0{,}102$ (pure horizontal force load). The pile head deflection after 1 000 load cycles is:

$$y_{cyc} = 1{,}31 \cdot 1.000^{0{,}102} = 2{,}65 \text{ cm} .$$

D3.2.2 Estimating deflection accumulation after D3.1.2

D.3.2.2.1 Objective

A tubular steel pile with an embedded length of 25,0 m, a diameter of 1,5 m and a wall thickness of 0,025 m is subjected to cyclic lateral loads with varying harmonic load ranges at the pile head as shown in Figure D3.4. The pile was driven into dense, saturated sand with a relative density D = 0,73.

The accumulated pile head deflections under cyclic lateral loads are estimated in accordance with [141].

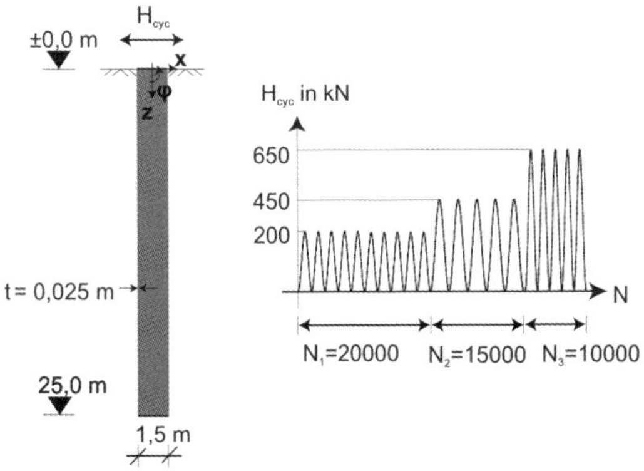

Figure D3.4 Dimensions of the pile subjected to cyclic lateral loads in dense, saturated sand

The parameters required for the analysis model are described below together with their experimental determination. Following a brief description of the calculation procedure the results are summarised.

D.3.2.2.2 Model parameters for describing the soil properties

The following material constants describing the initial density of the soil are requires as model parameters for describing soil properties:

- effective soil friction angle φ';
- constants K_a, K_{ur}, R_f and n after [25];
- cyclic exponent a;
- Poisson's ratio of the soil ν.

The required model parameters are described in more detail below.

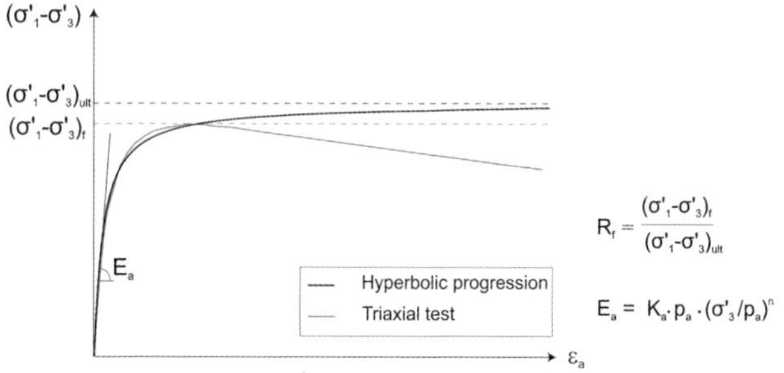

Figure D3.5 Hyperbolic soil stress-strain relationship profile compared to triaxial test profile

[25] proposes a hyperbolic function with the form:

$$(\sigma_1' - \sigma_3') = \frac{\varepsilon_a}{\dfrac{1}{E_a} + \dfrac{\varepsilon_a \cdot R_f}{(\sigma_1' - \sigma_3')_f}} \tag{D3.7}$$

to describe the non-linear soil stress-strain relationship. Here, σ_1' is the greater and σ_3' the smaller effective principal stress and ε_a the axial strain in the triaxial test. A failure ratio R_f defined below was introduced in Figure D3.5 to adapt the asymptotic value of the differential stress $(\sigma_1' - \sigma_3')_{ult}$ to the stresses in the failure state of the soil $(\sigma_1' - \sigma_3')_f$:

$$R_f = \frac{(\sigma_1' - \sigma_3')_f}{(\sigma_1' - \sigma_3')_{ult}} . \tag{D3.8}$$

The following approach, used in [51], is adopted to describe the initial tangent modulus E_a as a function of stress as shown in Figure D3.5:

$$E_a = K_a \cdot p_a \cdot \left(\frac{\sigma_3'}{p_a}\right)^n \tag{D3.9}$$

Where p_a is the atmospheric pressure. The dimensionless stiffness factor K_a and the exponent n in Eq. (D3.9) are material constants which are experimentally determined.

To describe unloading and reloading processes it is assumed after [25] that the soil behaves linearly on unloading and reloading. The unloading or reloading modulus E_{ur} in Figure D3.6 is formulated by the following relationship analogous to the initial tangent modulus E_a:

$$E_{ur} = K_{ur} \cdot p_a \cdot \left(\frac{\sigma_3'}{p_a}\right)^n \tag{D3.10}$$

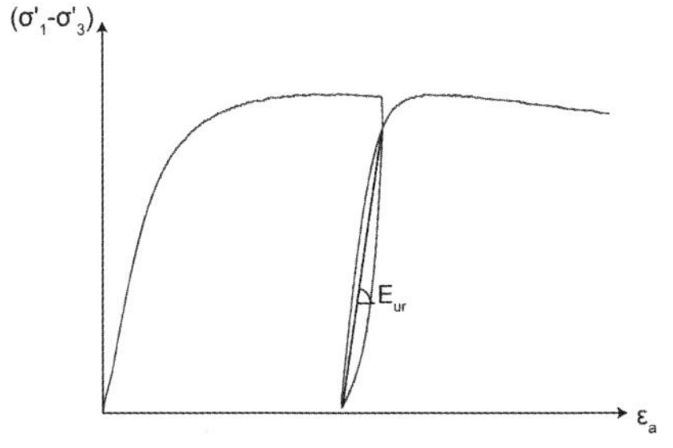

Figure D3.6 Definition of the unloading and reloading modulus E_{ur} in the triaxial test

In Eq. (D3.10) K_{ur} is a dimensionless stiffness factor, which generally has a greater value than the factor K_a.

Soil behaviour under cyclic loads is characterised in the analysis model by the exponent a, which is used to describe the development of the plastic strain of the soil over the cycle number. In Figure D3.7 the cyclic exponent a for the sand with an initial density D = 0,73 is shown. It was determined from a cyclic, drained triaxial test with a cyclic load ratio X = 0,37. The cyclic load ratio is defined as the ratio of cyclic deviator stress q to the deviator stress in the static failure state $q_{stat,f}$. When determining the exponent it is assumed that it is independent of the stress state of the soil and differs only in its dependence on the physical condition of the soil, e.g. its density and saturation.

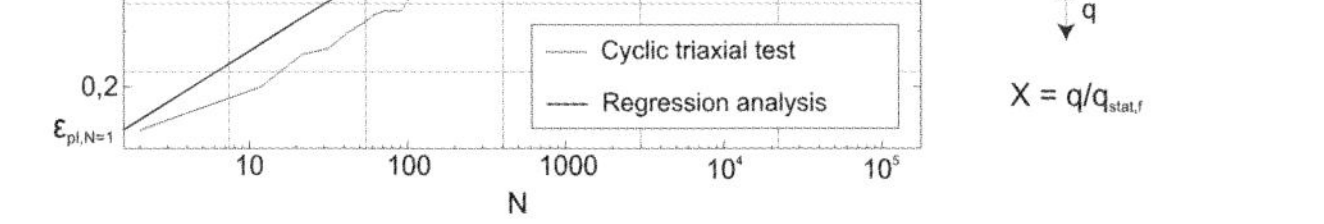

Figure D3.7 Cyclic exponent a for sand with an initial density D = 0,73 from a cyclic, drained triaxial test

D.3.2.2.3 Experimental determination of the model parameters

The effective friction angle φ′ of the soil can be determined to DIN 18137-2:2011-04.

Static, drained triaxial tests on characteristic soil samples are required to determine the material constants K_a and K_{ur}, the failure ratio R_f and the exponent n.

The soil samples are subjected to a monotonous static load to determine the initial tangent modulus E_a and the asymptotic value of the differential stress $(\sigma_1' - \sigma_3')_{ult}$ after Figure D3.5. The vertical initial stress of the soil sample in question is adopted as the triaxial, isotropic consolidation stress σ_3'.

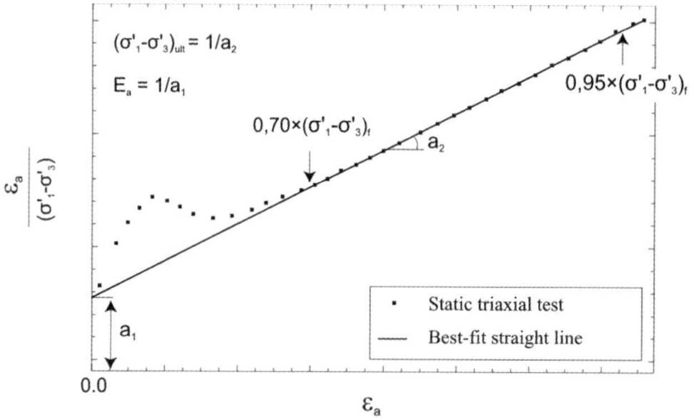

Figure D3.8 Determining the initial tangent modulus E_a and the asymptotic value of the differential stress $(\sigma_1' - \sigma_3')_{ult}$ from the transformed stress-strain profile of a triaxial test

The stress-strain relationship of the samples from static triaxial tests are shown in Figure D3.8, transformed and linearised with a best-fit curve through the stresses at 70% and 95% of the failure stress $(\sigma_1' - \sigma_3')_f$. The reciprocal of the determined parameters a_1 and a_2 corresponds to the initial tangent modulus E_a or the asymptotic value of the differential stress $(\sigma_1' - \sigma_3')_{ult}$.

The failure ratio R_f can then be calculated from the triaxial test results after Eq. (D3.8) using the known asymptotic value of the differential stress $(\sigma_1' - \sigma_3')_{ult}$ and the failure stress $(\sigma_1' - \sigma_3')_f$.

It is necessary to determine the stiffness factor K_a and the exponent n to describe the stress-dependence of the initial tangent modulus E_a using Eq. (D3.9). For this purpose, the determined initial tangent moduli E_a are drawn in a double-logarithmic graph as a function of the triaxial consolidation stresses σ_3' as shown in Figure D3.9. By linearising the results using a best-fit curve the stiffness factor K_a and the exponent n are derived as shown in Figure D3.9.

Linear-elastic soil behaviour after [25] is assumed for unloading and reloading processes. Triaxial tests with at least one soil unloading and reloading stage respectively are required to determine the unloading and reloading modulus E_{ur} of a soil sample (Figure D3.6). The dimensionless stiffness factor K_{ur} in

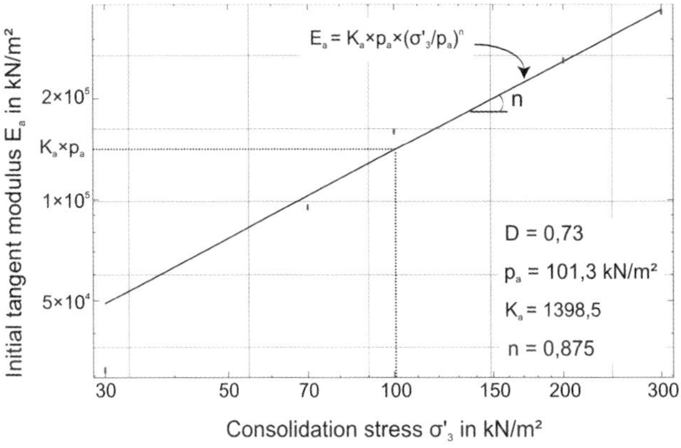

Figure D3.9 Determining the stiffness factor K_a and the exponent n for an example of 5 triaxial tests with differing respective isotropic consolidation stress σ_3' for sand with an initial density D = 0,73

Eq. (D3.10) is calculated analogous to the stiffness factor K_a with the known unloading and reloading moduli E_{ur} and the triaxial consolidation stresses σ_3'.

The cyclic exponent a is used to describe the development of the plastic strain of the soil over the cycle number. To determine this exponent the plastic strain profile is drawn over the cycle number in double-logarithmic scale as shown in Figure D3.7. The exponent a is given by the plastic soil strain after load cycle $\varepsilon_{pl,N=1}$ from the gradient of the plastic strain profile.

The Poisson's ratio of the soil is required as an input parameter for the analysis model. The influence of the Poisson's ratio of the soil on the results of the model is limited [141]. For granular soils a constant Poisson's ratio $\nu = 0{,}3$ is assumed in the calculation model.

Table D3.1 summarises the determined model parameters for the dense sand with an initial density $D = 0{,}73$ used in the calculation model.

Table D3.1 Model parameters for describing the soil properties of the dense sand with an initial density $D = 0{,}73$

φ [°]	K_a [–]	K_{ur} [–]	R_f [–]	n [–]	a [–]	ν [–]
40,4	1 398,5	1 950,7	0,90	0,875	0,31	0,3

D.3.2.2.4 Calculation procedure

To determine the pile deflections and action effects it is necessary to perform iterative calculation steps on the solution to an elastically supported beam. The relationship between the subgrade stress and the pile deflection is derived from the three-dimensional stress-strain behaviour of the soil.

The calculation flow diagram pile for determining the deflections and action effects of a pile subjected to cyclic lateral loads is described in [141].

D.3.2.2.5 Results of the calculation example

The estimated accumulated pile head deflections as a function of the cycle number are shown in Figure D3.10.

The results show that the horizontal pile head deflections at the end of cyclic loading reach values of approx. 2,6 cm.

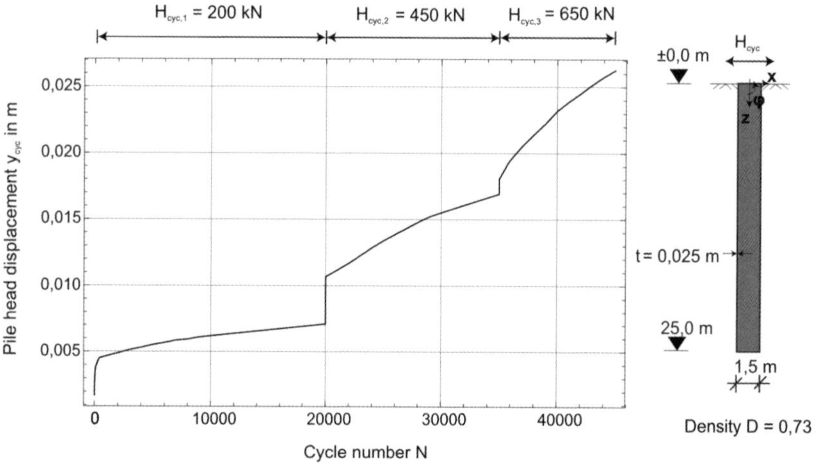

Figure D3.10 Accumulated pile head deflections as a function of the cycle number

D3.2.3 Subgrade degradation adopting the p-y method after D3.1.3

D3.2.3.1 Objective

A driven, tubular steel pile (D = 5,7 m, yield strength f_y = 335 N/mm²) in dense sand is considered. The system and the loads are shown in Figure D3.11. The subgrade is described by non-linear resistance-deflection curves (p-y curves) after D3.1.3. With regard to the given effect, it is assumed that it occurs within a design event with a limited cycle number (N < 100), see D3.1.3 (1) and (2). A modification, for example after D3.1.3 (3), is necessary for larger cycle numbers. The pile shall be analysed for the given load in the ultimate limit state after 13.8.1.

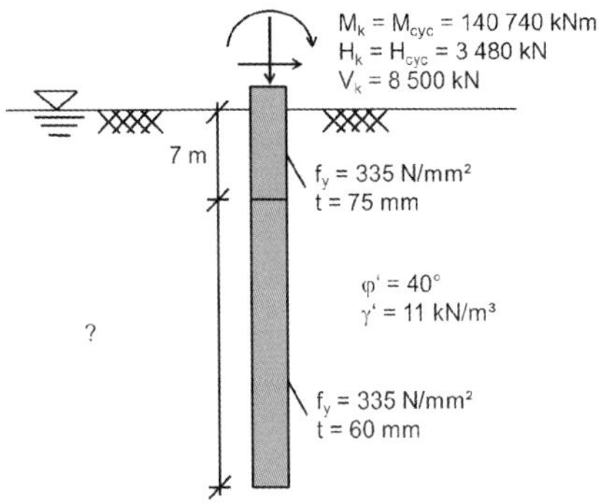

Figure D3.11 Pile system and loads

D3.2.3.2 Determining the required embedment depth

The parameters necessary to construct the p-y curves are determined from Figure D3.1 for $\varphi' = 40°$ as $C_1 = 4{,}62$, $C_2 = 4{,}38$, $C_3 = 104{,}2$ and $k = 45$ MN/m³. The empirical factor A after D3.1.3 (2) is adopted as 0,9 for the ultimate limit state analysis. The pile head rotation as a function of the pile length is shown in Figure D3.12 for the characteristic maximum load from M_k and H_k. In accordance with 13.4.3 (9) the required pile length is adopted such that any additional pile length has no additional impact on the deformation properties of the pile. The embedment depth is therefore adopted as 26 m.

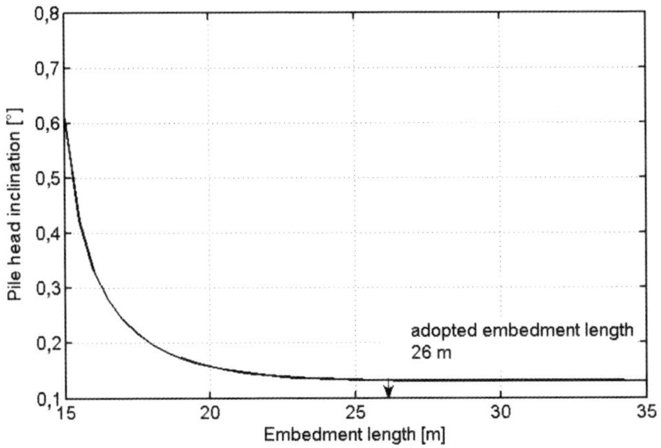

Figure D3.12 Calculated pile head rotation as a function of pile length

D3.2.3.3 Analysis of the soil reaction

Figure D3.13 shows the calculated deflection curve, the maximum and the mobilised subgrade resistances, and the moment profile under characteristic load. The integral of the mobilised subgrade stress from grade to the deflection curve's zero transition point at z = 14,19 m is $B_{h,k}$ = 14 721 kN. The integral of the maximum subgrade stress $A \cdot p_u$ to this depth is $R_{ult,k,cyc}$ = 68 618 kN. Analysis of the soil reaction block using the partial factors γ_Q = 1,50 and $\gamma_{R,e}$ = 1,4 gives:

$$B_{h,d} = B_{h,k} \cdot \gamma_Q = 14.721 \times 1,50 = 22\,082 \text{ kN} < 49\,013 \text{ kN}$$
$$= 68\,618/1,40 = R_{ult,d,cyc}$$

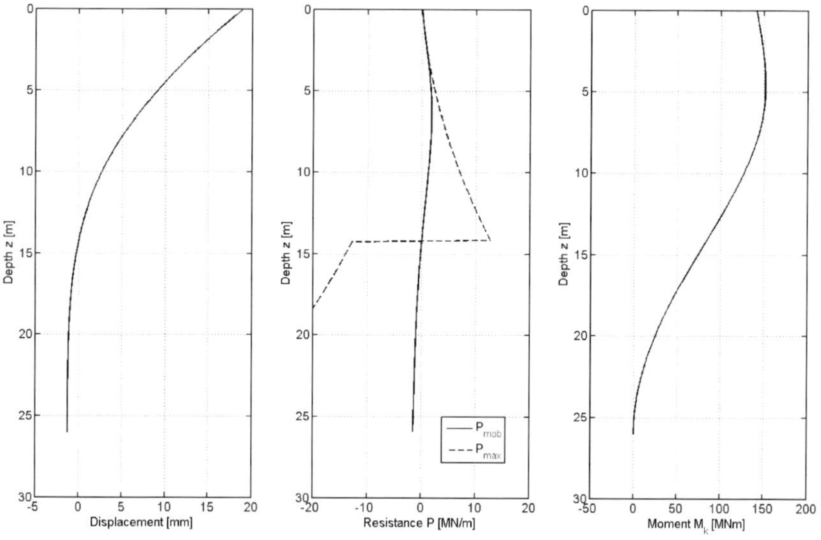

Figure D3.13 Deflection curve with cyclic p-y curve (A = 0,9) calculated (left), mobilised and maximum subgrade resistances (centre) and moment profile (right)

D3.2.3.4 Analysis of internal capacity

The characteristic design moment is calculated for a depth of z = 4,55 m as M_k = 151 490 kNm. The characteristic axial load taking the self-weight of the pile under buoyancy into consideration to this depth is 8 905 kN. The characteristic resistance moment at this depth is determined as follows:

$$W = \frac{\pi}{32}\frac{(D^4 - (D-2t)^4)}{D} = \frac{\pi}{32}\frac{(5,7^4 - (5,55)^4)}{5,7} = 1,84\,\text{m}^3.$$

Using $\gamma_Q = 1{,}50$ and $\gamma_G = 1{,}35$ the largest design value of the compressive stress in the cross-section is:

$$\sigma_{com,Ed} = \frac{M_k \gamma_Q}{W} + \frac{V_k \gamma_G}{A} = \frac{151\,490 \cdot 1{,}50}{1{,}84} + \frac{8\,905 \cdot 1{,}35}{1{,}325} = 132{,}6\,N/mm^2.$$

The factor $\varepsilon = \sqrt{235/f_y}$ may be increased to DIN EN 1993-1-1 as follows:

$$\varepsilon_{increa.} = \sqrt{\frac{235}{fy}} \cdot \sqrt{\frac{f_y/\gamma_{M0}}{\sigma_{com,Ed}}} = \sqrt{\frac{235}{335}} \cdot \sqrt{\frac{335/1{,}0}{132{,}6}} = 1{,}33.$$

With $D/t = 5\,700/75 = 76 < 90\varepsilon^2_{increa.} = 159{,}5$ the cross-section falls into cross-section class 3 and may be analysed using elastic-elastic methods. With $\gamma_{M0} = 1{,}0$ the "internal" capacity of the pile is given:

$$\sigma_{S,d} = \sigma_{com,Ed} = 132{,}6\,N/mm^2 < 335\,N/mm^2 = f_y/\gamma_{M0}$$

The stability calculations are not reproduced here.

D3.2.3.5 Estimating pile head deflection

Adopting the static p-y curves after D3.1.3 (2) returns a pile head deflection of 16,9 mm. The deflection increases to 18,9 mm using the cyclic p-y curves (Figure D3.13, left).

D4 Procedure for determining an equivalent single-stage load spectrum

D4.1 Calculation method

D4.1.1 Method for determining an equivalent load cycle number for piles subjected to cyclic axial loads

(1) The majority of the methods available for serviceability and ultimate limit state analyses of piles subjected to cyclic axial loads require the definition of an equivalent single-stage load spectrum.

(2) Hereby, the equivalent single-stage load spectrum is defined as the cyclic action of N_{eq} load cycles with a mean value of $F_{mean,eq}$ and a cyclic load amplitude $F'_{cyc,eq}$, which has the same impact with regard to the cyclic effect as the true, generally stochastic, load profile.

(3) The principle for determining an equivalent load cycle number for a reference spectrum for axially loaded piles is analogous to the procedure for piles subjected to cyclic, lateral loads (D4.1.2). The example in D4.2.1 explains the procedure. It can generally be assumed that the governing case with the greatest total impact of the cyclic load is when the reference spectrum with the greatest individual impact is adopted.

D4.1.2 Method for determining an equivalent load cycle number for piles subjected to cyclic lateral loads

(1) Existing calculation approaches allow the deflection accumulation resulting from a specified cycle number under a uniform load or from an action combination consisting of horizontal force and moment to be estimated (D3.1.1). If the magnitude of the prevalent loads is variable a cyclic equivalent load with a corresponding equivalent load cycle must be defined.

(2) [78] proposes the strain superposition method described by [139] for calculation of the pile deflections under cyclic loads of changing magnitude, see Figure D4.1. For the total head deflection of a pile under n different action combinations E_k (horizontal force and moments at the pile head) with N_k cycles respectively, Eq. (D4.1) gives.

$$y_{cyc} = y_{1,1}\left[1 + t \ln\left(N_1 + \sum_{k=2}^{n} N_k^*\right)\right] \qquad \text{(D4.1)}$$

where:

$y_{1.1}$ (static) pile head deflection as a result of a single (N = 1) loading by E_1;
N_1 number of cycles of the load or action combination E_1;
N_k^* equivalent cycle number from the action combinations E_2 to E_n;
t parameter for system behaviour as a result of cyclic loads (D3.1.1).

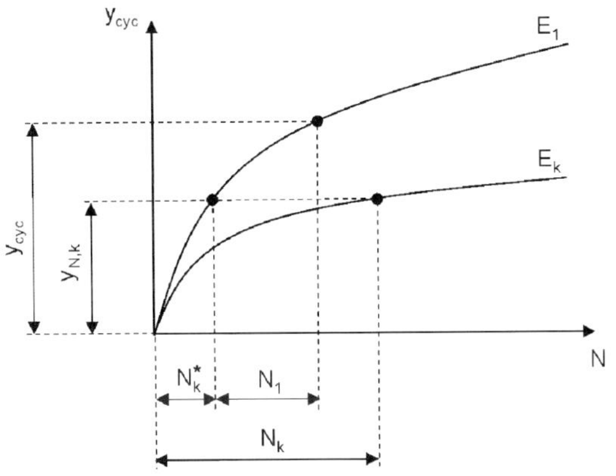

Figure D4.1 Determining the resulting pile deflections for varying cyclic action combinations (horizontal force and moment) E_k at the pile head

(3) The action combination E_1 with the load cycle number N_1 represents the equivalent load $E_{cyc,eq}$. The equivalent cycle number is given by:

$$N_{eq} = N_1 + \sum_{k=2}^{n} N_k^* \tag{D4.2}$$

where:

$$N_k^* = e^{\frac{1}{t}\left(\frac{y_{1,k}}{y_{1,1}}(1+t\ln N_k)-1\right)}$$

$y_{1,k}$ static pile head deflection as a result of action combination E_k;
N_k number of cycles of the action combination E_k.

The respective load or action combination is therefore weighted for the size of the corresponding pile deflection.

(4) It should be noted that the result is not independent of which action combination is adopted as the reference variable for which the equivalent load cycle number was determined. Adopting the combination with the greatest static pile head deflection as the reference is recommended. The equivalent load cycle numbers for the remaining action combinations can then be respectively determined by directly relating them to this reference variable.

Alternatively, the load cycle number of the combination with the smallest static deflection can be converted to the equivalent load cycle number of the combination with the second-smallest static deflection. This, in turn, can then be converted to the equivalent cycle number of the combination with the next largest static deflection, etc.

(5) For uniform cyclic loads the method assumes the validity of a linear relationship between the increase in deformation and the natural logarithm of the load cycle number (D3.1.1):

$$y_{N,k} = y_{1,k} \left[1 + t \ln N_k\right] \tag{D4.3}$$

(6) The method initially only applies to pulsating loads acting in one direction. Any additional, opposing loads act to reduce the accumulated pile deformation. Based on a proposal by [74] this can be approximately taken into consideration by a corresponding reduction in the load cycle number of the loads occurring in the direction of the accumulation.

D4.2 Calculation examples

D4.2.1 Determining an equivalent load cycle number for piles subjected to cyclic axial loads after D4.1

D4.1 proposes the method after [78] for determining an equivalent load for piles subjected to cyclic horizontal loads. This method can also be adopted for use with piles subjected to cyclic axial loads, by applying it to the reduction in capacity instead of the respective deflection.

Table D4.1 Multi-stage load spectra

Load spectrum	Cyclic action [kN]	Mean static action [kN]	Cycle number	Skin friction reduction [kN/m²]
1	700	700	200	9,31
2	500	900	100	2,50
3	700	100	10 000	9,24
4	400	400	1 000 000	0,37

A pile with a diameter of 19 m, a diameter of 46 cm and an axial pile shaft resistance of $R_{ult} = 2\,500$ kN, subjected to a tensile load is considered. The multi-stage load spectra in Table D4.1 are investigated.

The reference spectrum with the largest characteristic skin friction reduction $\Delta\tau(N) = \Delta q_s(N)$ can be adopted as the most conservative reference spectrum. Spectrum 1 is the governing reference spectrum here and the shear stress and skin friction reductions are now simulated together with the remaining spectra by adapting the cycle number and adopting this reference spectrum. This gives:

Table D4.2 Determining the equivalent cycle numbers

Spectrum	Equivalent cycle number
1	200
2 (1)	5
3 (1)	200
4 (1)	1,25
Total	**406**

The equivalent single-stage load spectrum is thus formed by $F'_{cyc} = 700$ kN, $F_{mean} = 700$ kN and $N_{eq} = 406$.

D4.2.2 Determining an equivalent load cycle number for piles subjected to cyclic lateral loads after D4.1

A foundation pile for an offshore wind turbine under a pure pulsating load is formed by a monopile with D = 5 m and L = 20 m, see Figure D4.2.

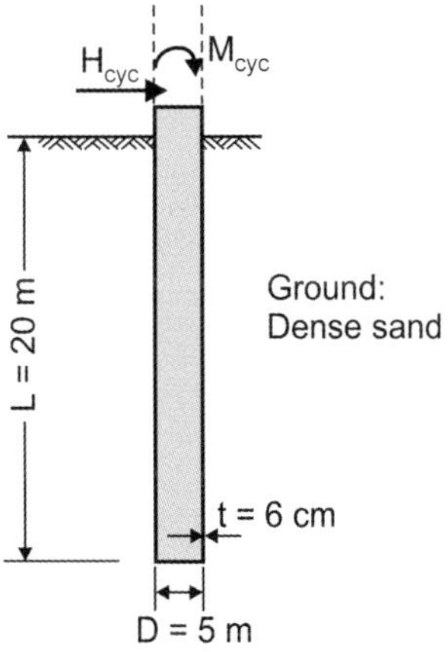

Figure D4.2 Monopile in dense sand; system and dimensions

Seven loading classes as shown in Table D4.3 were identified on the basis of an analysis of the prevalent wind and wave loads. Each load class consists of an action combination E_i (horizontal force H and head moment M) and the corresponding number of load cycles N_i.

In an analysis of the (static) bearing behaviour of the pile under the individual action combinations the static pile head deflections $y_{1,i}$ also given in Table D4.3 are received.

Table D4.3 Loading classes, load cycle numbers and static pile head deflections

Class	Action combination		Number of cycles	Static pile head deflection $y_{1,i}$
	H_{cyc} [MN]	M_{cyc} [MNm]		
1	2,14	118,75	$N_1 = 5$	$y_{1,1} = 3{,}7$ cm
2	1,91	106,25	$N_2 = 40$	$y_{1,2} = 3{,}3$ cm
3	1,69	93,75	$N_3 = 250$	$y_{1,3} = 2{,}8$ cm
4	1,46	81,25	$N_4 = 1\,800$	$y_{1,4} = 2{,}3$ cm
5	1,24	68,75	$N_5 = 12\,000$	$y_{1,5} = 1{,}9$ cm
6	1,01	56,25	$N_6 = 80\,000$	$y_{1,6} = 1{,}5$ cm
7	0,79	43,75	$N_7 = 550\,000$	$y_{1,7} = 1{,}1$ cm

The action combination (load class 1) leading to the largest static pile head deflection is adopted as the equivalent load. The parameter t is estimated at $t = 0{,}20$. Evaluation of the above equations gives the load cycle numbers N_k^* summarised in Table D4.4 for each load class. The resulting equivalent load cycle number for the total load is $N_{eq} = 73{,}3$. The variable loads can therefore be replaced by a harmonic load $H_{cyc} = 2{,}14$ MN and $M_{cyc} = 118{,}75$ MNm where $N_{eq} = 73{,}3$ cycles. The resulting pile head deflection y_{tot} for the example is $y_k = 6{,}9$ cm.

Table D4.4 Equivalent load cycle numbers

Class	Action combination		Number of cycles	Equivalent load cycle number N_k^* of the respective load class
	H_{cyc} [MN]	M_{cyc} [MNm]		
1	214	118,75	$N_1 = 5$ (design load)	5,00
2	1,91	106,25	$N_2 = 40$	15,63
3	1 69	93,75	$N_3 = 250$	19,34
4	1,46	81,25	$N_4 = 1\,800$	15,92
5	1,24	68,75	$N_5 = 12\,000$	10,92
6	1,01	56,25	$N_6 = 80\,000$	4,97
7	0,79	43,75	$N_7 = 550\,000$	1,52
Equivalent load cycle number from total load N_{eq} = 73,30				

If, alternatively to the procedure described above, the class 7 load cycle numbers are converted to class 6 equivalent load cycle numbers, the resulting load cycle number then to class 5 equivalent load cycle numbers, etc. (solve from the bottom to the top – cf. D4.1.2 (4)), the result in this case is a class 1 equivalent load cycle number of $N_{eq} = 52{,}1$, and therefore a resulting pile head deflection of $y_k = 6{,}6$ cm.

Literatur

[1] Alizadeh, M., Davisson, M. T.: Lateral load test on piles-Arkansas River project. Journal of the Soil Mechanics and Foundations Division, ASCE, Vol. 96, No. 5, 1970, pp. 1583–1604.

[2] API RP 2A-WSD: Recommended Practice for Planning, Designing and Constructing Fixed Offshore Platforms – Working Stress Design; 21st Edition, American Petroleum Institute, Washington, 2007.

[3] Ashour, M., Norris, G.: Lateral Loading of a Pile in Layered Soil using the Strain Wedge Model. Journal of Geotechnical and Geoenvironmental Engineering, Vol. 124(4), 1998, pp. 303–315.

[4] Baumbach, J., Schwarz, P.: Axial zyklisch belastete Mikropfähle in bindigen Böden. Schlussbericht Forschungsvorhaben. Zentrum Geotechnik, Technische Universität München, 2010 (unveröffentlicht).

[5] Bender, M., Mark, P., Stangenberg, F.: Querkraftbemessung von bügel- oder wendelbewehrten Bauteilen mit Kreisquerschnitt. Beton- und Stahlbetonbau 105, Heft 7, 2010, S. 421–432.

[6] Bergfelder, J., Schmidt, H. G.: Zur Planung und Auswertung horizontaler Pfahlprobebelastungen, Geotechnik 1989, Heft 2, S. 57–61.

[7] Bergfelder, J.: Zur Wechselwirkung zwischen Pfahlgründung und Tragwerk. Pfahl-Symposium. Mitteilungen des Institutes für Grundbau und Bodenmechanik, TU Braunschweig, Heft Nr. 60, 1999, S. 29–48.

[8] Blum, H.: Wirtschaftliche Dalbenformen und deren Berechnung, Bautechnik 10, 1932, S. 50–55.

[9] Brandl, H.: Stützbauwerke und konstruktive Hangsicherung. In: Grundbautaschenbuch, 7. Auflage, Teil 3, Kapitel 3.9, Verlag Ernst & Sohn, Berlin, 2009, S. 747–901.

[10] Brandtzaeg A., Harboe E.: Buckling tests of slender piles in soft quick clay; Proc. of the fourth Intern. Conference on Soil Mechanics and Foundation Engineering Vol. II, p. 19, London 1957.

[11] Briaud, J. L., Jeong, S., Bush, R.: Group Effect in the Case of Downdrag, ASCE GSP 27, Geotechnical Engineering Congress, 1991, S. 505-518.

[12] Brinch Hansen, J. (1961): The Ultimate Resistance of Rigid Piles Against Transversal Forces, Geoteknish Institute, Copenhagen, Bulletin No. 12.

[13] Broms, B. B.: Negative Skin Friction, Proceedings of the 6th Asian Regional Conference on Soil Mechanics, Singapore, Vol. 2, 1979, S. 41–75.

[14] Bustamante, M.: Current french Design Practice for axially loaded Piles, Ground Engineering, 1999, p. 38–44.

[15] Cooke, R. W.: The settlement of friction pile foundations. In: Proc. Conf. Tall Buildings, Kuala Lumpur, Vol. 3, 1974, S. 1–16.

[16] Coyle, H. M., Reese, L. C.: Load transfer for axially loaded piles in clay. Journal of the Soil Mechanics and Foundation Division, Vol. 92, No. 2, 1966, pp. 1–26.

Editors: German Geotechnical Society (*Deutsche Gesellschaft für Geotechnik e.V.*)

[17] De Beer: Piles subjected to static lateral loads. State-of-the-Art-Report, Specialty Session no 10 Proceedings, ICSMFE, Tokyo, 1977.

[18] DGGT AK 1.4: Empfehlungen des Arbeitskreises „Baugrunddynamik": Deutsche Gesellschaft für Geotechnik e. V. (Hrsg.), Berlin 2002, 2012.

[19] Dührkop, J.: Zum Einfluss von Aufweitungen und zyklischen Lasten auf das Verformungsverhalten lateral beanspruchter Pfähle in Sand. Veröffentlichungen des Instituts Geotechnik und Baubetrieb, Technische Universität Hamburg-Harburg, Band 20, 2009.

[20] Dührkop J., Grabe J.: Monopilegründungen von Offshore-Windenergieanlagen - Zum Einfluss einer veränderlichen zyklischen Lastangriffsrichtung. Bautechnik, 85(5), 2008, S. 317–321.

[21] Dürrwang, R., Otto, U.: Die Sanierung von Rutschungen mit Verdübelungen aus Großbohrpfählen und Schlitzwandscheiben, Tagungsband zur 8. Donau-Europäischen Konferenz über Bodenmechanik und Grundbau, Nürnberg, 1986.

[22] Dürrwang, R., Schulz, G.: Sanierung einer Rutschung mittels Großdübel, Bericht der 8. Nationalen Tagung für Ingenieurgeologie, Berlin, 1991.

[23] Dürrwang, R.: Pfahltragfähigkeiten im Grenzbereich Lockerboden-Fels. Geotechnik 20(3), 1997, S. 168–176.

[24] Dürrwang, R., Hecht, T., Johann, S., El-Mossallamy, Y.: Die Gründung der Weida-Talbrücke: Pfahlprobebelastung mit der Osterberg-Zelle. In: Pfahlsymposium, Mitteilungen des Institutes für Grundbau und Bodenmechanik, Technische Universität Braunschweig, Heft 80, 2005, S. 377–388.

[25] Duncan, J. M., Chang, C. Y.: Nonlinear Analysis of Stress and Strain in Soils. In: ASCE, Vol. 96, No. 5, 1970, pp. 1629–1653.

[26] EAB: Empfehlungen des Arbeitskreises 2.4 „Baugruben" der Deutschen Gesellschaft für Geotechnik e. V. (DGGT), 4. Auflage, Verlag Ernst & Sohn, Berlin, 2006.

[27] EAU: Empfehlungen des Arbeitsausschusses „Ufereinfassungen", 10. Auflage, Hrsg. Arbeitsausschuss Ufereinfassungen der Hafenbautechnischen Gesellschaft und der Deutschen Gesellschaft für Geotechnik e. V. (DGGT), Verlag Ernst & Sohn, Berlin, 2005.

[28] EBGEO: Empfehlungen für den Entwurf und die Berechnung von Bewehrungen aus Geokunststoffen. 2. Auflage, Verlag Ernst & Sohn, Berlin, 2010.

[29] Elborg, E.-A.: Verbesserung der Vorhersagbarkeit des Last-Setzungsverhaltens von Bohrpfählen auf empirischer Grundlage, Dissertation: Technische Hochschule Darmstadt, D 17, 1993.

[30] Ellner, A., Floom, K. J.: Mantelreibung an einem Großbohrpfahl im verwitterten Fels. Veröffentlichungen des Grundbauinstitutes der LGA Bayern, Heft 39, 1980.

[31] Endo, M. et al.: Negative Skin Friction acting on Steel Pipe Piles in Clay, Proceedings of the 7th International Conference on Soil Mechanics and Foundation Engineering, Vol. 2, 1969, S. 85–92.

[32] Fahey, M., Carter, J. P.: A finite element study of the pressuremeter test in sand using a non-linear elastic-plastic model. Canadian Geotechnical Journal, Vol. 30, No. 2, 2003, pp. 348–362.

[33] FGSV: Merkblatt über Felsgruppenbeschreibung für bautechnische Zwecke im Straßenbau, Forschungsgesellschaft für das Straßenwesen, 1980.

[34] Fedders, H.: Seitendruck auf Pfähle durch Bewegung von weichen bindigen Böden – Empfehlung für Entwurf und Bemessung. Geotechnik, 1978, Heft 2, S. 100–104.

[35] Fedders, H.: Zuschrift zu Seitendruck auf Pfähle, Geotechnik, 1980, Heft 4, S. 209–210.

[36] Fellenius, B. H.: Negative Skin Friction on long Piles driven in Clay, Swedish Geotechnical Institute, Proceedings No. 25, 1971.

[37] Fellenius, B. H.: Down-Drag on piles in clay due to negative skin friction. Canadian Geotechnical Journal, 1972, S. 323–337.

[38] Franke, E.: Probebelastung an Großbohrpfählen. Die Bautechnik, Heft 1, 1973, S. 7–19.

[39] Franke, E., Garbrecht, D.: Spitzendruck und Mantelreibung von Großbohrpfählen, Geotechnik, 1980, Heft 1, S. 1–51.

[40] Franke, E., Gollub, P.: Zur Berechnung von Pfahlgruppen, insbesondere von Zugpfahlgruppen. Bautechnik 73, 1996, Heft 9, S. 605–613.

[41] Guideline: Guideline on the interpretation of Rapid Load Testing on Piles. Cur report 230, Cur Commission H410, Netherland, 2010.

[42] Gruber, N., Koreck, H.-W., Schwarz, P.: Beiträge zum Tragverhalten axial zyklisch belasteter Pfähle, Schriftenreihe Heft 5, Lehrstuhl und Prüfamt für Grundbau, Bodenmechanik und Felsmechanik der Technischen Universität München, 1985.

[43] Gudehus, G., Leinenkugel, H.-J.: Fließdruck und Fließbewegung in bindigen Böden: Neue Methoden, Vorträge der Baugrundtagung 1978 in Berlin, Deutsche Gesellschaft für Erd- und Grundbau, Essen, 1978, S. 411–429.

[44] Handbuch EC 7-1: Handbuch Eurocode 7 – Geotechnische Bemessung, Band 1 Allgemeine Regeln. 1. Auflage, Beuth Verlag, Berlin, 2011.

[45] Handbuch EC 7-2: Handbuch Eurocode 7 – Geotechnische Bemessung, Band 2 Erkundung und Untersuchung. 1. Auflage, Beuth Verlag, Berlin, 2011.

[46] Hanisch, J., Katzenbach, R., König, G.: Kombinierte Pfahl-Plattengründungen, Verlag Ernst & Sohn, Berlin, 2002.

[47] Hettler, A.: Verschiebungen starrer und elastischer Gründungskörper in Sand bei monotoner und zyklischer Belastung. Veröffentlichungen des Instituts für Bodenmechanik und Felsmechanik, Universität Karlsruhe,1981, Heft 90.

[48] Hinz, P.: Beurteilung des Langzeitverhaltens zyklisch horizontal belasteter Monopile-Gründungen, Mitteilungen Fachgebiet Grundbau und Bodenmechanik, Universität Duisburg-Essen, Heft 37, VGE Verlag, Essen, 2009.

[49] Hölscher, P., Brassinga, H., Brown, M., Middendorp, P., Profittlich, M., van Tol, F.: Rapid Load Testing on Piles: Interpretation Guidelines, CRC Press – Tylor and Francis Group, the Netherlands, 2011.

[50] Horch, M.: Zuschrift zu Seitendruck auf Pfähle, Geotechnik, 1980, Heft 4, S. 207.

[51] Janbu, N.: Soil compressibility as determined by oedometer and triaxial tests. European Conference on Soil Mechanics & Foundations Engineering, Wiesbaden, Germany, Vol. 1, 1963, pp. 19–25.

[52] Kase, E., Ross, T.: Seismic Imaging as a means to evaluate foundation structures, In: DiMaggio, J. A. & Hussein, M. H. (Hrsg.): Current Practices and Future Trends in Deep Foundations, ASCE Geotechnical Special Publication No. 125, 2002, S. 361–369.

[53] Kempfert, H.-G., Lauffer, J.: Probebelastungen in wenig tragfähigen Böden unter statischer und wechselnder Belastung. Geotechnik 14, 1991, Heft 3, S. 105–112.

[54] Kempfert, H.-G.: Negative Mantelreibung bei Pfahlgründungen nach dem Teilsicherheitskonzept, Vorträge zum 12. Darmstädter Geotechnik-Kolloquium, Mitteilungen des Institutes und der Versuchsanstalt für Geotechnik der Technischen Universität Darmstadt, Heft 71, 2005, S. 21–31.

[55] Kempfert, H.-G., Gebreselassie, B.: Excavations and Foundations in Soft Soils, 1. Auflage, Springer Verlag, Berlin, Heidelberg, 2006.

[56] Kempfert, H.-G., Thomas, S.: Zum axialen Pfahltragverhalten unter zyklisch-dynamischer Belastung. VDI-Bericht 1941. VDI-Verlag Düsseldorf, 2006, S. 521–535.

[57] Kempfert, H.-G., Becker, P.: Grundlagen und Ergebnisse der Ableitung von axialen Pfahlwiderständen aus Erfahrungswerten, Bautechnik 84, 2007, S.441–449.

[58] Kempfert, H.-G., Raithel, M.: Bodenmechanik und Grundbau, Band 1 und 2. Bauwerk Verlag, 2007, 2012.

[59] Kempfert, H.-G.: 3.2: Pfahlgründungen. In: Grundbautaschenbuch, 7. Auflage, Teil 3, Kapitel 3.2. Verlag Ernst & Sohn, 2009, S. 73–277.

[60] Kempfert, H.-G., Thomas, S., Gebreselassie, B.: Observation of Pile-Soil-Interaction during Cyclic Axial Loading using Particle Image Velocimetry; Proceedings of the GeoShanghai 2010 International Conference, Deep foundations and geotechnical in situ testing, 2010, pp. 67–72.

[61] Kempfert, H.-G., Lüking, J.: Untersuchung zur Ableitung von Verbundspannungen bei Verpressmörtelpfählen. Abschlussbericht zum F+E Projekt, Universität Kassel, 2010 (unveröffentlicht).

[62] Kempfert, H.-G., Lüking, J., Mardfeldt, B.: Zum Ansatz von Verbundspannungen bei Verpressmörtelpfählen, Bauingenieur, Band 86, 2011, S. 464–474.

[63] Kempfert, H.-G., Thomas, S.: Untersuchungen zum Pfahltragverhalten unter zyklisch axialer Belastung in nichtbindigen und bindigen Böden; DFG-Abschlussbericht, Universität Kassel, 2011 (unveröffentlicht).

[64] Kempfert, H.-G.: Pfahlgründungen. In: Kommentar zum Normenhandbuch Eurocode 7 – Geotechnische Bemessung, Allgemeine Regeln, Verlag Ernst & Sohn, Berlin 2011.

[65] Kirsch, F., Richter, Th.: Ein analytisch-empirischer Ansatz zur Bestimmung der Tragfähigkeit und der Verformungen von axial zyklisch belasteten Pfählen, Veröffentlichung des Instituts für Bodenmechanik und Felsmechanik, KIT Süd, Workshop „Offshore- Gründungen von Windkraftanlagen“, Heft 172, 2010, S. 151–164.

[66] Kirsch, F., Richter, Th.: Ein einfaches Näherungsverfahren zur Prognose des axial-zyklischen Tragverhaltens von Pfählen, Bautechnik 88, Heft 2, 2011, S. 113–120.

[67] Kirsch, F., Richter, Th., Mittag, J.: Zur Verwendung von Interaktionsdiagrammen beim Nachweis axial-zyklisch belasteter Pfähle. Bautechnik 88 (2011), Heft 5, S. 319–324

[68] Klingmüller, O.: Dynamische Pfahlprüfung als Optimierungsproblem, In: Pfahlsymposium, Mitteilung des Instituts für Grundbau und Bodenmechanik, TU Braunschweig, Heft 38, 1991, S. 149–176.

[69] Klüber, E.: Tragverhalten von Pfahlgruppen unter Horizontalbelastung, Institut für Grundbau, Boden- und Felsmechanik, TH Darmstadt, Heft 28, 1988.

[70] Kluckert, K., Placzek, D.: Aktueller Stand der Methoden zur Tragfähigkeitserhöhung und Setzungsminimierung bei Großbohrpfählen, Baugrundtagung Mainz, 2002.

[71] Kraft, L. M., Cox, W. R., Verner, E. A.: Pile load tests: Cyclic loads and varying load rates. Journal of the Geotechnical Engineering Division, Vol. 107, No. 1, 1981, pp. 1–7.

[72] Krieg, S., Goldscheider, M.: Bodenviskosität und ihr Einfluss auf das Tragverhalten von Pfählen, Bautechnik (75), 1998, Heft 10.

[73] Latotzke, J., Triantafyllidis, Th.: Horizontale Stoßbelastung von Pfählen – Modellversuche in der geotechnischen Großzentrifuge in Bochum. Bautechnik 79, Heft 3, 2002, S. 144–159.

[74] LeBlanc, C: Design of Offshore Wind Turbine Support Structures, Ph.D. Thesis, Aalborg University, Denmark, DCE Thesis No. 18, 2009.

[75] Lesny, K.: Foundations for Offshore Wind Turbines – Tools for Planning and Design. VGE Verlag Essen, 2010.

[76] Lesny, K., Hinz, P.: Design of Monopile Foundations for Offshore Wind Energy Converters. In: Contemporary Topics in Deep Foundations. Geotechnical Special Publication ASCE, No. 185, 2009, pp. 512–519.

[77] Likins, G. et al.: Monitoring Quality Assurance for Deep Foundations, In: DiMaggio, J.A. & Hussein, M.H. (Hrsg.): Current Practices and Future Trends in Deep Foundations, ASCE Geotechnical Special Publication No. 125, 2004, S. 222–238.

[78] Lin, S. S., Liao, J. C.: Permanent strains of piles in sand due to cyclic lateral loads. ASCE Journal of Geotechnical and Geoenvironmental Engineering, Vol. 125, No. 9, 1999, pp. 798–802.

[79] Linder, W.-R., Richter, Th.: Die Gründung einer Experimentierhalle mit höchsten Anforderungen an die Setzungsbegrenzung. Tagungsband 7. Österreichische Geotechniktagung, Wien, 2009, S. 71–82.

[80] Linder, W.-R., Siebke, H.: Bohrpfähle: Kommentar zu DIN EN 1536. 1. Auflage, Beuth Verlag, Berlin, 2004.

[81] Lippomann, R.: Ingenieurgeologische Kriechhangsicherung durch Dübel, Veröffentlichungen des Institutes für Boden- und Felsmechanik, Universität Karlsruhe, Heft 111, 1988.

[82] Little, J. A.: Downdrag on Piles: Review and Recent Experimentation, Vertical and Horizontal Deformations of Foundations and Embankments, Proceedings of Settlement 94, 1994, S. 1805–1825.

[83] Little, R. L., Briaud, J.-L.: Full scale cyclic lateral load tests on six single piles in sand. Miscellaneous paper GL-88-27, Geotechnical Division, Texas A&M University, College Station, Texas, 1988.

[84] Long, J. H.; Vanneste, G.: Effects of cyclic lateral loads on piles in sand. ASCE Journal of Geotechnical Engineering, Vol. 120, No. 1,1994, pp. 225–244.

[85] Lüking, J.: Tragverhalten von offenen Verdrängungspfählen unter Berücksichtigung der Pfropfenbildung in nichtbindigen Böden. Schriftenreihe Geotechnik, Universität Kassel, 2010, Heft 23.

[86] Lüking, J., Kempfert, H.-G.: Untersuchung der Pfropfenbildung an offenen Verdrängungspfählen (Stahlrohrpfählen). Bautechnik, 2012 (in Vorbereitung).

[87] Lunne, T., Robertson, P. K., Powell, J. J. M.: Cone Penetration Testing, Blackie Academic & Professional, London, 1997.

[88] Lutz, B., El-Mossallamy, Y., Richter, Th.: Ein einfaches, für die Handrechnung geeignetes Berechnungsverfahren zur Abschätzung des globalen Last-Setzungsverhaltens von Kombinierten Pfahl-Plattengründungen. Bauingenieur 81, 2006, S. 61–66.

[89] Masing, G.: Eigenspannungen und Verfestigung beim Messing. Proceedings of the 2nd International Congress of Applied Mechanics, Zürich, 1926, pp. 332–335.

[90] Matlock, H.: Correlations for Design of Laterally Loaded Piles in Soft Clay; Tagungsband 2nd Offshore Technology Conference, Houston, 1970, S. 577–594.

[91] Matlock, H., Foo, S. H. C.: Axial analysis of piles using a hysterics and degrading soil model. Proceedings of the 1st International Conference Numerical Methods of Offshore Piling, London, 1979, pp. 127–133.

[92] Mazurkiewicz, B.: Einfluss von Rammgeräten auf die Tragfähigkeit von Stahlbetonprofilen, Beitrag in Tagungsband zum Symposium Pfahlgründungen TH Darmstadt, 1986, S. 31–36.

[93] Meek, J. W.: Das Knicken von Verpresspfählen mit kleinem Durchmesser in weichem, bindigem Boden; Bautechnik 73, Heft 3, 1996.

[94] Meek, J. W.: Sind Kleinverpresspfähle knickgefährdet?, Pfahlsymposium Braunschweig, Mitteilung des Instituts für Grundbau und Bodenmechanik der TU Braunschweig, Heft Nr. 60, 1999, S. 221–234.

[95] Meinhardt, G., Hoefsloot, F., Bakker, K. J., de Jong, E.: Statnamic Probebelastungen von Großbohrpfählen, Ausführung und Beurteilung. Tagungsband 31. Baugrundtagung 2010, DGGT, S. 27–33.

[96] Melzer, K.-J., Bergdahl, U., Fecker, E.: Baugrunduntersuchungen im Feld. In: Grundbautaschenbuch, 7. Auflage, Teil 1, Verlag Ernst & Sohn, Berlin, 2008, S. 43–122.

[97] Meyerhof, G. G.: General Report. Proc. European Symposium on Penetration Testing, Stockholm, Band 2.1, 1974, S. 41–48.

[98] Mittag, J., Richter, Th.: Beitrag zur Bemessung von vertikal zyklisch belasteten Pfählen, Symposium „Verkehrswegebau und Tiefgründungen", Schriftenreihe Geotechnik, Universität Kassel, Heft 18, 2005, S. 337–354.

[99] Moormann, C.: Zur Tragwirkung und Beanspruchung von Gründungspfählen beim Baugrubenaushub. In: Pfahlsymposium, Mitteilungen des Institutes für Grundbau und Bodenmechanik, Technische Universität Braunschweig, Heft 71, 2003, S. 351–378.

[100] Moormann, Chr., Rumpelt, Th., Schmidt, H.-H., Jud, H.: Bemessungs- und Optimierungsansätze für Tiefgründungen in verwitterten Halbfestgesteinen. Beiträge zum 19. Christian Veeder Kolloquium, „Tiefgründungen – Bemessung und Ausführung", 15./16. 04. 2004, Gruppe Geotechnik Graz, Technische Universität Graz, Heft 21, 2004, S. 43–71.

[101] Moormannn, C., Jud, H., Keysberg, J.: Single- und Multi-Level Tests nach dem Osterberg-Verfahren – Erfahrungen bei Probebelastungen an Großbohrpfählen in Zentralamerika. Pfahlsymposium, Mitteilungen des Institutes für Grundbau und Bodenmechanik, Technische Universität Braunschweig, Heft 80, 2005, S. 391–428.

[102] Moormann, Chr.: Pfahltragverhalten in festen und veränderlich festen Gesteinen. Symposium „Verkehrswegebau und Tiefgründungen", Schriftenreihe Geotechnik, Universität Kassel, 2005, S. 249–273.

[103] Nendza, H., Placzek, D.: Die Erhöhung der Tragfähigkeit durch gezieltes Nachverpressen – Stand der Erfahrungen. Baugrundtagung 1988, Hamburg.

[104] Norris, G.: Theoretically Based BEF Laterally Loaded Pile Analysis. 3rd International Conference on Numerical Methods in Offshore Piling, Nantes, France, 1986, pp. 361–386.

[105] Österreichische Vereinigung für Beton und Bautechnik: Richtlinie Bohrpfähle, Ausgabe März 2005, Wien 2005.

[106] Ofner, R., Wimmer, H.: Knicknachweis von Mikropfählen in geschichteten Böden. Bautechnik (84), 2007, Heft 12, S. 881–890.

[107] Osterberg, J. O.: New device for load testing driven piles and drilled shafts separates friction and end bearing. Int. Conf. On Piling and Deep Foundations, London, Balkema, 1989, S. 421–431.

[108] Osterberg, J. O.: The Osterberg load test method for bored and driven piles: the first ten years. 7th Int. Conf. On Piling and Deep Foundations – DFI 98, 15.–17. June 1998, Wien, 1.28.1–1.28.11.

[109] Osterberg, J. O.: Load testing high capacity piles – what have we learned? Proceedings of the 5th Int. Conf. on Deep Foundation Practice, Singapore, April 2001.

[110] Placzek, D.: Vergleichende Untersuchungen beim Einsatz statischer und dynamischer Sonden. Geotechnik, Heft 2, 1985, S. 68–75

[111] Placzek, D., Schmidt, H. G., Oetjeng, D.: Zum Tragverhalten von Großbohrpfählen mit Fußaufweitung. Bautechnik 71, Heft 10, 1994, S. 626–632.

[112] Placzek, D.: Mantel- und Fußverpressung zur Tragfähigkeitserhöhung von Bohrpfählen – aktueller Stand der Erfahrungen. In: „Symposium Verkehrswegebau

und Tiefgründungen". Schriftenreihe Geotechnik, Universität Kassel, Heft 18, 2005, S. 275–290.

[113] Plaßmann, B.: Zur Optimierung der Messtechnik und der Auswertemethodik bei Pfahlintegritätsprüfungen; Mitteilungen d. Inst. f. Grundbau und Bodenmechanik, TU Braunschweig; Heft Nr. 67, 2001.

[114] Peralta, P., Achmus, M.: An Experimental Investigation of Piles in Sand Subjected to Lateral Cyclic Loads. 7th Int. Conf. on Physical Modeling in Geotechnics, ICPMG Zürich 2010.

[115] Poulos, H.: Spannungen und Setzungen im Boden. In: Grundbautaschenbuch, 6. Auflage, Teil 1, Kap. 1.6. Verlag Ernst & Sohn, 2001, S. 255–305.

[116] Quark-Vonscheidt, J., Walz, B.: Die Grenztragfähigkeit von Zugpfählen und Zugpfahlgruppen in Sand. Bautechnik 79, Heft 1, 2002, S. 42–47.

[117] Rackwitz, F.: Numerische Untersuchungen zum Tragverhalten von Zugpfählen und Zugpfahlgruppen in Sand auf Grundlage von Pfahlprobebelastungen, Veröffentlichung des Grundbauinstitutes der TU Berlin, Heft 32, 2003.

[118] Raithel, M., Kempfert, H.-G., Quick, H.: Statische und zyklische Pfahlprobebelastungsergebnisse als Grundlage für Fahrwegsgründungen. 4. Österreichische Geotechniktagung, Österr. Ingenieur- und Architektenverein, 2003, S. 395–414.

[119] Raithel, M.: Lärmschutzwände auf horizontal belasteten Pfählen bei zyklischen/ dynamischen Einwirkungen – Versuchsergebnisse und Berechnungsansätze. Symposium Geotechnik, Schriftenreihe Geotechnik, Universität Kassel, Heft 18, 2005, S. 21–34.

[120] Randolph, M. F., Wroth, C. P.: Analysis of deformation of vertically loaded piles. ASCE J. GE Div. 104, 1978, pp 1465–1488.

[121] Randolph M. F., Houlsby G. T.: The limiting pressure on a circular pile loaded laterally in cohesive soil. Geotechnique, Jg. 34, Nr. 4, 1984.

[122] Richter, Th.; Kirsch, F.; Mittag, J.: Bemessungskonzepte für axial-zyklisch belastete Pfähle – Ein Überblick und neue Ansätze. In: DGGT (Hrsg.): Vorträge zur Baugrundtagung München, 2010, S. 263–270.

[123] Rollberg, D.: Zur Bestimmung der Pfahltragfähigkeit aus Sondierungen. Bauingenieur 60, Springer Verlag, Berlin, Heidelberg, 1985, S. 25–28.

[124] Rudolf, M.: Beanspruchung und Verformung von Gründungskonstruktionen auf Pfahlrosten und Pfahlgruppen unter Berücksichtigung des Teilsicherheitskonzeptes, Schriftenreihe Geotechnik, Universität Kassel, Heft 17, 2005.

[125] Rudolf, M., Kempfert, H.-G.: Setzungen und Beanspruchungen bei Gründungen auf Pfahlgruppen, Bautechnik 83, Heft 9, 2006, S. 618–625.

[126] Savidis, S., Vrettos, C.: Baugrunddynamik. In: K. Zilch, C. J. Diederichs, R. Katzenbach (Hrsg.): Handbuch für Bauingenieure, Springer Verlag, Berlin, Heidelberg, 2011.

[127] Schallert, M.: Faseroptische Mikrodehnungsaufnehmer für die Bewertung der Struktur von Betonpfählen; Mitteilungen des Instituts für Grundbau und Bodenmechanik, TU Braunschweig, Heft Nr. 93, 2010.

[128] Schiel, F.: Statik der Pfahlwerke, Springer Verlag, Berlin, 1960.

[129] Schmidt, H. G.: Horizontale Gruppenwirkung von Pfahlreihen in nichtbindigen Böden, Geotechnik, Heft 1, 1984, S. 1–6.

[130] Schmidt, H. G.: Großversuche zur Ermittlung des Tragverhaltens von Pfahlreihen unter horizontaler Belastung. Mitteilungen des Instituts für Grundbau, Boden und Felsmechanik der TH Darmstadt, Heft 25, 1986.

[131] Schmidt, H. G.: Großbohrpfähle im Übergangsbereich Boden – Fels. Bautechnik 67, Heft 1, 1990, S. 361–366.

[132] Schultze, E., Muhs, H.: Bodenuntersuchungen für Ingenieurbauten. 2. Auflage, Springer-Verlag, Berlin–Heidelberg–New York, 1967.

[133] Schuppener, B., Hrsg: Kommentar zum Normenhandbuch Eurocode 7 – Geotechnische Bemessung, Allgemeine Regeln, Verlag Ernst & Sohn, Berlin 2011.

[134] Schwarz, W.: Verdübelung toniger Böden, Veröffentlichungen des Institutes für Boden- und Felsmechanik, Universität Karlsruhe, Heft 105, 1987.

[135] Schwarz, P.: Beitrag zum Tragverhalten von Verpresspfählen mit kleinem Durchmesser unter axialer zyklischer Belastung. Schriftenreihe Lehrstuhl für Grundbau, Bodenmechanik und Felsmechanik, TU München, Heft 33, 2002.

[136] Seitz, M., Schmidt, H.-G.: Bohrpfähle, Verlag Ernst & Sohn, Berlin, 2000.

[137] Smoltczyk, U.: Pfahlgründungen (Abschnitt 8). In: Grundbautaschenbuch 6. Auflage, Teil 3, Kapitel 3.2. Verlag Ernst & Sohn, Berlin, 2001, S. 207–228.

[138] Stahlmann, J., Fischer, J., Zahlmann, J.: Theoretische Grundlagen und Untersuchungen zur Eindeutigkeit dynamischer Pfahltests, Geotechnik 2012 (in Vorbereitung)

[139] Stewart, H. E.: Permanent strains from cyclic variable-amplitude loadings. ASCE Journal of Geotechnical Engineering, Vol. 112, No. 6, 1986, pp. 646–660.

[140] Studer, J. A., Laue, J., Koller, M. G.: Bodendynamik; 3. Auflage, Springer Verlag, Berlin, 2007.

[141] Taşan, H. E.: Zur Dimensionierung der Monopile-Gründungen von Offshore-Windenergieanlagen. Veröffentlichungen des Grundbauinstitutes der Technischen Universität Berlin, Heft 52, 2011.

[142] Thomas, S.: Zum Pfahltragverhalten unter zyklisch axialer Belastung. Schriftenreihe Geotechnik, Universität Kassel, 2011, Heft 25.

[143] Thomas, S., Kempfert, H.-G.: Untersuchung des Pfahltragverhaltens infolge zyklisch axialer Einwirkungen in einer Spannungszelle; Beiträge zum Pfahl-Symposium 2011, Mitteilung des Instituts für Grundbau und Bodenmechanik, TU Braunschweig, 2011, S. 139–157.

[144] Thomas, S., Kempfert, H.-G.: Ein Beitrag zum Tragverhalten zyklisch axialer belasteter Pfähle (in Vorbereitung).

[145] Tomlinson, M. J.: Adhesion of Piles in Stiff Clays, Construction Indian Research Inf. Association CIRIA, report No. 26, 1970.

[146] Vermeer, P.A., Wehnert, M.: Numerische Simulation von Pfahlprobebelastungen an Großbohrpfählen. Tagungsband zum 4. Kolloquium „Bauen in Boden und Fels“ in Ostfildern, Technische Akademie Esslingen (TAE), 2004, S. 555–573.

[147] Vogler, M., Katzenbach, R.: Ergebnisse einer Pfahlprobebelastung in den Frankfurter Kalken mit 78 MN Grenzlast. Symposium „Messen in der Geotechnik“ 2004, 09./10. September 2004, Mitteilungen des Institutes für Grundbau und Bodenmechanik, Technische Universität Braunschweig, Heft 77, 2004, S. 53–66.

[148] Vogt, N.: Vorschlag für die Bemessung der Gründung von Lärmschutzwänden. Geotechnik 11, 1988, S. 210–214.

[149] Vogt, N., Vogt, S., Kellner C.: Knicken von schlanken Pfählen in weichen Böden, Bautechnik 82, 2005, Heft 12, S. 889–902.

[150] Vrettos, Ch.: Bodendynamik. In: Grundbau-Taschenbuch, Teil 1, 7. Auflage, Verlag Ernst & Sohn, Berlin, 2008, S. 451–500.

[151] Vucetic, M.: Cyclic Threshold Shear Strains in Soils, Journal of the Geotechnical Engineering Division, Vol. 120, No. 12, pp. 2208–2228.

[152] Waas, G.: Pfahlgründungen unter dynamischer Belastung. In: Haupt W. (Hrsg.): Bodendynamik, Grundlagen und Anwendung, Vieweg Verlag, 1986, S. 189–224.

[153] Wallrauch, E.: Verwitterung und Entspannung bei überkonsolidierten tonigschluffigen Gesteinen Südwestdeutschlands, Dissertation, Tübingen, 1969.

[154] Weiß, K., Hanack, S.: Der Einfluss der Lagerungsdichte des Bodens und der Herstellung von Großbohrpfählen auf deren Tragfähigkeit. Mitteilung der Deutschen Gesellschaft für Bodenmechanik (DEGEBO) an der TU Berlin, Heft 35, 1983.

[155] Wenz, K.-P.: Über die Größe des Seitendruckes auf Pfähle in bindigen Erdstoffen, Institut für Boden- und Felsmechanik der Universität Karlsruhe, Heft 12, 1963.

[156] Wenz K. P.: Das Knicken von schlanken Pfählen in weichen bindigen Erdstoffen, Mitteilungen Institut für Bodenmechanik und Felsmechanik, Universität Karlsruhe, Heft 50, 1972.

[157] Winter, H.: Fließen von Tonböden. Eine mathematische Theorie und ihre Anwendung auf den Fließwiderstand von Pfählen, Veröffentlichung des Instituts für Boden- und Felsmechanik, Universität Karlsruhe, Heft 82, 1979.

[158] Witzel, M.: Zur Tragfähigkeit und Gebrauchstauglichkeit von Verdrängungspfählen in bindigen und nichtbindigen Böden, Schriftenreihe Geotechnik, Universität Kassel, Heft 15, 2004.

List of Advertisers

Recommendations on Piling – EA Pfähle.
Editors: German Geotechnical Society (*Deutsche Gesellschaft für Geotechnik e.V.*)